Mind From Body

MIND FROM BODY

Experience From Neural Structure

Don M. Tucker

Illustrations by Anne Awh

2007

OXFORD
UNIVERSITY PRESS

Oxford University Press, Inc., publishes works that further
Oxford University's objective of excellence
in research, scholarship, and education.

Oxford   New York
Auckland   Cape Town   Dar es Salaam   Hong Kong   Karachi
Kuala Lumpur   Madrid   Melbourne   Mexico City   Nairobi
New Delhi   Shanghai   Taipei   Toronto

With offices in
Argentina   Austria   Brazil   Chile   Czech Republic   France   Greece
Guatemala   Hungary   Italy   Japan   Poland   Portugal   Singapore
South Korea   Switzerland   Thailand   Turkey   Ukraine   Vietnam

Copyright © 2007 by Don M. Tucker

Published by Oxford University Press, Inc.
198 Madison Avenue, New York, New York 10016

www.oup.com

Oxford is a registered trademark of Oxford University Press

All rights reserved. No part of this publication may be reproduced,
stored in a retrieval system, or transmitted, in any form or by any means,
electronic, mechanical, photocopying, recording, or otherwise,
without the prior permission of Oxford University Press.

Library of Congress Cataloging-in-Publication Data
Tucker, Don M.
Mind from body : experience from neural structure / Don M. Tucker ;
illustrations by Anne Awh.
p. ; cm.
Includes bibliographical references and index.
ISBN 978-0-19-531698-8
1. Cognitive neuroscience.   2. Mind and body.   3. Neural networks
(Neuroscience).   I. Title.
[DNLM: 1. Psychophysiology—methods.   2. Mental Processes—
physiology.   3. Neural Pathways.   WL 103 T891m 2007]
QP360.5.T83 2007
612.8'233—dc22   2006032313

Illustrations by Anne Awh, Early Stuff Productions
www.earlystuffproductions.com

9 8 7 6 5 4 3 2 1
Printed in the United States of America
on acid-free paper

Preface

This book is about the brain and its function, the mind. As we learn to use science to analyze the structure of the brain's networks, we are finding new opportunities for insight into the structure of experience. The mind is embodied. We now recognize that by studying the patterns of the brain's anatomy, we can look upon a roadmap to the mind.

Interpreting the roadmap, however, is not easy. In this book, we will first study the general functional systems of the brain: how psychological abilities emerge from the major dimensions of cortical and subcortical organization. Next, we will examine principles of cognitive representation and control: how ideas are held within brain tissue and how they are regulated to meet real-world needs. Just like navigating in a strange country, finding our way with the brain's roadmap depends on the principles of interpretation that we use to read the signs. As we learn how neural connectivity implies psychological function, the map comes to life.

The Fundamental Opacity of Introspection

When we apply scientific evidence to an analysis of mind, we find that the mind which we always took for granted turns out to be, well, unavailable. We naïvely assume a state of mental agency, a state in which we are in control of the mind. But when we study its regulatory mechanisms—especially in their biological forms—we find the mind is only partially transparent to conscious inspection, and only partly under volitional control.

Consider the example of emotion. In Western philosophical tradition, we assume we have volitional control of the mind, and that whereas emotions may influence us, they do so as we consciously evaluate them, as we appraise our feelings. This is wrong. When people suffer from depression, for example, they don't just feel bad, recognizing that the feeling is caused by the depression.

Rather, once in the grip of depression, the world looks terrible. The self becomes terrible. For those who are depressed, reality seems to change before their very eyes, but they don't see how the depression made it change. They just find themselves in a mean, cold world. Even if they have bipolar disorder and were manic just days before, depressed people have no memory of the good mood and the good world. With no subjective insight into their emotional bias, they may commit suicide, eager to end this terrible self and terrible world.

As we come to understand such integral adaptive mechanisms of emotional and motivational processes, science teaches us that the mind is organic, and embodied. This biological way of thinking is foreign to our naïve assumptions about human experience. Although often difficult to accept, the implications of brain research offer us new perspectives on subjectivity, and on the limits of introspection. If the only result of psychological study is to recognize the scope of our ignorance, that in itself provides a kind of enlightenment.

Principles for Young Minds

As a university professor for many years, I have learned to recognize the ability of people, particularly young people, to master difficult concepts. Young adults are still flexible enough to give up naïve assumptions about the nature of mind. Whereas older people have such a long history with their mental assumptive matrix that they cannot be expected to give it up, young people have only been conscious for a short while, so they can try new ways of thinking. I have therefore oriented this book toward adolescents and young adults, in the hope that it might be useful for those students who want to understand the psychological and philosophical implications of neuroscience and computational science. To the extent that older people remember how to study, and have not become so comfortable with subjectivity that they cannot approach it from a novel vantage, this book may be useful for more mature readers as well.

Visceral and Somatic Boundaries of Mind

The central theme of the book is that we can discover basic patterns in the organization of subjective experience by examining the structures of the brain's networks. In the classical studies of the mammalian nervous system, it became clear that there were two major divisions. One division is the *somatic* nervous system, which organizes the skin surface and the skeletal muscles, and thus mediates the bodily engagement with the world. As the special senses (sight, hearing, smell) evolved, they extended the interface with the world beyond the primitive somatic sense of touch.

Equally important to the body's somatic function is the *visceral* nervous system, which regulates the internal state. This regulation is achieved at the most basic level through homeostasis, making internal changes in response to environmental variation in order to maintain bodily integrity. From this primitive basis, the visceral function of the nervous system evolved to provide a rich substrate of motivational and emotional controls to direct the somatic function, and therefore to guide both experience and behavior.

As we study the neural structures of mind, we find that the visceral networks at the limbic core of each cerebral hemisphere are integral to the consolidation of experience. These play a remarkable counterpoint to the interface with the world mediated by the somatic networks of the sensory and motor functions at the lateral surface of the hemisphere. It is within the dialectical interplay—of internal, visceral motives with external, somatic contact with the world—that memories are formed. In this process, structures of mind arise. The goal of this book is to understand this process of organizing the neural structures of experience that give form to the subjective life of the mind.

Acknowledgments

In the Notes section I acknowledge some of the many works that form the basis of both the basic studies and the speculative ideas of this book. Two of my colleagues have been particularly influential on the theory of visceral and somatic psychological structures. Phan Luu has been the original thinker in many of our joint theoretical papers that established this line of reasoning. Doug Derryberry was important in directing my thinking to the role of interoceptive, visceral mechanisms in organizing subjective experience. I am grateful to the several editors at Oxford who helped me work this material into readable form. Catharine Carlin has been especially supportive of my efforts to frame complex scientific notions in an informal style. Finally, I thank my wife, Susan, and my children, Corin and Ethan, for their patience with my theorizing over many years. They were directly influential on this book, challenging me from the outset to apply neuropsychological theory to real world issues.

Contents

1. Mind in the Information Age — 3
2. Structures of Intelligence — 30
3. Principles of Representation and Control — 94
4. Motivated Experience — 143
5. Visceral and Somatic Frames of Mind — 187
6. Subjective Intelligence — 231
7. Objective Experience — 254
8. Information in the Age of Mind — 281

Notes — 301

Index — 313

Mind From Body

1
Mind in the Information Age

This must be the Information Age. As new vistas of knowledge open up before us, we are continually overwhelmed by more data than we can comprehend. Yet a short while later, there we are again, like unrepentant addicts, searching out more news. What is valuable is not the raw data but information that is processed and organized in a way that helps us get something done or, more important, understood. We struggle to use informatic tools—communication links, computers, and software—to organize and process the information. In just a few decades, these tools have become so integral to both work and play that computer literacy has become a requirement for participation in society. Just a generation ago, computer skills were restricted to specialized scientists and engineers, the nerd prophets of the new order. Soon computer skills will be as fundamental to participating in modern culture as language itself.

The Information Paradox

Asked what information is, one might say it is "what tells me something." This commonsense meaning was captured in the 1940s in the field of communications engineering by Claude Shannon's definition: Information is what reduces uncertainty. A message is informative to the extent that it reduces the uncertainty of the person who gets it.

The interesting implication of Shannon's definition—and our commonsense understanding—is that information is not an absolute thing. Rather, it gains meaning only within a context of knowledge. The idea of uncertainty is important because it implies the existence of a particular context of the knowledge that gives meaning to the message. Raw data do not constitute information.

We often think of intelligence as the possession of information or at least the capacity to understand it. If so, it is interesting to take Shannon's definition literally. Maybe information can be meaningful only to those who are uncertain. This reciprocity of information and uncertainty causes what we might call the "information paradox." Information has meaning only in the context of need and incomplete knowledge. This contradiction may be difficult to contemplate in abstract terms, but most of us are familiar with it in practical terms. We know that getting an answer is easy once you know to ask the question.

Recognizing this, we can see that if you have no question and no sense of uncertainty, then information, even if presented, remains meaningless. Data without need are like pearls before swine. We think of information as an objective quantity, so that information processing is an objective task. However, in a fundamental sense, information is subjective. The stuff in a computer is data. The stuff that is meaningful to a person is information.

In abstract terms, we can see how data become information when they have a relation to a context and complete a pattern. In experiential terms, we can see how information appears only when we understand its significance and when it fulfills a need we have. Otherwise it is just data—the raw, useless facts of the world.

Once we recognize the subjective context of information, the information paradox explains why new knowledge leads to hunger for more. When truly uninformed, we have no questions. The more we learn, the more we know we do not know. Uncertainty becomes shaped and prepared. An initial uncertainty grows into a more defining knowledge context.

The Information Age thus draws us into a state of continual knowledge hunger and expansion. Because information is fundamentally subjective, this knowledge expansion gives us an opportunity to expand subjective experience. Is there a point at which a more explicit knowledge of the information structure of experience would lead to a more explicit awareness of uncertainty? Perhaps a more differentiated state of conscious experience?

Memory and Vision

To understand the course of today's knowledge expansion, we can consider its roots. If communication is the flow of information, then the evolution of language gave a powerful new informatic capacity to humans who could speak. Language then evolved to transcend evolution. It was only a matter of time before the development of stories and myths allowed the retention of information beyond the immediate, present context. Soon writing, at first a business accounting tool, was used to make records of events in general. Words then enabled knowledge to extend beyond the memory of the storyteller, the evolved biological structure. With digital information technology, we now capture not only language and the sensory records of art and music but also the varied forms of enumerated raw data that provide potential information for future uncertainties. Our data technology can now mirror the reality of events with a precision limited only by sensor quality and digitization rate.

Through a history we only vaguely understand, we humans became the master data technologists. In the last century we discovered that calculating machines could be powerful vehicles for those essential accounting practices at the foundation of our commercial culture. In the last half-century we recognized the general use of Turing's programs of stereotyped machine code algorithms, which could be executed on data ordered in systematic addresses. With our natural technical sense, we knew that electronic circuits would be an efficient way to actualize these machine programs. By the last quarter of the century we had built digital circuits that were physically matched to the logical algorithms of the Turing machine. Etching silicon for digital circuits, we spawned a new culture of hardware data technologists. This culture rapidly delivered successive new generations of tools to the data technologists of software, who, in turn, are eagerly supplying information tools to satisfy society's increasing hunger for information.

Importantly, the information demands that transformed this new age came not from the office but the home. Of course, commerce nurtured the new computer technology, just as it nurtured reading, writing, and arithmetic in previous millennia. But the roots of the information revolution were established by the home computer. At first this personal fascination with data technology was sensed only by the nerd prophets, the high-functioning savants of data technology. Then, however, expanding from the vague initial attraction, the digital information seduction took hold of the rest of us.

Fumbling and inarticulate, compromising wherever possible, avoiding upgrades until the pain of obsolescence could be borne no longer, we integrated computation into work, recreation, and family interactions. Perhaps most important, we offered the information pipelines to our children. Generations now come of age in a culture in which video games, instant messaging, and processor upgrades are woven in the fabric of daily experience. Those of us who became parents in this era are like immigrants to a new world. We look on in awe, recognizing that we do not really understand how young people use this new language. We watch, with a bittersweet mixture of pride and alienation, as our children become the natural masters of a new world.

But children do not have the awareness to understand what is happening. As the Information Age unfolds, it is only those with memory and the perspective of history who can achieve vision. And with just the barest vantage of history, we can see that data technology is exploding.

In the last several decades we have measured the progress in data technology by Moore's law. A founder of the Intel Corporation, Gordon Moore remarked in 1965 that microprocessor capacity was doubling every 18 months and that—incredible as it might seem—he predicted that this unprecedented expansion could continue until 1975. We are now three decades past the reach of Moore's vision, and we still watch, transfixed, as the progression continues.

Nevertheless, this narrow span of history may not be sufficient for us to understand the significance of what is happening. Some of the most concrete prophecy of the cybernetic future has come from the imagination of Ray Kurzweil. An engineer who made significant advances in the digital synthesis of both speech and music, Kurzweil paused from his work to reflect on the momentum of the digital revolution and on where it is taking us. He looked into the new millennium by studying the history of data technology over the last century. Kurzweil found that, if we consider computing power in its early forms, such as mechanical calculators at the end of the 19th century and the first electronic computers in the middle of the 20th century, we see that data-processing inflation has been going on for at least a hundred years. This is not a phenomenon of silicon transistors in the 1960s. Rather, it is a fundamental phenomenon of data technology, expanding in power and complexity at a multiplying rate.

If we think beyond the development of machine intelligence in the last century and consider the development of biological intelligence in general, we can reflect on information complexity in the most relevant context, that of evolutionary time. In this frame of at least ten millennia, human culture itself has been exploding at a comparable rate.

In his book *Dragons of Eden*, Carl Sagan recognized that something remarkable happened when intelligence was no longer restricted to the biological mechanisms of genetic mutation and natural selection. Rather than being limited by the rate of the rare fortunate mutations in the blueprints of individual brains, intelligence became capable of cultural transmission. Evolution was no longer the same.

On those occasions in the history of our species when children were educated effectively, ideas and technologies were maintained across generations, spanning the gaps left by the passing of individual brains. When reading and writing were mastered effectively, ideas and technologies could be maintained by anyone with access to books, even the books of a different culture or era. With literature, the transmission and perhaps the evolution of information took new form.

With the digital computation of the last few decades, Sagan's extracorporeal intelligence was given an even more abstract medium. Taking digital form and thriving with the appropriate caretakers, information could now expand on an unprecedented scale. The data capacity seduces new levels of uncertainty to promise new forms of information. Fascinated by its power and speed, we are well prepared to serve as caretakers of information in its new digital form. We organize its code structures into more efficient patterns, causing each new generation of silicon to grow these patterns with increasing speed. It is now easy to think that the real potential for intelligence on this planet is held not by our brains and not by our children but by our machines.

Computational Mind

It is natural to identify artificial intelligence with digital logic, as if information processing really were a property of the familiar microcomputer chips of our daily experience. However, the real advances in the science of information are giving us insight into the nature of information in the abstract. The more theoretical, general models of information processing are not restricted to a particular machine code but can be related to general methods of intelligence, including how the human brain handles information.

Remarkably, the models of information organized in distributed networks of processing units, termed "artificial neural networks" or "connectionist networks," are turning out to behave in an inherently brainlike fashion. In fact, there is now a growing convergence of the science of brains (neuroscience) and the science of computers (information science), which is sometimes called "computational neuroscience." It is a productive exchange for understanding both biological and artificial forms of intelligence. In most cases the computation is still carried out with the familiar digital machines. Yet the models of computational neuroscience can now simulate the actual stuff of brains: neural membranes, synapses, and chemical transmitters. Although this is the physical substrate of the brain, it is not really hardware. As neuropsychologist Karl Pribram recognized, the physical stuff of our brain is *wetware*.

Computational models of neural activity are giving us new insight into the mechanisms of brains, including human brains. At the same time, the science of biological neural networks—the study of wetware—is teaching us how to build more interesting and effective artificial neural networks. A key realization is that artificial neural networks are showing novel forms of machine intelligence that we never saw in digital computer circuits.

For example, an appropriately constructed artificial neural network can recognize a human face. Of course, even with conventional digital logic it is possible to identify a specific photograph of a face by digitizing it into pixels and contrasting the pixels one by one with those of a known photograph. However, digital codes are inherently rigid. Present a photo of the same face in a different perspective, and digital logic fails. In contrast, with the distribution of information across the many connections of an artificial neural network, the patterns of the face are represented in a fundamentally different, more holistic manner. It generalizes readily from one photograph to another of the same face or even *similar* faces. As we study artificial neural networks, we are seeing the first evidence of intelligence that seems emergent, more general and novel than what we expect from a linear and logical addition of the component algorithms.

These properties of artificial neural networks might never have been discovered if we had simply kept grinding out greater bandwidth with digital microprocessors. It was the effort to understand brainlike networks and to simulate them with functional circuitry (even if still mostly emulated in digital logic) that created these distributed representations with their surprising properties.

Computational science will continue to benefit from brain science as we learn more about how the brain's cognition—the mind—arises from bodily processes. In nature, information gains meaning because it is mapped in some way to physical reality. For the nervous systems of animals, information gains meaning as it maps to bodily functions. For the human mind, science is showing us how bodily functions are integral to the basic operations of the mind, both in evaluating experience and in forming ideas. As we learn of this physical embodiment of mind, we may gain insight into how information processing may be controlled more generally, even in machines.

Nonetheless, even in machines and with massive distributed networks, information processing is not just representation. It requires control. The capacity for representation and for mapping or mirroring aspects of reality in some way must be motivated and regulated. Control theory may be more important to computational science than representation theory. In fact, many limitations of parallel distributed models of computation arise because connectionist representation capacities are so difficult to control for a useful purpose. As we understand the control of human intelligence within a physiological framework—within the wetware of the brain's bodily mechanisms—we may gain insights that will lead to improved control of information in any of the new distributed architectures of intelligence now emerging.

The science of artificial neural networks is still in its infancy and struggling in many ways, yet the underlying science has already created a revolution in machine intelligence. Perhaps because we have made them in our own image, we look on the performances of the artificial neural networks with a particular delight. In discussions of connectionist simulations at scientific meetings, you often hear the excitement and pride of young scientists as they describe the unexpected behavior of their new simulations. Without knowing the context of the conversation, you might think they were describing their children.

Maybe they are. Maybe this is our place in the order of things—to be the transitional forms, the last disposable biogenetic wetware, as we give birth to machine intelligence. Computational capacity continues to double every 18 months, even as our education system falters. Without advancing education, all we give our children is the human brain, which saw its last upgrade with the genetic mutations of the Stone Age. With these odds, one would have to bet on the machines.

Even a cursory review of human accomplishments in the last few centuries will show that, if there is a choice between developing insight into human nature and developing a new, more clever and powerful technology, we choose the technology. Although some may hope that we are creating a future to actualize human potential, there can be little doubt that, over the last century at least, the practical progress has been with machines.

The Lost Mind

Splitting off from the philosophy departments at the end of the 19th century, the new discipline of psychology embraced the scientific method as the way to achieve cumulative progress in the knowledge of the mind. As one of the first scientific psychologists, William James was an original thinker. He was also a careful student of the 19th-century literature on many topics, from philosophy to physiology. But what is most impressive from today's perspective is the breadth of the questions that James considered as essential to the new science of the mind. He saw it as psychology's task to explain the mind as it appears to people, as emotions arise and we reflect upon them, as religious experiences change our lives, and as we try to understand the memories of a lifetime.

Since James wrote, there has been considerable progress in the objective science of psychology. What has not progressed is the translation of the science into terms that are meaningful for personal experience. In the early 20th century we made significant advances in clarifying the mechanisms of animal learning, with sophisticated experiments that could be applied to study human learning as well. The Gestalt psychologists of Germany showed us the principles of pattern formation in perception. Social psychologists applied systematic observation and experiments to show how people think in social situations. Jean Piaget showed how to analyze the development of intelligence in a child's actions and explained how this could give us clues to the foundations of adult intelligence.

In the mid-20th century many psychologists recognized the importance of abstract principles of information processing developed from communications engineering. Here psychology could become highly objective for the first time, with formal mechanisms and even mathematical descriptions of mental processes. Concepts such as feedback were applied not only to electronics but also to real-world issues such as social interactions and mental disorders. As data-processing machines became capable of greater and greater feats of memory and calculation, psychologists took the metaphors of information processing from the machines and applied them to a detailed experimental analysis of the mechanisms of human cognition, including perception, language, memory, and attention.

The result of these contributions was a scientific discipline of human information processing that can proceed on an objective basis. We now have limited, testable theories that are closely linked to experimental studies of how people perceive, speak, remember, and concentrate under carefully controlled laboratory conditions.

However, the researchers drawn to the information-processing approach to psychology were mostly the disciplined, hard-nosed types, who were uncomfortable with the vague humanism of the broader areas of psychology. As a result, in the 1970s they began to separate themselves from university psychology departments. Much as the first psychologists had done with philosophy a century before, these scientists moved out on their own to create departments of cognitive science. Because discrete and testable experiments are best done with specific aspects of cognition, cognitive science became a discipline of specific components or "distinct modules" of objective cognition, such as word form perception in reading, spatial visual attention, and verbal memory.

With human cognition clearly organized and specified in this distinct way, cognitive science could proceed as "normal science," with agreement among a community of researchers on what constituted acceptable questions for experiments and methods of addressing those questions.

When neuroscientists developed new technologies for imaging the activity of the human brain, it soon became clear that they needed the help of cognitive psychologists. It was not possible to simply put someone's head into a PET scanner or an fMRI magnet and scan the brain while the person did some interesting mental thing like reading a book or listening to music. What they got was a complex mixture of many different mental operations, often at different points in time. It was therefore necessary to separate the specific mental operations and trap them precisely in time before these operations could be located in the brain. As neuroimaging technologies (made possible of course by advances in computation) emerged in the 1980s, cognitive science gradually morphed into cognitive neuroscience. With this new discipline we are learning to map the functions of the brain with the modules of cognitive science, and we are using the technology for imaging neural activity to improve our specification of the modules of human information processing.

Yet somewhere along the way we may have lost the mind. Cognitive scientists (and now cognitive neuroscientists) have objectified the mind's operations as information-processing capacities and distinct mental modules that can be measured in controlled experiments. In the process, many of the personal, experiential questions of psychology that were so obvious at the end of the 19th century have been abandoned. Science never did get a grip on subjective experience.

In my research on the brain, as I have studied the neural structures that give form to the mind, I have come to the realization that information does not exist without value. For humans, this means that information processing must be understood in relation to personal significance, which in turn becomes the engine of subjective experience.

In practical terms, every college student knows that subjectivity is essential to having a mind. Without experience, the mind's information-processing functions become empty academic abstractions. If science cannot be meaningful to subjective experience, it will leave us with naïve and superstitious ways of understanding our lives. Science will then fail to guide our culture. It will remain a restricted technical discipline, suitable only for technological applications such as creating new generations of information-processing machines.

Chasm of the Two Cultures

In the 1950s C. P. Snow considered the split between science and the humanities in his famous "two cultures" lecture. He once remarked that he simply intended to discuss problems in the English university system, the academy of scholars. Instead, he uncovered what many now recognize as an inherent defect of Western intellectual progress. Snow's analysis became emblematic of the chasm between science and the humanities in society at large.

The two cultures of the academy persist because those who choose the discipline of science cannot seem to speak the same language as the "literary intellectuals" that Snow described in the humanities. Within their circles, many of the literary intellectuals have formulated their own approach to the subjective mind. While the scientists write off subjectivity as unapproachable (and naïvely assume that objectivity can be gained simply by taking up the trappings of science), many of the literary intellectuals have decided that objectivity itself is an illusion. In the more extreme forms of the critical or deconstructionist analysis popular at universities in recent years, a kind of left-wing, histrionic political correctness has leaked into the dialogue, with no one able to stand up to it. Deconstructionists hold that the cultural bias of the European civilization has achieved an unnatural hegemony, blinding and deluding all those who accept the premises of the societal establishment. Especially politically incorrect are the scientists, who may think they are creating objective knowledge when in reality they are perpetrating their European, dominant, white, male hegemony.

This is because, according to the deconstructionists, objectivity, even when framed in the equations and logic of science, is a sham. Reality is not a material fact to be revealed to all through observation, analysis, and insight. Rather, it is a cultural construction that is valid by definition for a particular culture. Not only can it not be objectively questioned, but the received reality of a culture cannot even be understood by someone outside that culture. In the deconstructionist philosophy of modern literary intellectuals, all opinions are equally valid.

Dead-End Subjectivity

Deconstructionist analysis has offered important insight into biases that may go unrecognized in the assumptions of a dominant culture. However, it exacts a steep price for its lessons in that it has caused many of the literary intellectuals of our age to lose hope.

It would not be so bad if it were just the old literary intellectuals, but it includes the young ones, too. Certainly energy and fervor characterize the struggles for political correctness by college students and liberal young intellectuals, but these are often negative qualities that find fault but little promise. As a result, we may need to recognize the dead end faced when subjectivity is held as the criterion for knowledge. In the deconstructionist analysis, in which every view is strictly subjective, every opinion is by definition equally valid. Then objective truth, even as an ideal, must be abandoned.

And if one seeks truth, to abandon evidence is to abandon hope. When subjective experience is affirmed as the only basis for knowledge, it cannot be validated by external evidence, and we face life with a solipsism, the conviction that reality is an opinion and a manufactured, biased illusion. Intellectual experience becomes a nightmare of narcissism in which the mind cannot approach reality and can see only itself. Instead of an objective reality to be approached through evidence and logical inference, the mind creates reality to manifest its own ethnocentric bias. No wonder hope fades.

Bodily Engines of Experience

Thus, for curious students and for intellectuals generally, it is science or humanism; take your pick. The nerdy, analytical types that can hack math but shiver when contemplating the ambiguity of humanistic questions opt for science. Or, if forced into the real world, they choose engineering. The sociable, argumentative types who get bored with numbers and would rather entertain clever ideas than struggle with dry evidence take up their lives as literary intellectuals. Alternatively, if pushed into the real world, they end up in marketing or management. As science produces new tools that could be used for understanding the human mind, the tension between science and humanism creates not progress but a chasm. The dialectic dead-ends.

The chasm remains because neuroscientists cannot quite reach the conceptual scope required for comprehending the mind as it appears in human experience. For their part, humanists cannot quite manage the discipline and study required to comprehend the principles of information processing, brain anatomy, and physiology that allow reasoned inquiry into the physical mechanisms of experience.

Sometimes I wonder whether certain faculties of the human intellect allow us to deal with data (the objective specification of reality) and others allow us to deal with need (the subjective basis of meaning). These faculties are so fundamentally differentiated that the corresponding capacities of the mind are polar opposites and impossible to reconcile. As a result, we cannot gain real insight into the capacities of data and need—the actual stuff of the mind.

In this book I explore the hypothesis that the mind evolved from the brain's mechanisms of bodily control. In the mind, information takes form only when it is personally, biologically meaningful. At the deepest level, subcortical mechanisms of homeostasis integrate experience and behavior with the body's internal physiology. Control processes of arousal and motivation respond to ongoing needs and urges. At the surface level, networks of the cortex interface with the environment, patterning both the sensory data of perception and the motor data of actions. Between these boundaries, spanning internal need and the external interface, information is created. Experience emerges—more or less.

In this way, the evolution of information is not just a curiosity of our biological history. More complex forms of information, as well as the associated capacity for intelligence, must emerge from the struggle between internal needs and the data of reality. This occurs not just in brains in general but also in each person's developing experience.

The approach of this book is straightforward. We can learn about the nature of experience through literal, functional interpretations of brain anatomy and physiology. Once we know how the brain's networks are connected, we can draw inferences about how the mind works. The reasoning is taken directly from connectionist or artificial neural network modeling, which teaches how function follows form. The function is the cognitive ability of a network; the form is the architecture, that is, the physical pattern of connectivity.

It is necessary to study the anatomy of the brain in some detail in order to learn its connectional forms. This is the discipline required in science before evidence will give up its secrets. When we study the brain's connectional architecture, we must be looking at the structure of the mind. It may not even be important for the theory to be right if we learn to appreciate the evidence. The function we can see emerging from the brain's architecture is not just information processing as some abstract capacity; it is the organization of experience.

In the last few decades pioneering neuroanatomists have worked out the architecture of the mammalian brain and specifically the primate form. Close behind this, other researchers are giving us increasing insight into the subtle variations that make up the distinctive features of the human brain. The implications are striking. We can see the modular networks of the cortex that represent sensory systems such as vision, hearing, and smell. We can see the linked networks that achieve motor control, all the way from the general attitudes that set posture to the fine articulation of the fingers and mouth.

At the center of the brain (in the limbic system), we find integrative networks forming a base for these modular networks of bodily control. The limbic system develops a border or limbus that separates the inner core of the cortex from the primitive subcortical structures. As we review the clinical evidence, we will see that these limbic networks are essential for integrating the sensory and motor patterns with memory. As we further consider the connectivity of these limbic networks, it will become evident that their function, through the extended course of vertebrate evolution, has been visceral control. These networks regulate the sensations from and the control of the internal organs. We will find that the visceral functions are not in the least vestigial or obsolete in humans but are fully operational. Indeed, they form the basis of the emotional qualities of experience.

What does this mean? Can the networks of the human cortex be so closely identified with bodily functions? At the base of the cortical architecture are the visceral functions of the body's internal milieu; at the apex of each cortical pathway are the finely differentiated networks of the somatic sensory and motor interface with the environment. Bounding the brain's function at both internal and external interfaces are networks specialized for the representation of the body. Where is cognition?

A similar question arises when we look for the neural substrate of the executive functions. As we identify components of the brain that carry out specific tasks, such as expressive language or spatial memory, it becomes important to understand how these components are coordinated to function in the complex decision-making contexts of daily life. We look then to the issue of control and to the executive functions that must regulate the various mental components.

What we will discover is that this search for executive control has pointed not to some higher, uniquely human regions, such as uncharted domains of the frontal lobe (even though we have searched there for many decades). Rather, the search for executive control is now leading to brain circuits that we know carry out the emotional and motivational control of behavior. In fact, these are the same core-brain limbic regions that originally evolved for the visceral functions. Here again we start by assuming that parts of the brain must be reserved for higher cognitive functions. But we find only bodily functions. Where then is cognition?

The answer may be that there are no brain parts for cognition, at least not separate from the brain mechanisms pertaining to bodily functions. In our cognitive tradition we have attempted to develop a science of mental processes in the abstract, a disembodied analysis of mental faculties, now reformulated as information-processing capacities. In a peculiar paradox these objectified abstractions may be as far removed from the actual mechanisms of human experience—and the biological meaning of information—as the politically correct narcissism of the deconstructionists. We will learn that information processing in the human brain evolved through coordination between the mammalian visceral networks of the limbic cortex and the mammalian somatic sensory and motor networks of the neocortex. For humans, just as for other mammals, the cortex is bounded by the body at both the internal core and the external shell. Intelligence can emerge only from networks that are interposed between the limbic visceral control (the internal core) and the neocortical somatic control (the external shell). There is nowhere else.

Sweet Naïve Subjectivity

From the point of view of scientific theory, philosophers and neuroscientists alike are now recognizing the bodily mechanisms of the mind's operations. In this way, the anatomy of the brain makes sense. We do not need to ignore the embarrassing parts. If we assume that the brain's anatomy gives us the road map to its function, we can see how bodily control mechanisms must generate human intelligence. As British neurologist John Hughlings Jackson recognized more than a century ago, the bodily control mechanisms are the operational systems of the vertebrate brain. These were the only basis for evolving the more complex functions of human intelligence.

As a scientific exercise, the present analysis of bodily mechanisms, I am confident, is a productive way of understanding the mind. However, from the point of view of subjective self-understanding, I am not so sure. It may not be safe for the more literal readers to take the hypothesis of this book seriously. To look into experience and see the operations of animal urges and quasi-vestigial evolutionary forms may be too much. Seeing the actual mechanisms of one's own mind in such objective, clinical terms could have unexpected consequences. I worry that, perhaps for the more sensitive reader, the shock of this exercise could dislodge the fiction of naïve agency. This is a fiction of assumptions that appears invalid but is nonetheless necessary for maintaining the illusion of a coherent self.

Maybe there is a reason that generations of intellectuals have divided themselves between subjective humanism and technical science. They knew better than to mix this stuff. Maybe the chasm is therefore unavoidable. Maybe we need to cultivate a naïve, narcissistic subjectivity in order to maintain the fragile mental balance. Maybe this is why Freud's threat—to bring biological objectivity into the private sanctuary of the naïve mind—was so great.

Wielding Wetware

There must be a reason that scientists can think only technically and humanists struggle to think objectively. Now we are taught in school what we should do but not how to do it. No one teaches children how their minds work. If we began laying the groundwork for physical principles of the mind, maybe some future generation would make use of them. Bored adolescents might take them up, perhaps as a novel conceptual game or even as implements of mischief, and learn to use them to think more effectively.

We can already see that the physical principles of the mind need to manifest not only information theory but also evolutionary biology. In the brain, at the core, information capacities arise from biological functions. In the following pages we will see that full participation in making a decision (even in a complex domain of expertise) requires emotions that arise from primitive brain regions. We may need to comprehend the emotions that arise even at deep preconscious levels before we can understand how to take charge of the unfolding decision process.

At the other end of the representational hierarchy (at the shell or sensorimotor interface with the world), we will see how creativity, when it remains a latent psychological talent, is virtually worthless. Real creativity requires differentiating nascent insights into concrete actions. The articulation—in words or deeds—of the form of an idea may be necessary before a creative idea becomes not only realized but also fully conscious.

In the following pages we will search for the mind across multiple psychological levels, each one of which is embedded in a context of physical, bodily mechanisms. The principles of information that arise from this study will be bodily ones, the operational principles of the visceral and somatic divisions of the central nervous system. These will prove foreign not only to the assumptions of naïve subjectivity but also to the abstracted computer metaphors of information processing in cognitive neuroscience. When we examine the mind's neural structure, we will see how it is formed by the inherent arbitration of sensorimotor data and visceral significance. The neural basis of information is thus a framework of evolved bodily functions and the venerable representation and control systems of the vertebrate plan. To find the mind, we must look to the body. To find the meaning of information in the Information Age, we must learn how data become transformed by uncertainty and need.

Overview of the Theory

The basic idea of this book is that the neural structures of the mind—the networks of the brain—exist to construct information. They do so by creating concepts that relate internal, personal need to external, environmental reality. In its basic form, personal need is represented within the emotional and motivational networks of the visceral nervous system. These neural networks are organized at the limbic core of both the left and right cerebral hemispheres. In a basic form, the data of the world are interfaced by the sensory and motor networks of the somatic nervous system in each hemisphere's neocortex. Between these boundaries—between the internal, visceral core and the external, somatic interface with the world—the human brain constructs the information of mind through linked patterns of meaning, patterns woven across the hierarchy of each hemisphere's corticolimbic (core-to-shell) networks. Each concept is then linked across these distributed network representations through waves of meaning recursively engaging each network in turn. Meaning is thus formed by neural network patterns traversing both the visceral (personal significance) and somatic (reality interface) structures of experience. The scientific question is, how does this happen? The philosophical question is, what does it mean for the nature of experience?

This overview is guaranteed to be frustrating on the first read. It simply states the main ideas without explaining them. However, it will familiarize you with the theory that can guide the development in the later chapters. After reading those chapters, you might return to this overview to review the theoretical statements in their simple form.

The Philosophical Approach

Let us start with the implications for subjective experience. We are now learning how each fact about neural structure can be translated directly into new insights into psychological structure. An essential starting point is the realization that we have little conscious insight into the organization and workings of the mind. This is undoubtedly a pathetic state of affairs, considering the mind is all we have to cope with the slings and arrows of daily life.

Teaching psychology courses to university students, I have long faced the limits of introspection and the scientific failure to address subjectivity. Students ask straightforward questions about how the mind works and why we dream. They want to know how we can understand the problems of the people in our lives and why relationships turn out the way they do. However, the coursework in psychology deals with the objective research of psychology, not the personal meaning of what we could learn from that research. Psychology is taught as a technical, academic discipline and does not deal with the subjective question of what it means to have a mind.

Unfortunately, when psychologists do adopt a humanistic approach, they often seem to lose the capacity for rational thought altogether. In the would-be popular approaches to psychology, the arguments often degrade to personal impressions, biases, and polemics and demonstrate little concern for evidence. The two cultures schism between technical science and the humanities is apparent not only between scientific and humanistic disciplines but also squarely within the one discipline, psychology, that should integrate them.

The problem of integrating technical, factual ways of thinking with subjective, humanistic ways must be faced as we develop the neuropsychological theory of this book. The mind's visceral core provides evaluative significance in relation to personal needs; when the mind is drawn to this pole of the brain's organization, it seeks roots in the subjective, humanistic perspective. In contrast, the somatic specification of the evidence of reality is the function of the sensorimotor networks of the neocortex; being drawn to this pole frames the mind in the cognitive mode of technical, scientific specification of evidence. It is difficult to organize concepts that bridge these poles.

Maybe this is why the scientific understanding of subjective experience has proven so intractable. To achieve complex forms of mind, we must transcend the dialectical opposition of the subjective, visceral base and the objective, somatic instantiation in reality. Facing this key challenge, Western intellectual culture seems to collapse under the tension. Ideas then degrade structurally, defaulting to one pole or the other—and information is lost.

Not being a philosopher, I take a somewhat naïve approach to this problem: I simply translate from neuropsychological structure to experiential structure, as if we can read the scientific evidence as a road map to explore this territory of experience. As a psychologist, I am well aware of the limited insight people have into the functions of their minds. The opacity of introspection is a fundamental limitation. We must understand that the mind is not subjectively transparent. We have to construct our knowledge of the mind carefully and systematically through scientific analysis of the evidence. At the same time, the questions naturally held by subjective minds are general ones about how we can understand the process of experience.

Adolescent Philosophy

While teaching university students for many years, I have often tried to relate the study of psychology to the lives of young people. Through their questions and challenges, each new class of students has taught me about how fresh young minds are awakening to understand life in the culture in which they find themselves immersed.

It is the nature of the adolescent mind to be intensely self-conscious. At the same time, the young person gains a new capacity for abstract thought that is simply not possible for the child. As a result, reflection on subjective experience is a novel operation of the adolescent mind. As we move into adulthood, we are more likely to take the mind's operations for granted and to become preoccupied with the more concrete demands of trying to get our needs met.

I will therefore illustrate how the neural structures of the mind may manifest in experience through fictional examples of adolescents reflecting on their lives. These examples show how the neural structures of the mind shape both the conscious and unconscious domains of mental life. My hope is to make this neuropsychological theory meaningful not only to the open minds of intelligent young people but also to the minds of those more senior readers who find themselves still wondering about the meaning of experience.

Scientific Analysis of Neural Structures

Because I want the result to be a clear scientific theory, I address the scientific questions somewhat more systematically and formally than the philosophical questions. At the same time, I want the ideas to be accessible to college-educated readers who are willing to study the evidence. "Accessible" here does not mean easy; it means one can eventually get the point after carefully studying the evidence and arguments.

The first job, in Chapter 2, will be to examine the neural structures of the mind—the functional networks of the brain. These can be understood as organized along three primary dimensions: left/right, front/back, and inside-out (core-shell). The best-known dimensions are the lateral dimension (left and right hemispheric specialization) and the front/back dimension (with action in the frontal networks and perception in the posterior ones). The dimension that is the least understood yet the most important to the present theory is the one running from the inner core (the visceral, limbic core of each cerebral hemisphere) out to the shell (the somatic, sensorimotor shell interfacing the brain with the world).

By studying these functional dimensions of the brain's networks, we can begin to see how the mind (the brain's function) is not an indivisible whole. Rather, it can be broken down into functional, psychological systems. The primary questions for Chapter 2 are these three dimensions of the networks of cerebral hemispheres, with each hemisphere made up primarily of the cortex (layer of gray matter composed of nerve cell bodies) and the interconnecting axon fiber tracts (white matter made up of nerve connections between regions of the gray matter).

The Challenge of Vertical Integration

As the embedding context for studying the cerebral hemispheres, the most complex and theoretically challenging dimension of the human neuraxis is more fundamental than the three dimensions of the cortex described earlier. This is a neural form we share with all vertebrates. It is the vertical dimension, extending from the most recently evolved *telencephalon* (primarily the cerebral hemispheres and associated subcortical gray matter) to the stacked structures of the brain stem, each of which is the residual of a more primitive stage of vertebrate evolution. These residuals of neural systems, stacked one on top of the other, show that evolution could build more complex brains only progressively by stacking new features on top of old—and still functioning—neural architectures.

The vertical dimension of neural evolution provides an essential basis for understanding the key principle of *encephalization,* the taking over of basic functions by more recently evolved brain networks. As each more recent brain level (e.g., telencephalon) evolved, it provided more complex ways of elaborating functions (sensing, evaluating, acting) than the existing levels of neural architecture (e.g., diencephalon, mesencephalon). Within the central theoretical dimension of the core-shell organization of the cerebral hemispheres, encephalization becomes a key principle for understanding how the mechanisms of the mind have evolved by elaborating on two primitive bodily structures: internal visceral control and external somatic interface. As we work to infer the structure of the mind from the scientific evidence, this evolved, vertical structure will perhaps be the most disturbing, yet interesting, constraint that we uncover as we apply scientific insights to analyze the latent structure of subjective experience.

Machine Information

Once Chapter 2 outlines the neural structures within which the mind must be found, then we have to look for it. To explain the theory of the multileveled cortical networks of the mind (from internal core to external shell), we need to know how information can be organized in neural networks. Chapter 3 develops the basic principles of cybernetics—principles of information representation and control in distributed networks. These are the essential tools for inferring a neural theory of mind and for bringing the theory to life.

We need to know how the mind could arise from patterns of neural network architecture. In the Introduction we saw that information implies relation; data become significant only within the relational context that gives them meaning. In Chapter 2 we will study the patterns of the brain's networks—the neural connectional structures—that help explain its function. The computational principles in Chapter 3 then show how information (the stuff of the mind) can be read directly from the patterns of connections in the neural networks. Confidence that this is the case comes from modern connectionist computational models, which demonstrate that biologically meaningful concepts can be distributed across widespread networks of simple processing units (neurons).

For the theoretician, the study of distributed information representation leads to some surprising realizations. A distributed information system is inherently holistic: Once it has learned a concept, the system cannot learn a new idea without disrupting the existing one. Now that we are learning how to make the literal translation from information systems to the mind, we must consider the possibility that the limitations of distributed networks in machine intelligence may hold true for the distributed networks of human intelligence. New learning may always require sacrificing old assumptions. Information is change. Becoming informed turns out to be an act of self-transformation, and the first step is a kind of self-destruction. Learning, when the information is personally significant, requires abandoning the old self.

Animal Information

One way to test the ideas of cybernetics derived from machine learning is to apply them to animal learning. Of course, most people would define the mind as a uniquely human thing, which implies that animals therefore cannot have one.

Moreover, even if we want to admit that more complex mammals have minds, these must certainly be primitive compared to our own. Yet, in considering the challenges faced by neuropsychological theory, most of what we seek to explain about the human brain would also inform our understanding of the brains of our more complex mammalian cousins. Certainly the differences are intriguing. Just as we gain essential tools for reasoning about the mind by studying the representation of information in machine learning, we also gain another set of theoretical tools by examining the nature of information in animal learning.

What we find is that mammals hold expectations for their lives. They are cognitive creatures and use their particular forms of intelligence to predict what will happen next. Animal predictions are valued expectancies based upon and continually directed by the animals' internal need states. Starting with their motivated expectancies, mammals respond to information that signals discrepancies from their predictions in order to adapt their world models to the realities they face. This is the way they learn.

The evidence shows that mammals experience two kinds of information. One is a minimal, gradual form that arises from experience that is more or less consistent with their expectations. This kind of learning (feedforward control) maximizes stability. The second is a strong, disruptive form of information that changes their current memories more radically when the events of the world are at variance with expectancies. Learning with this discrepant form of information (feedback control) allows plasticity at the expense of stability.

What we learn from studying animal information turns out to be highly complementary to what we learn from studying machine information. Animals are expectant, predictive engines that are ready to anticipate their worlds and yet are capable of adapting rapidly when the world violates their expectations. In both modes of animal learning (feedforward and feedback), information is the vehicle of adaptive self-regulation. It arises only in the context of adaptation in a carefully modulated confluence of uncertainty and need. The theory of visceral and somatic structures of the nervous system explains the neural, physical structure of this confluence.

Conscious Information

In scientific research on animals and machines, we discover the nature of information by studying its functional role. We see information created in certain machine architectures that yield the remarkable properties of intelligent concepts. We see information created in animals through their efforts to hold properties of the world in memory as they bid for increasing mediation (and increasing intelligent agency) between stimulus and response.

But is information to be known only through the objective specification of its function? Do we see only its tracks, its causal residuals, as its substance remains fully opaque in the subjective life of the mind?

That's it, pretty much. The rational, scientific analysis of the subjective mind must recognize, first and foremost, the limited grasp of naïve introspection. We know the mind's workings largely through careful analysis of their effects, not through any simple inspection of them in action.

This realization is one of the most important, as well as frustrating, lessons of psychology. Scientific discipline leads us to realize the facile, presumptuous, even delusional assumptions we hold about conscious control of the mind. Nonetheless, the scientific analysis of the mind's informatic mechanisms is intriguing enough. We will conclude our study in Chapter 3 by questioning whether the fundamental principles of cybernetics—of information representation and control—could give us direct insight into the workings of the mind.

Can scientific analysis shed light on the subjective capacity that appears to consciousness? We are largely conscious of the products of the mind, not its process. We can know the unconscious generative mechanisms only by applying scientific inference. However, through a careful application of this inference, drawing on the principles of information in animals and machines, we may learn to reflect upon the constituent mechanisms of subjectivity with a new clarity of understanding.

By studying machines we will see that distributed representations are holistic by nature. The inevitable result of this holism is that new information (allowing plasticity and change) disrupts the prior information (the stability of existing knowledge). So it is with the subjective mind. The mind is not neatly modular, with components or faculties of specific cognitions. Rather, it is *of a piece*, such that new learning disrupts old knowledge. This is a dialectical balance in which information is not free. It requires transformation. In order to find a new and improved self, one must sacrifice one's old self.

The principles of Chapter 3 are basic ones that deal with the nature of information representation and control. Yet, even before we apply these principles to the specific architectures of mammalian neural networks, we can take them to heart. What we are learning about information from scientific research must teach us about the substance of experience—the stuff of the mind. Even if the mind's processes cannot be grasped directly but can be approached only through inference, the principles of biological information may offer new ways of participating in the subjective process. We will entertain some of these new ways of participating, methods of grasping the implicit, generative structures that yield conscious experience.

Consolidating Experience in the Linked Structures of the Mind

The principles of information representation and control in Chapter 3 will give us the theoretical tools for analyzing the neural structures of the mind. In Chapter 4 we will put these tools to work. We will use them to take apart the neural structures, these linked networks of the cerebral hemispheres, to look for clues to a fundamental question: How is memory organized adaptively? For animals, it is obvious that memory must be motivated. So it is with us as well.

Perhaps the most important insight into the mechanisms of the mind is that the formation of cognitive representations—through the consolidation of memories—is achieved by the same limbic networks that provide the motivational control of behavior. The simple implication is that all experience is motivated.

These are exciting times for psychologists. We now can examine and analyze the architecture of human memory, which extends from the networks of the limbic (visceral) core out to the networks of the neocortical (somatic) shell of the hemisphere. This anatomy shows how the visceral significance of events provides the motive engine for selecting and consolidating events in memory. Studying the consolidation process therefore becomes the central theoretical discussion of Chapter 4. We will use the principles of information representation and control to interpret the functional properties of the core-shell architectures of the cerebral hemispheres.

Because our goal is a subjective, phenomenological understanding, in this work we will confront the familiar problem of the opacity of introspection. The influence of motivational controls is not particularly apparent in conscious experience. When a mood state, such as depression, shapes your experience of the world, you have no feeling of being biased by the motivational influence. You just see the world as a negative place. As we examine the subjective perspective on the mechanisms of adaptive memory, this fundamental limitation of introspection will again present an essential starting point for building careful inferences about the latent content of experience.

Memory is organized with a visceral base, its foundation resting on internal motive controls that resonate to personally significant information. Memory is articulated in relation to the sensory and motor interfaces we construct to mirror the external world. We are aware of the sensory and motor patterns: These are the structures of conscious sensation and voluntary action. At the same time we are largely unaware of the motive processes that bias our interpretation of events: These form implicit structures of the mind that can be known largely through careful psychological (and, in the present approach, neuropsychological) analysis.

Chapter 4 therefore begins the search for the latent motive structures of the mind through a rational, psychological analysis the demonstrates that the base structures of the mind are inherently unconscious. We will examine not only the integrative networks at the limbic core of the hemisphere but also the hypothalamic and brain stem mechanisms of motivation underlying those limbic networks. Together, these limbic and subcortical networks shape the unconscious context of experience.

During this search we will examine remarkable evidence of the neural structures of the mind. It is essential that this evidence be understood clearly in both psychological and neural terms before we move on to developing a more speculative theory in later chapters. A good way to examine the theoretical development of Chapter 4 is to consider its topics in reverse order. The goal at the end will be to understand the network architecture of memory, where the functional control (and the consolidation of memory) is provided by motivational controls at the limbic core of the hemisphere. A key neuroanatomical insight is that this core-shell (somatic) architecture of the hemisphere is emergent from its continuity with the structure of the subcortical brain. This continuity is especially apparent in relation to the hypothalamic (visceral) and thalamic (somatic) divisions of the diencephalon (interbrain) at the top of the brain stem. Given this functional continuity with the brain stem substrate, it is then clear why, before studying the architecture of memory, Chapter 4 needs to review the mechanisms of motivation that control the formation of memories.

And then, because the eventual goal is to understand the body's motive influences in psychological, as well as neural, terms, we need to consider motive influences as they manifest not only objectively (in behavior) but also subjectively (in experience). Here, starting the chapter with a psychological analysis, is where we face the elusive manifestation of motivation in conscious awareness. As we saw in the question of conscious information at the close of Chapter 3, the opacity of subjective introspection to motive influences is a fundamental fact of the unconscious, bodily structure of the mind. If we are to bring a subjective perspective to the analysis of neural mechanisms of memory, this must also be the starting point for the analysis in Chapter 4.

Visceral and Somatic Engines of Experience

Once we understand both the exquisite connectional architecture representing memory and the dynamic neurophysiological process that brings it to life, we can begin to appreciate memory's motivational control. It is then time to consider possible ways in which the more complex properties of the mind could arise from these principles of embodied cybernetics. Chapter 5 organizes a theoretical approach to three fundamental issues.

First, we study in detail the physiological engines of memory consolidation, which are the neural mechanisms that weave the tapestry of meaning across the corticolimbic architecture. Memory consolidation is the mind's ever-present substrate of physiologic excitability. It should be understood in detail by anyone who thinks, all those who remember episodes of their own life, and everyone who fantasizes and wonders why.

Second, we consider the psychological context of the consolidation process, which is a context in which experience is embedded in a lifelong matrix of interpersonal relations. Because knowledge is personally contextual—that is, a function of the historical self—to participate effectively in the task of knowledge requires understanding the historical self that one brings to the occasion.

Third, we consider a novel theory of how the visceral and somatic structures could yield the abstract patterns of the mind. Abstract thought is the most psychological stuff of the mind, and we naïvely think that it must be far removed from bodily mechanisms. Yet we now have scientific ways of reasoning about how abstract thought could emerge directly from bodily mechanisms.

The psychological description of abstract concepts in Chapter 5 is a structural one. Abstract concepts are holistic, relating to a general class of things (such as mammals) without losing the differentiated meaning of their specific referents (having body hair, nursing the young). We will see first that holistic meaning is naturally opposed to differentiated specifics. It is intrinsically difficult to entertain both of these levels of meaning (generality and specifics), and yet both are captured by the abstract idea. By this point we will have seen that these two forms of psychological structure are emergent directly from the differing forms of network architecture at the core of the hemisphere (syncretic holism) and at the sensorimotor shell (differentiated specificity). The theoretical question we then face is how the holistic and differentiated structures of meaning can be combined.

To address this question we will borrow liberally from theories of psychological development that propose that conceptual integration can be *hierarchic*. The abstract concept is formed to integrate meaning at a higher level while at the same time maintaining links to the more differentiated level. To put this theory of psychological structure into neuroanatomical terms, we simply look to the architecture of the brain's networks, where we see holism at the core and specificity at the shell. It is then a simple task of the theory to formulate the patterns of arbitration between core and shell that can achieve effective information linkage and consolidate the waves of meaning across the linked corticolimbic network architecture.

In this way we will theorize about the way in which significant qualities of the mind can be explained by interpreting the capacities of the body. Although this particular theory will probably eventually prove wrong in important respects, it may illustrate an important philosophical point: Science is on the verge of explaining the mind in physical terms. We do not need to hold on to an artificial dualism and claim that the mind is one thing and the body is another. Even the more complex psychological qualities of the mind can be understood to emerge directly from the functional properties of the body's organized network structures.

Intuition at a Gut Level

For the more personal question of understanding subjective experience, this theoretical work has straightforward implications. We can recognize the holistic patterns of meaning—intuitions—that are based at the visceral core of the brain. In common language, this is experience at the gut level. A more fully developed neuropsychology could teach us about the foundations of the mind at this adaptive base. The result could be an appreciation of the mind's preconscious structures. This appreciation could then allow us to use intuition more effectively and achieve a more disciplined and refined application of the preconscious, developmental structures of thought.

Similarly, the study of the brain's somatic networks could teach us how to apply the implicit patterns of meaning at the visceral core to achieve conscious realization at the shell. We could understand the articulation of thoughts as we bring them toward action through the somatic processes of differentiation and specification. We may learn that certain forms of realization in sensorimotor networks are not limited to concrete actions and sensations but may be necessary for the structural operation of differentiating conscious experience.

According to this theory (developed in Chapter 5), we can now begin to interpret patterns in the flow of conscious experience in relation to those of information taking form in the linked corticolimbic networks. We may examine not only the mind's structure but also its generative process. By studying the physiology of consolidation—the excitable electrical dynamics of corticolimbic interaction—we may gain a better appreciation of the mind's ongoing, unconscious, generative mechanisms. We could learn not just to recognize the opposing poles of visceral and somatic influences but also to combine these influences in hierarchic patterns. Subjective experience may at times be concrete and at other times abstract. It is remarkable enough that we can now infer patterns of the brain as they manifest in the flow of experience. Even more than this, we might learn how subjective experience itself can take on abstract form.

Subjective Intelligence

The basic theory is laid out in Chapter 5, but we will spend another chapter exploring its breadth and depth. Like a good solution to any puzzle, a theory works best when it opens up new explanations we did not expect. In Chapter 6 we will see how the theory of consolidating concepts across visceral and somatic networks explains other structures of intelligence in new ways. In fact, this theory helps to explain one of the fundamental questions of neuropsychology: the psychological specialization of the left and right hemispheres. The holistic right hemisphere seems to achieve its unique conceptual structure by elaborating the dense, diffuse interconnection pattern of corebrain limbic networks. In contrast, the analytic left hemisphere may develop its skill in differentiating the elements of ideas—and then logically sequencing them—by elaborating the cognitive networks of the shell, which allow modules of meaning that remain close to the sensorimotor interface with the world.

Theorizing in this way could help us explain both cognitive and emotional aspects of hemispheric specialization: The right hemisphere's organization of nonverbal communication of emotion is not only holistic and syncretic but emotional as well. It is strongly rooted at the limbic core. The left hemisphere's contribution to rational thought is not just a product of the structural differentiation of ideas; rationality may be supported by a greater distance of left hemisphere networks from the embedding base of limbic, emotional meaning.

The theory of abstract intelligence can thus be extended to enable us to imagine hierarchic patterns not only within each hemisphere, connecting the hemisphere's core and shell networks, but also across hemispheres, linking the dual hemispheres' specialized elaborations of core and shell patterns into a coherent assembly of mind. In a similar fashion, we will use our theoretical framework to gain a new perspective on the front (motor) and back (perceptual) networks of the brain. Consolidation requires a recursive, back-and-forth processing of information (patterns weaving back into themselves) between core and shell networks. However, the primary direction of the relation between core and shell differs between the back of the brain, which gathers sense data into memory, and the front of the brain, which organizes cognition to direct actions. By understanding these patterns, we will see that the core and shell theory provides a new explanation for the different structures of attention required for comprehension (perception) and expression (action). In perception, we gain a holistic capacity for integration afforded by the shell-to-core (articulate-to-diffuse) direction of elaboration. In action, we achieve a focusing of attention as we proceed from the broad scope of the core toward selection and articulation of specific motor patterns at the somatic shell.

The structures of intelligence in brain networks are thus complex: Each one is not independent of the others but intrinsically related to them. In Chapter 6 we will see that for both the lateral (hemispheric) and front/back (action/perception) dimensions of neural architecture, the core-shell organization of corticolimbic networks gives us new insight into the adaptive base of the mind. We will find important opportunities for theoretical integration, seeing how hemispheric specialization has elaborated on different aspects of the core-shell organization of each hemisphere. The right hemisphere seems specialized for the holistic cognitive structure and the emotional responsiveness of the core-brain networks. The left hemisphere seems to articulate the more differentiated organization of the sensorimotor shell. The core-shell dialectics are then organized at a higher level, in which intrahemispheric processing elaborates one pole of the dialectic and the productive arbitration must be between the left and the right sides.

In a similar way, we will gain new insight into the front/back dimension of the brain's networks by understanding the organization of the core-shell dimension. The posterior brain's perceptual networks have corebrain holism as their target; this will explain not only important properties of perceptual attention but also the qualities of aesthetic experience. Conversely, the anterior brain's organization of motor sequences must emerge from the limbic core and become progressively articulated out toward discrete specification in the motor cortex of the shell. We will see that the integration of value with action forms a bodily basis for ethical decisions. Both perception and action thus emerge as memory operations consolidated upon a motivated base, a foundation for experience that emerges from personal, emotionally significant representations of meaning.

In this way, intelligence is subjective. Without personal meaning, there is no basis for forming memories. Subjectivity is the first approach to the depth of the mind that we will examine. The subjective, visceral organization of personal history forms the basis for the mind's operations. This is much the same perspective that Freudians gained with their depth psychology: ideas of how each of us organize a mind from the emotionally significant events of development in childhood and adult life.

The Objective Perspective

In Chapter 7 we take up a complementary perspective on the organization of the mind. This is the question of objective intelligence. Of course, the operations of most minds in everyday life are directed by egoistic and thus narrowly subjective motives. However, the most important product of core-shell consolidation is the capacity for abstract thought. As we see in Chapter 5, both the visceral (core) and somatic (shell) constraints on the mind's consolidation are inherently concrete. On the one hand are biological and personal motives, and on the other are the immediate realities of sensation and action. The interesting theoretical insight has to do with how hierarchic and abstract concepts may be generated that transcend both forms of concrete constraints (at the core and the shell). As we build upon the subjective basis of the mind in Chapter 7, we will work through an understanding of how objective intelligence could arise from these adaptive control mechanisms as well. The primary example we will consider is perhaps the most fundamental one: the abstract perspective on the self that is gained from understanding the perspectives of others.

Social intelligence may be the mind's most challenging and at the same time rewarding capacity. Although there are many domains of objective intelligence—in school, work, or other interests—it is in abstract ways of understanding social perspectives that we learn to balance the natural egocentrism that is inherent to the embodied mind. In examining the capacity for perspective-taking in social interaction, we will see how the visceral and somatic structures of the mind have become capable of organizing complex, abstract representations of the minds of other people. For the children who learn to represent social perspectives in these flexible and complex ways, the result is a more highly organized, abstract, perspective on the self. In this way, as the process of experience becomes more abstract, it becomes more objective and less egocentric. The narcissism of subjectivity can be transcended as experience develops beyond its primitive, childhood form.

Mind Evolving

In the concluding chapter I use the theory in these pages as a springboard to look toward future developments. The basic theory involves biological structures that have evolved over thousands of millennia and that generate the mind. As we look to the future, we may need to take this history of the body seriously. Only then can we project from the trends we identify in our historical analysis to imagine what happens next. Historians know that civilization has brought profound changes in the structure of human information. Only historians, knowing the momentum of our past, may be audacious enough to achieve real vision for the future that awaits us.

2
Structures of Intelligence

We begin our scientific analysis with a study of the brain's physical architecture—its structure. Structure refers to the way in which something is organized, how the parts are put together. A house has structure. We can study the organization of the walls, beams, and rafters to understand how the house will hold up to stress, such as from a storm. But the mind's structure is not readily apparent. From the naïve view—the one that we all begin with—the mind seems not to be physical, not to have real substance. We have thoughts and feelings, but nothing seems to change in the external world. The mind's qualities therefore appear elusive, even spooky.

The naïve view of the mind makes it difficult for us to make decisions about mental qualities such as intelligence, even when we face these decisions on a daily basis. A teacher must determine whether a child is able to deal with a new lesson. A supervisor on the factory floor must evaluate whether a new worker will understand the problems that appear in the work flow. With no real training in how the mind works, people must discover for themselves how to judge the qualities of intelligence. Because we are unaccustomed to grasping the mind's qualities in the abstract, we are more likely to understand its achievements in specific domains of knowledge.

An experienced heavy equipment operator, for example, has a finely honed intelligence for dealing with bulldozers. A contractor knows from observations on past jobs that this operator will be able to complete the work effectively, even when unexpected problems arise. What is it about the operator's mind that makes this happen? If asked, the contractor might say something like, "He certainly knows bulldozers" or "He's the smartest excavator I've seen." Or, more likely, in the colorful language of the trades, "He rides that Cat like it's glued to his butt."

When we recognize these structures of intelligence, what exactly do we see? Is there a quality of mind that experts have acquired? When skilled operators move the bulldozer onto a difficult piece of ground, they are making better decisions on many levels than would a novice. Does this occur because of a more refined form of subjective experience? A more organized form of consciousness?

Research on the development of expertise in many domains has suggested that an expert gains a form of structure in mental operations that provides speed and efficiency in both formulating problems and finding solutions. Importantly, the expert's mind gains a hierarchic structure: Concepts become sufficiently organized at a basic level so that they operate automatically at that level, thereby allowing the expert to pay attention at a higher level.

In contrast, the novice mind is fully occupied with the elementary problems. If we could listen in, it might sound like this: "To go left, back off on the left track and speed up the right. Oh, s#%!, I meant to push the *right* track forward."

The novice must use all of the mind's control capacities in each elementary motor skill. The expert, on the other hand, is fully unaware of the motor tasks, operating as if the machine is an extension of the body. The expert mind automatically subordinates the motor tasks in the service of more complex goals of attention, like taking a good bite of dirt, keeping a safe center of gravity, and planning where the dirt should go. The expert's consciousness seems to take on a more complex, organized structure. This structure allows a more abstract form of awareness, perhaps because it can relegate certain mental operations to the unconscious domains of the mind. When athletes, musicians, and other expert performers describe mental states of peak performance, they often say they were "unconscious." The structure of intelligence is often latent, so we see the results but not the generative mechanisms.

Even if the mechanisms become unconscious, an expert's experience with applying intelligence seems to provide the capacity for a more differentiated perception. A skilled gardener makes fine discriminations about the unique advantages and limitations of various soils. I can understand how these things can be different in the abstract, yet to me it just looks like dirt. There must be a different reality (conscious at least to a degree) appearing to this gardener, a quality to each kind of soil accruing to senses sharpened by the discipline of careful observation.

With experience, the events and facts of the world become differentiated. They can then be apprehended through more automatic categorizations, allowing more integrative concepts to emerge as the contents of attention. How is this conceptual structure achieved? Could we see how a more differentiated structure is represented in the brain's anatomy? Could we describe such a thing with principles of information structure from computational science? What does psychological theory tell us about conceptual structure?

By considering these questions, we will see how it may be possible to relate the mind's psychological structures to the brain's physical architecture. This is an architecture formed by the pattern of connections among neurons.

Intelligence and the Brain

Studying the brain's architecture leads us into the territory of neuroanatomy. This territory will seem foreign to the uninitiated. Much of its landscape has been known to humans for hundreds of years. Its place names are still the Greek and Latin terms used by early physicians and scientists. Yet the classical anatomy was crude in that it was based on superficial appearances. Accurate maps of the brain's connections are just now being worked out. As they are becoming clearer, we are starting to realize the functional implications, gaining new insights into the structures of the mind. Exactly how experience emerges from the brain's architecture is still a matter of theoretical speculation. However, we have important clues. From the anatomy of the human cortex we can see patterns in the network architecture that place constraints on the patterns of experience.

In this chapter we will study the connectional anatomy of the human cortex in order to understand the patterns of the neural architecture. These patterns offer insights into the structural basis of intelligence.

Consider Intelligence

Some of the best evidence that psychologists have provided on the structure of the mind has come from research on the measurement of intelligence. Over the years, this measurement has had a practical, clinical focus and led to the use of standardized tests of mental ability to assess a person's intelligence. The goal is usually to help with tasks such as knowing when a child needs special education or determining whether an adult will be afflicted with dementia. But the theoretical analysis has also been important in providing indications of the way intelligence is organized.

Key early results were obtained with simple tests of a child's abilities, including reasoning, vocabulary, pattern recognition, memory, and problem solving. Scores on these tests were added together to provide an intelligence quotient (IQ). Each of the tests is fairly simple, and it would be difficult to argue that any one of them defines intelligence as an abstract quality. However, when taken together, the scores have proved highly successful in predicting which children would do well in school. Even though it may not have provided much insight into the nature of intelligence, the scientific measurement of the qualities of mind thus received an important practical validation.

A novel theoretical insight was gained as researchers examined the pattern of relations among specific intelligence subtest scores. A consistent finding was that a person who did well on one subtest, such as vocabulary, was likely to do well on others, such as abstract reasoning. In terms of the theory of measurement, the implication of this finding was clear. Whereas each test taps a specific facet of intelligence, it also measures a general quality or factor that is shared with other tests. This "general intelligence factor" was inferred to be the underlying structure that causes diverse ability tests to be correlated. Although any one test was an incomplete measure, the set of tests taken together appeared to be a very reliable measure of the general factor of intelligence. By the mid-20th century, general intelligence became accepted as a quality of mind that could be indexed by an IQ test.

The theory of general intelligence thus holds that intelligence is a unitary personal quality, something a person has more or less of. Although this idea was accepted as a fact of life throughout most of the 20th century, it did not fit well with the more liberal sensitivities of the U.S. and European education systems in the later decades of the century. Rather than a single quality that would mercilessly divide people into the gifted, the normal, and the retardates, intelligence was reinterpreted by modern liberal academics as respecting diversity, with every person having a unique and equally admirable form.

In a popular culture that readily fixates on the most beautiful, the extremely talented, and the highly intelligent, it is certainly important to recognize that human abilities come in many forms. However, from a scientific perspective, the complexity and variety of the dimensions of intelligence must be evaluated on the basis of evidence. Do mental abilities cluster together as components of a single general intelligence? Or are there separate dimensions that are truly unrelated? As we make progress in brain research, do we find brain systems that support unique dimensions of mental ability? Could the structures inherent to these systems explain the appearance of multiple forms of intelligence?

Examine the Brain

What brain systems might be responsible for specific intelligences? We know that intelligence depends on a healthy brain. When a person's brain is damaged, intelligence is invariably diminished. There are, of course, many factors determining the nature of the deficit, including the degree of damage and the part of the brain that is injured. But in any community, the loss of intelligence due to head injury, stroke, or brain tumor is not only apparent to trained neuropsychologists but also becomes all too obvious to brain-injured patients themselves and their family members. The mind depends on its underlying neural tissue.

Unfortunately, many physicians, including neurologists and neurosurgeons, have only rudimentary training in the psychological capacities of the brain. They are well trained in medical or surgical methods for dealing with the brain's pathologies, but they are taught only basic rules of thumb for how brain function is related to psychological functions, such as language or personality. It is therefore not uncommon for a well-meaning physician to tell a family member that, even after a substantial brain injury, a patient has a good chance for a full recovery. Such a prediction was made by the physicians in charge when James Brady, press secretary to President Reagan, was shot in the head during the 1981 attempt to assassinate Reagan.

Brady, of course, never did fully recover. His tortured efforts to communicate and maintain emotional stability were soon accepted as graphic evidence of the need for gun control. For many who saw them, Brady's television appearances were also graphic evidence that the intelligence and personality that we usually take for granted are dependent in every waking moment on the healthy functioning of an intact brain.

How does the brain achieve intelligence? Centuries ago, early scientists and physicians cut open the heads of cadavers to examine the brain. They found that the human brain is large and extensively wrinkled. However, when they cut further to examine the brain's internal composition, they discovered it has the same mammalian architecture that is found in familiar animals such as dogs and raccoons. In fact, for our closer genetic relatives such as the large apes, it is clear that our brains are only minor variants on the same plan.

I believe that this is one of the most fundamental discoveries of science—the fact that the human brain shares the common structure of all vertebrate brains. Yet, with a few exceptions, this discovery was simply integrated as a technical fact in medicine and brain research rather than acknowledged as a discovery with important theoretical and philosophical implications. It seems that scientists and physicians alike assumed that these qualities of brain anatomy must be properties of the body, not the mind.

Yet, if we consider the anatomy and physiology carefully, the mystery of brain function is more than just those qualities—like expansive frontal lobes or asymmetric cerebral hemispheres—that appeal to our hominid pride. Rather, the real scientific mystery is how the vertebrate brain functions at all in any of its mammalian, avian, or even reptilian forms. Even if we put aside fundamental intellectual curiosity to focus on practical human problems that require a specific knowledge of the human brain, careful scientific analysis eventually demonstrates that the uniquely human qualities can only be understood as elaborations on the functional modes of the vertebrate neuraxis, the venerable neural algorithms of adaptive self-control.

Few of today's biologists emphasize the continuity in brain evolution across species. Almost as if they have a modern politically correct respect for diversity, many of today's biologists emphasize that every species has adapted to a unique environment with its own unique brain features. Although the diversity of brains across species is certainly important, the biologists of the 19th century were more impressed with the continuity of neural structures that are common across all vertebrate species. I think this continuity is also true, perhaps in a more profound sense.

The continuity of vertebrate neural evolution can be traced by examining the levels of neural organization in any mammal's brain. Although textbooks usually present the brain in the abstract, it might be useful to consider it more concretely, as the physical mechanism of an actual person's intelligence. Consider the appearance of the human brain shown in Figure 2.1, which shows the brain of Jared, an 18-year-old high school student. Jared is above average in his schoolwork and does well on tests, but his intelligence is most acutely developed in relation to the thing he is most interested in now: automobile engines. He has rebuilt the engine of his car, studied shop manuals and car enthusiast magazines, and gone to the public library to read up on the principles of engine operation and ways to increase their performance. He can afford only a few performance modifications to his car, but he chose these with knowledge of how to get the most performance increase for his money. Although Jared seldom thinks about his brain, each of his considerable intellectual skills is achieved through the physical actions of his brain tissue.

The obvious feature of Jared's brain is the large, wrinkled cortex. In this chapter we will follow the typical approach of relating the mind's functions to the structures and mechanisms of the cortex. Although the terms may seem difficult at first, they reflect a fairly simple descriptive scheme. The lobes of the brain are labeled in relation to the bones of the skull, which cover them. There

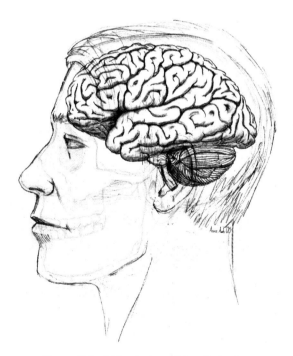

Figure 2.1. Lateral view of Jared's brain.

Structures of Intelligence

is thus no profound meaning to the division of the lobes, but it gives us a first crude mapping of the cortex. For each hemisphere (both left and right sides), there is a frontal lobe, a temporal lobe, a parietal lobe, and an occipital lobe. Within each hemisphere, the front of the brain is termed frontal or anterior, and the back is termed posterior. The medial surface of the hemisphere (Figure 2.3) is along the midline of the body, whereas the lateral surface is toward either the left or right side. The top surface is described as dorsal or superior, and the bottom surface is termed ventral or inferior.

Normally, "dorsal" describes the back surface of an animal, and "ventral" the front surface. Because humans often stand on their hind legs, the front-back axis of the brain is often perpendicular to the front-back axis of the rest of the body. However, we retain the dorsal/ventral terms and often use them to describe the superior (dorsal) and inferior (ventral) surfaces of the cerebral hemispheres.

The wrinkles of the cortex include a convex surface (called a gyrus) and a concave furrow or groove (termed a sulcus) in the crease. The central sulcus (Figures 2.2 and 2.8) is a notable landmark because it divides the motor (action control) from the somatosensory (body sensation) cortex.

As we study the complex operations of the cortex, we will find that its functions can be understood only through careful attention to the simpler subcortical circuits and structures that are common to all mammals. In architectural terms, this dependence of more complex levels on simpler levels can be described as *vertical integration*. Each higher structure of the vertebrate nervous

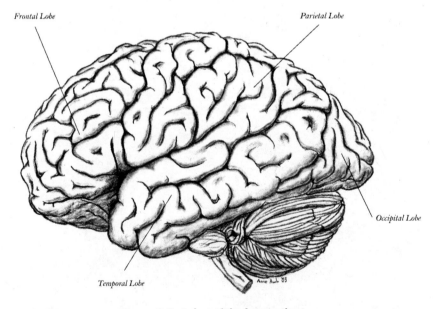

Figure 2.2. Lobes of the human brain.

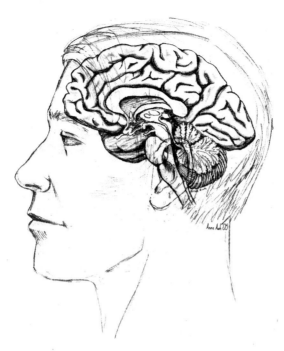

Figure 2.3. Medial (middle) view of Jared's right hemisphere, with the left hemisphere removed.

system, such as the cortex, has evolved to extend and modify the function of the lower structures, such as the brain stem and diencephalon. (In the next several chapters we will study this vertical organization in detail.) Even for this introduction to the brain's anatomy, however, it is important to examine the vertical dimension of the phylogenetic order, the levels of brain organization that have appeared successively in vertebrate evolution. Every function of the brain, from the simplest reflex to the most complex thought, is affected by this process of vertical integration.

The evolutionary order of the brain's levels starts with the divisions of the brain stem. This is the primordial brain at the top of the spinal cord. When anatomists first preserved brains in jars, the brain stem got its name because it protruded from the center of the cerebral hemispheres, like the stem of an apple or a pear. Figure 2.3 shows Jared's right hemisphere from the medial (or middle) view, with the left hemisphere removed to reveal this midline surface of the hemisphere. Figure 2.4 (which shows this same view of his right hemisphere) labels the major divisions of the neuraxis. At the lowest level of the brain stem, just above the spinal cord, is the *hindbrain* or *rhombencephalon*. This can be further divided into the medulla of the lower brain stem *(myelencephalon)* and pons and cerebellum *(metencephalon)*. Evolving from this, stacked on top of it, and subordinating its functions to some extent is the *midbrain* or *mesencephalon* of

Structures of Intelligence 37

the upper brain stem. At the top of the brain stem is the *interbrain* or *diencephalon,* composed of the *thalamus* (which is the input/output conduit and activation regulator for much of the neocortex) and the *hypothalamus* (which monitors and regulates the body's internal, visceral state). Finally, we find the *endbrain* or *telencephalon,* whose structures form the most complex circuits at the end of the neural tube. The telencephalon includes the *basal ganglia* and the limbic nuclei (*amygdala* and *hippocampus*) at the core of the cerebral hemisphere, the limbic or inner border cortex next to this core, and finally the large wrinkled cortex of the cerebral hemisphere.

The cortex of each hemisphere is made up largely of the neocortex, called "neo" or "new" in contrast to the more primitive or "old" limbic cortex at the core or inner border of the hemisphere (around the diencephalon). The limbic cortex appears to have evolved from some of the early telencephalic structures (the evolutionary origins are hotly debated by some modern neuroscientists). More specifically, the *archicortex* (anterior and posterior cingulate cortex forming the dorsal or top border of the limbic cortex) seems to have evolved from the hippocampus and its associated circuits through the septum and thalamus. This dorsal limbic circuitry was first identified as the "limbic system" by Papez in his classic studies of limbic seizures.

In addition, the *paleocortex* (comprising the perirhinal, anterior temporal, and caudal orbital frontal regions at the bottom or ventral limbic border of the

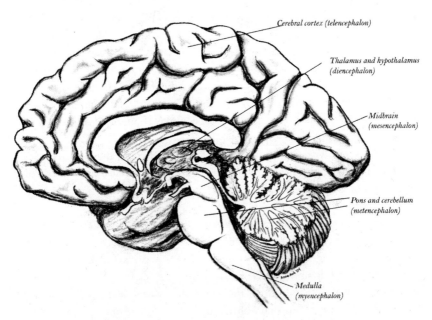

Figure 2.4. Medial view of the right hemisphere, showing the major phylogenetic divisions of the brain stem.

hemisphere) appears to have evolved from ventral limbic (amygdala and olfactory) circuits, which are closely connected to the basal ganglia.

The expansion of the mammalian cortex was a remarkable event in the evolution of general-purpose neurocognitive systems. Recognizing this and keeping with the tradition of human brain research of concentrating on the cerebral hemispheres, I focus this chapter on the networks of the cortex that provide cerebral structures of intelligence.

Still, it is essential from the outset to understand the overall evolutionary order, the fundamental blueprint of vertebrate neural architecture. The mammalian cortex is an elaboration on top of the stacked, vertical hierarchy of evolved systems of the vertebrate neuraxis (nerve pathways). The primitive structures of the brain stem evolved initially in simpler creatures, such as fish and lizards, and we therefore tend to discount them as obsolete or vestigial. Nevertheless, as we continue our studies, it will become clear that the evolved anatomy of vertebrates is an essential perspective even when considering the functions of the cerebral cortex in humans. As later chapters will demonstrate, the structures of mind taking form within this cortex owe their existence at several critical junctures to the balance of the regulatory influences—ranging from memory consolidation to motivational arousal—that are applied by subcortical systems.

Structure of Mind

Even a cursory introduction to the human brain must thus begin with the remarkable fact that the outlines of its structure parallel the history of several hundred million years of evolution. Yet how can the mind be found in this hierarchy of ancient nerve nets? We have seen that physical examination shows the brain to be an organ with a historical, hierarchic structure built upon multiple levels of evolutionary organization. But when we consider the mind, even in the most objective scientific assessment, we find nothing with levels.

Certainly when people, including philosophers and ordinary people alike, have reflected upon the mind's appearance in experience, evolutionary structure was not the first thing they noticed. In fact, when we take the time to examine the brain as the bodily organ that it is, we find things—the vestiges of primitive animal brains, the elementary visceral and motivational circuits—that barely seem appropriate for polite company. No wonder humanists, psychologists, and philosophers have been quick to skip the brain classes at college, concluding with relief that these vaguely repulsive substances of vertebrate neurophysiology are unnecessary for their studies of the nature of mind.

Au contraire, as we shall see in the chapters ahead. The evolutionary architecture of the vertebrate brain, all the way down to the brain stem, is an unavoidable fact in the scientific understanding of the mind. Indeed, it may be necessary for understanding the immediate structure of subjective experience.

Without doubt, understanding the vertical integration of the neuraxis is a daunting task. To avoid becoming overwhelmed too quickly, we can begin with the conventional approach—by considering the functional roles of the cortical networks. These are the obvious residences of mental representation, the massive cortices of the cerebral hemispheres.

From at least 200 years of systematic medical observations of the effects of brain lesions and at least 20 years of research on functional neuroimaging methods, the primary evidence we have of the psychological functions of the human brain is biased toward the cortex. This is true for the lesion (brain damage) evidence for a simple reason: Damage to subcortical structures is usually fatal, so that the specific psychological effects of partial brain damage to the subcortex cannot be easily studied. Subcortical structures include the basal ganglia, limbic nuclei, and brain stem. Although neurologists understand the function of these structures in broad outlines, psychologists have largely ignored them and have organized classical neuropsychological theory around the functions of the cortex.

Modern neuroimaging methods—positron emission tomography (PET), functional magnetic resonance imaging (fMRI)—give pictures of blood flow and metabolism. Although these imaging methods often reveal interesting patterns of subcortical activity, they show cortical activity so clearly that it has been natural for researchers to continue the tradition of lesion studies by focusing on cataloging maps of mental tasks and activities in the cortex. The result is that modern cognitive neuroscience has succumbed to the "cortical chauvinism" of the neurology and neuropsychology that preceded it.

As the mammalian cortex evolved, it elaborated and extended the simpler algorithms of the vertebrate subcortical systems. And as it did, the cortex gave rise to new, more complex functional systems. These cortical systems superseded the more rigid behavior-control circuitry of subcortical systems, which at the simplest level is reflexlike. As we have seen, this evolution of superordinate levels of representation and control, resulting in the emergence of new levels that subordinate the previous levels, is termed *encephalization.* The primate cortex is so extensively developed that many of the functions of perception and action are no longer carried out exclusively in the subcortical structures but are encephalized within the cortex.

A classic example of encephalization is the cortical control of motor function. A cat is a fairly advanced mammal, with much the same gross cerebral anatomy as humans. Yet if a cat's motor cortex is damaged in one hemisphere, the cat will show loss of movement ability (hemiplegia, on the opposite side of the body) only temporarily. After a few months of recovery, the cat's motor capacities return to apparently normal ability.

But if a human's motor cortex is damaged (by a stroke, for example), the hemiplegia may be permanent. Although the motor cortex is important to the cat, it doesn't fully supersede the subcortical motor systems. In humans it does, to the extent that the subcortical motor systems can no longer take over motor function with cortical damage.

As we examine the major systems of the human cortex, three divisions are instructive. The first is that between the left and right hemispheres. Each hemisphere includes its half of the cortex and the associated limbic circuits, basal ganglia, and thalamic nuclei. A second major division is that between input and output, between the sensory modules of the posterior brain (posterior to the postcentral gyrus in Figure 2.8) and the motor modules of the anterior brain (anterior to the precentral gyrus in Figure 2.8). A third division is not so apparent at first, but it becomes critical as soon as we consider the patterns of connectivity within each hemisphere. This is the separation between the limbic networks at the core of the brain and the neocortical networks forming the sensory and motor modules of the lateral convexities (hemispheres) of the cortex. By understanding these major divisions, we can begin to formulate a functional architecture of the mammalian cortex and examine the features that are uniquely human. In considering this architecture, we see first that the functions of the mind have structure—patterns of organization. Next we will discover how we can relate this structure to the physical structure of the cortical networks.

The Left and Right Sides of Intelligence

The body of a vertebrate animal is more or less bilaterally symmetrical. This means that the left and right halves are mirror images of each other. When the telencephalon differentiates at the end of the neural tube in embryogenesis, giving rise to the large cortical bulbs, it forms the two cerebral hemispheres, one left (Figure 2.2) and one right (Figure 2.3). As they develop in the womb, the hemispheric bulbs first bulge out, and then, as they expand still further, the hemispheres have to wrap themselves, dragging the attached limbic and basal ganglia circuits with them, around the diencephalon and brain stem. The circuits that look strange and arbitrary in the adult brain turn out to make sense if we visualize the organ as wrapping and stretching as needed to expand the cortical networks, all the while keeping the major subcortical circuits connected to maintain the essential continuity of the vertebrate plan. These are the remarkable structural transformations of the primordial neuraxis that each of us went through during the course of morphogenesis, the embryonic differentiation of form.

A remarkable event in the morphogenesis of each brain is *decussation*, the crossing of the nerve tracks from left to right, such that the left hemisphere is then connected to the right side of the body (and the right hemisphere is connected to the left side). This is one of the basic mysteries of vertebrate evolution, and it applies to sensory input, as well as motor output. Visual input, for example, from the right visual field (the area to the right of the center of Jared's visual fixation in Figure 2.5) is connected to the left hemisphere (mediated through the temporal field of the right eye and the nasal field of the left eye). As he cruises through town, Jared, of course, has no awareness of this asymmetry of his visual field. The two hemispheres share their information about visual events quite thoroughly, and his experience is of a unified perceptual space.

However, if there is damage to one hemisphere, such as from a stroke or tumor, then the deficits in sensation, action, and attention will be seen in the opposite side of the body. I once examined a woman with a slowly developing tumor in her right hemisphere. At the time I evaluated her, she had a profound attentional neglect for the left side of space. Yet she had continued driving,

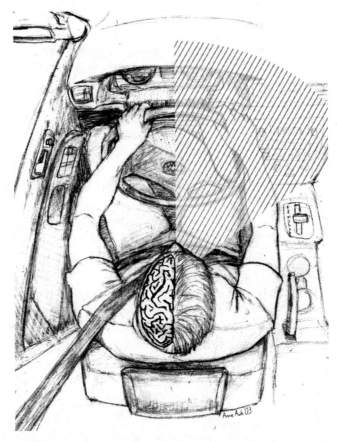

Figure 2.5. Crossing of connections from each cerebral hemisphere to the opposite side of the body. The right visual field of both eyes is projected to the left hemisphere.

even after her attention had been damaged for the entire left half of space. When I asked her how she managed this, she explained that she could not turn left very well, so she planned her driving trips to make only right turns.

Half a Mind

Each hemisphere, together with its associated brain stem circuits, includes all the functional brain parts. Once we recognize this, it becomes natural to wonder whether one hemisphere could manifest on its own the full properties of a mind. In the mid-19th century, A. L. Wigan, a physician, considered this possibility. Wigan reviewed several autopsy reports in which people with apparently normal behavior were found to have only one intact hemisphere. In one case, a man's left hemisphere, either from disease or a birth defect, was completely absent, leaving, in Wigan's terms, a "yawning chasm" on that side of the cranium. If these people could have a complete mind with one cerebral hemisphere, then what about people with two hemispheres? Could two minds emerge in the normal brain? If so, might they sometimes disagree? Intrigued with this possibility, Wigan published a book called *The Duality of the Mind*.

The questions that Wigan raised are reasonable ones and are not fully answered to this day. For many years, the relation between the hemispheres was explained by the concept of "hemispheric dominance." Based on the observation that the right hand is dominant for fine motor control (at least in the majority of people) and the left hemisphere is dominant for speech (also in most people), neurologists assumed that the left cerebral hemisphere was the true seat of the mind. It was the brain region that provides us with language. Neurologists then concluded that the right hemisphere must then be a dumb slave, carrying out elementary sensory and motor operations for the left side of the body, without presence of mind.

The Right Hemisphere Liberation Movement

The notion of left hemisphere dominance was never supported by convincing evidence, but it solved the dual brain problem. It allowed the existence of just one mind, which, from the naïve view of experience, seems to be an apparent fact that must be reconciled with the problem of dual brain organization.

Interestingly, the problem of two minds arose because of the implicit assumption that mind is brain. If two parallel brain systems were recognized and if mind is brain, then there is no getting around some duality of mind, unless, of course, one assumes that only one of these brains has the property of mind. For people with no training in psychology, language might seem to be the defining property of the human mind. It was therefore a convenient fiction for early physicians to assume that the hemisphere that contains language must be the residence of the mind and that the other one is not.

Even as Wigan was on tour with the dual brain theory, observant neurologists knew quite well that fundamental mental deficits are also caused by damage to the right hemisphere. In England in the late 19th century, John Hughlings Jackson observed that, whereas propositional reasoning (posing statements about reality in a structured, logical order) seems to require an intact left hemisphere, certain capacities (such as integrating perceptual patterns) require an intact right hemisphere.

Yet the concept of left hemisphere dominance seemingly solved the problem of dual minds. Even today physicians are reluctant to give it up. During neurosurgery, the surgeon is likely to be very careful not to cut the language centers of the left hemisphere but will be less concerned about cutting into the right hemisphere. This is in spite of the evidence that, over the months and years following brain damage, as much (if not more) social and occupational dysfunction occurs with right (rather than left) hemispheric damage.

As psychological testing developed in the 20th century, neuropsychologists realized that damage to the right hemisphere not only lowers a patient's scores on an IQ test but also does so in a revealing way. At first intelligence researchers recognized the general factor of intelligence implied by the tendency for all of the test scores to cluster together. Next, however, they identified subcomponents of intelligence, which were implied by clusters of tests that covary together. One consistent cluster represented various forms of verbal skills, suggesting a component of intelligence comprising language abilities. It was not too surprising that this verbal element was impaired by left hemisphere damage.

Also observed, however, was a separate cluster of intelligence made up of nonverbal concept-formation and problem-solving skills. This cluster was termed "performance" or nonverbal IQ. When IQ tests were applied to understand the effects of brain damage, the nonverbal performance cluster turned out to be differentially impaired by right hemisphere damage.

Clinical psychologists have often focused on the practical use of intelligence testing to help in the diagnosis and rehabilitation of brain damage, but some have recognized the theoretical implications of these results as well. Because of the underlying structure of the mind, the verbal and nonverbal factors must emerge in the quantitative analysis of intelligence scores. This structure reflects the functional differentiation—the psychological specialization—of the left and right cerebral hemispheres.

The two hemispheres must represent component subsystems of mind. If they were perfectly integrated, then there would be no tendency for their contributions to intelligence to separate into two clusters of intellectual skills. Rather, through some combination of genetic endowment and developmental process, each hemisphere seems to develop its own form of intelligence and to contribute uniquely to the general factor representing the capacity of the brain and mind as a whole.

For many years after psychological testing was applied to brain-injured patients, the bifurcation of intelligence into left and right hemispheric subsystems was a realization of specialists, primarily clinical psychologists and neurologists. The focus was mainly practical and clinical rather than theoretical or philosophical. But then in the 1960s, nearly a hundred years after Wigan, the theory of the dual brain reemerged in popular culture.

Around that time, a neurosurgeon in Los Angeles, Joseph Bogen, conducted a series of cerebral commissurotomy operations—cutting the commissures (nerve tracts) between the hemispheres. Performed on patients with severe epilepsy, the operation was intended to limit the spread of epilepsy from the affected hemisphere to the other, relatively normal one.

Bogen collaborated in evaluating these patients with a psychologist, Roger Sperry, who had conducted commissurotomy operations in monkeys. This was important scientifically because Sperry had developed experimental methods for testing the isolated hemispheres. These techniques involved separating the input to one hemisphere, such as with a visual stimulus to the opposite half of the visual space (Figure 2.5), and isolating the response from that hemisphere, such as with the hand on the opposite side of the body.

In a series of remarkable experiments, Sperry and his students were able to determine that each hemisphere of these "split-brain" patients was capable of sophisticated thinking and decision making. The presence of a more or less complete mind in each half of the brain was striking. In terms of specific results, these experiments largely confirmed what studies of brain injury had already shown about the verbal skills of the left hemisphere and the nonverbal skills of the right. What was remarkable was the capacity that each one showed for independent thought. Never before had the duality of hemispheric intelligence been so dramatically illustrated, and the phenomenon is compelling even today.

Maybe it was that these discoveries came out of California, or maybe it was that the 1960s were in full sway. Whatever the reason, many scientists and the popular press shared a fascination with the right hemisphere. The discovery of right hemisphere intelligence seemed to provide a kind of scientific validation for the creative and humanistic side of human nature. It was as if we all had been stifled by left hemisphere dominance for way too long.

A kind of right hemisphere liberation movement then swept through the popular culture, leading the two sides of the brain to become the metaphor for favorite dichotomies of human nature. The metaphor was freely applied, particularly in those disciplines that are traditionally less than rigorous intellectually: education, advertising, and New Age aromatherapies.

Of these, of course, only advertising has any financial support. Because of this, it soon became clear that the understanding of the two sides of the brain that would take root in the popular culture would not come from science, education, or even aromatherapy. Rather, most people know about hemispheric specialization from advertisements in magazines and on TV.

Because advertisers seem to have neither creativity nor shame about plagiarizing previous ad campaigns, the right hemisphere ads have gained a life in our culture and appear every few months or years hawking a new product. The format of the ad is predictable: It invites us to throw off the bonds of left hemisphere dominance and make a particular purchasing decision with the liberated right hemisphere. Or, in another variant, we are encouraged to realize how this new product appeals to both the logical and the creative sides of the mind.

Of course, from the beginning of the right hemisphere liberation movement, the more rigorous brain scientists wanted nothing to do with this loose popular thinking. In fact, with typical academic negativism, many experts were ready to downplay the significance of the discovery of the dual brains. Some supposedly knowledgeable neuroscientists went so far as to maintain that the differences between the left and right hemispheres were not really very important.

Fortunately, the technical research on brain lateralization has continued. In the later decades of the 20th century substantial progress was made in understanding hemispheric contributions not only to cognition and perception but also to emotion and personality. Yet there remained a gap between the technical scientific characterization of hemispheric differences and any general understanding of the meaning of hemispheric specialization for human experience and behavior.

What happened in this process? Are scientists incapable of more than technical thinking? Is our popular culture so vapid as to be unable to understand science? Are our insights to be limited to 30-second film clips and the glossy pages of magazine ads?

Asymmetric Mental Contents

The discovery of hemispheric specialization thus led first to some grand generalizations by the more enthusiastic scientists and journalists and then to rather unimaginative and shortsighted criticism by academic scientists. Certainly the fact that the duality of the mind resonated with many people could suggest the existence of a component structure of the mind that is related in some way to the specialized skills of the left and right hemispheres. For those who are not fully occupied by the unity of the naïve mind, this component structure might even impinge on awareness. To consider this possibility more closely, let us examine the evidence that the left and right hemispheres contribute unique cognitive subsystems.

We have seen that, in the 1950s, neuropsychological testing of brain-injured patients began with the observation that the verbal and performance IQ subscales were reflective of primary contributions from the left and right hemispheres, respectively. Many test results and experiments were consistent with the idea that the left hemisphere is important to verbal intelligence and the right hemisphere is important to nonverbal intelligence. This approach could be thought of as demonstrating hemispheric specialization for the *content* of the mind. Language is a privileged form of mental content. When thoughts are framed in words, we understand them as explicit and definite representations. These explicit representational forms invite processes of reflection and critical analysis that are not possible with more ephemeral forms of cognition, such as sensory impressions or images.

On the other hand, certain important forms of mental operation are nonverbal and are carried out in images, geometric concepts, and analog forms rather than symbolic representations. These contents of the mind are fully complete without requiring verbal mediation.

Thus, a distinction between hemispheric skills in terms of mental content—verbal versus nonverbal—provided an interesting generalization about the brain's organization of the mind. However, although we now understand that this generalization is mostly correct, in certain intriguing examples it didn't work.

In science, when a generalization fails, we see the anxiety and general hand-wringing from those who find themselves cut adrift. However, excitement and enthusiasm also occasionally occur if it becomes apparent that this new failure is a clue to a more fundamental truth. In the case of hemispheric cognition, the clues were apparent early on, and they led to intriguing insights into the neural basis of psychological structure.

The left hemisphere was found to be essential to language (in right-handers at least; left-handers require a somewhat more complex explanation). However, some patients with right hemisphere damage also show impairment in language tasks. They may have difficulty understanding the meaning of a story, even though they can answer specific questions about the specific facts that were related. They may not be able to understand a joke, even though they clearly understand all of the words and sentences. Something about the synthesis of meaning into larger concepts, the holistic organization of semantics, seems to require the right hemisphere, even when the mental contents are verbal.

In a similar way, neuropsychologists observed that, although the right hemisphere is important in nonverbal, visuospatial problems such as solving geometric puzzles or using maps, some visuospatial abilities were markedly impaired by left hemisphere damage. What could the verbal hemisphere contribute to visuospatial tasks that obviously require no language?

To explain these deficits in psychological test results, clinical observation of patients' behavior proved highly instructive. While watching brain-injured patients drawing pictures, French neurologist Henri Hecaen noticed stylistic differences between those with left and right hemisphere damage. The patients with an intact right hemisphere (but left hemisphere damage) would get the general pattern of the drawing but would ignore important details, so the drawing would look like a ghostly outline. On the other hand, patients with an intact left hemisphere (but right hemisphere damage) would get the details right but seemed to perceive them in a piecemeal fashion, such that they would draw the parts but be unable to assemble them in the correct configuration. The drawing would show recognizable elements, but they would be arrayed in a jumbled mess.

The implication of these observations was that the right hemisphere was able to synthesize the whole but was deficient in articulating the elements. In contrast, the left hemisphere was able to perceive and construct the elements but couldn't put the parts into a whole. Hecaen's insights proved highly useful in explaining the anomalous results of psychological tests and other clinical observations. Many findings soon made sense because we understood that the left hemisphere is not only verbal but also fundamentally analytic. The right hemisphere is not only spatial (or nonverbal) but also by its very nature holistic.

These fundamental structural differences explained the discrepancies of hemispheric contribution to mental content. The nonverbal tasks that required an intact left hemisphere were those that required analytic perception—breaking down the pattern into detailed elements. The verbal tasks that required an intact right hemisphere required synthetic perception—organizing the facts into a general meaning that could be understood for its gist or its humor. Whatever unity we may hope for in mental organization, the studies of hemispheric brain damage in the mid-20th century implied that the two sides of the brain contribute not only different but also opposing forms of psychological structure.

Independent Structures of Mind

The observations on hemispheric contributions to holistic and analytic cognitive structure in the 1950s were reaffirmed in several research settings in the last half of the 20th century. When Joseph Bogen, Roger Sperry, and Sperry's students examined the California split-brain patients in the 1960s, several of their observations were consistent with the differences in hemispheric contribution to mental structure that Hecaen and others had seen in the previous decade. For example, when performing a block-design task (an IQ task in which blocks with geometric shapes must be configured to create more complex geometric arrangements), the right hand (controlled exclusively by the left hemisphere in a split-brain patient) would attempt to organize elements of the design but seemed unable to fit them into the whole. The left hand (right hemisphere), on the other hand, was quickly able to assemble the correct whole.

The split-brain observations again confirmed an important distinction between the hemispheres' unique skills. They again revealed in dramatic fashion the possibility of independent function that some physicians and scientists had guessed at, but they had never been seen in such a remarkable form. For example, while skillfully performing the block-design task with the left hand, a commissurotomy patient's right hand was unable to restrain itself. It reached over to fight with the left hand in an effort to take control. Because the intrusive right hand (left hemisphere) was ineffective at this nonverbal, spatial task, the patient (or, we could assume, his right hemisphere) tried to sit on the right hand in an effort to keep it out of the way so he (it?) could complete the task without interference.

Once again, it was not the specific results from the split-brain studies that were so instructive but the general implications for human duality. Not only are holistic and analytic structures of the mind segregated in the brain, but sometimes they don't even get along.

Although not so dramatic as the initial split-brain observations, cognitive neuroscience research in the last several decades has confirmed the lateralization of holistic and analytic perceptual structure. Experimental psychology studies have shown that, when people first examine a scene, they tend to apprehend it globally. Then, oriented to the holistic form, they attend to the local features and analyze the discrete elements.

The degree of attention to global versus local features can be assessed with materials specially constructed to be recognizable at one level or the other but not both. An example is a large letter *H* made up of small *s* letters. When the composite letter is briefly shown, subjects usually see the *H* and may not even notice the little *s*'s. Several findings have suggested that this "global precedence" in perception requires an intact right hemisphere and that the left hemisphere, when left to its own devices, is drawn to the local features. This research suggests not only that there is a division in the brain/mind between modes of structural organization but also that there is an order to how these modes are applied, with global apprehension first and then local analysis.

The Meaning of Emotion

A patient with damage to the right hemisphere may have intact language function (because the left hemisphere is intact) but still have difficulty communicating. Several neurologists and clinical psychologists have observed that patients with right hemisphere damage seemed to have particular difficulty in understanding emotional communication. Systematic research then showed that these patients are specifically impaired in both understanding and expressing emotion. The implication seemed to be that, paralleling the left hemisphere's processing of verbal information, the right hemisphere is specialized for processing the nonverbal information of the speaker's emotional tone of voice, facial expression, and even bodily posture.

Research with normal subjects confirmed this finding. When nonverbal emotional information, such as that in a person's tone of voice or facial expression, is presented with experimental procedures that maximize its input to one hemisphere or the other, the information that is presented to the right hemisphere results in better emotional comprehension than that presented to the left (at least in right-handers; left-handers are not exactly opposite but often have more complex patterns of lateralization).

Even in the expression of emotion, the right hemisphere may show greater range or intensity of emotion than the left. When photographs of facial expression are divided down the middle and mirror images are added to each side to make whole-face composites out of each half, the composite made from the left half of the face (in right-handers) is typically judged as more emotionally expressive.

The research on emotional communication has raised important questions about how the intelligence of each hemisphere of the brain is applied to the cognition and experience of everyday life. Among the many that remain unanswered, there seem to be implications at many levels—biological, psychological, and even philosophical. Biologically, why does one cerebral hemisphere—the highest structure of the vertebrate brain—become specialized for emotional communication? We have always thought that emotion is a more primitive function, emerging from subcortical centers. Psychologically, what does it mean that emotion develops within the hemisphere that is *less* competent in language? Are language and emotion incompatible? Philosophically, does this specialization within the brain imply that there are separable and possibly incompatible domains of experience? In our everyday life, does one domain of the mind draw conclusions about logical content, while another forms impressions of emotional significance?

Mapping Mental to Neural Structure

An important clue to the physical basis of hemispheric intelligence has been structural organization. Language is able to parse meaning into components, and this may be related to the left hemisphere's analytic skills. Emotional experience is syncretic—with multiple elements fused into holistic constructs—and this form of experience may be related to the right hemisphere's synthetic skills. If each hemisphere applies a certain organization to perceptual information, how does this come about? It is one thing to map functions of the mind onto brain parts, but is there something about the organization of the neurons in each hemisphere that can explain its unique psychological functions? Something that can teach us how mind emerges from tissue?

An interesting clue came from what might be thought of as one of the first failures of brain-mapping research. In the 1960s Josephine Semmes studied the way in which brain lesions result in motor (action) and somatosensory (skin sensation) deficits. The idea was to uncover the map of the body that is laid out on the somatosensory and motor regions of the cortex. For the left hemisphere, the mapping went fairly well because a focal lesion (such as from a gunshot wound in World War II veterans) could be correlated with a localized loss of sensory or motor function on the right side of the body.

However, when looking at the data for the right hemisphere, Semmes found it was much more difficult to map localized function to a focal brain site. Although clearly regulating input and output for the left side of the body, the right hemisphere seemed to be more diffusely organized, without the mapping of specific functions to focal areas seen in the left hemisphere.

Rather than considering this just a failure of mapping, Semmes questioned whether the anomalous finding about the right hemisphere might present an opportunity for theoretical insight. Examining similar questions of mapping for spatial attention and motor strength, Semmes built a case for a general principle of hemispheric asymmetry in the mapping of function to the cortex. She proposed that a specific area in the left hemisphere is specialized for *like* elements (things that are similar or convergent), such as the input from a specific area of the body surface. Although this organization is not found in the right hemisphere, Semmes did not consider the right hemisphere as simply disorganized or diffuse. Rather, she proposed that a given region of the right hemisphere might be specialized for the integration of *unlike* elements (things that represent divergent functions). Figure 2.6 presents a schematic of how connections among the hemispheric regions might be organized in the way Semmes suggested.

Semmes's theory influenced many later theoretical efforts (including my own) to understand brain lateralization. It has proven to be a pioneering model for reasoning from mental capacity to brain tissue. With enhanced spatial accuracy due to modern neuroimaging and electrophysiological methods, we can now conduct tests of the Semmes hypothesis that are more specific than when the predominant measures were deficits due to brain injuries. Although new tests must now be conducted with the high-resolution measures, even

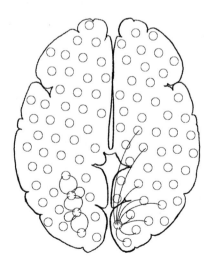

Figure 2.6. Schematic of differing left and right hemisphere architectures that might be consistent with Semmes's theory. The view is of a horizontal slice across the cerebral hemispheres. Each location within the left hemisphere (on the left of this image) is connected locally with just a few neighboring regions. Each location within the right hemisphere is connected more diffusely with both distant and adjacent regions.

before all of the evidence is in, we can see why Semmes's ideas have been so important for theoreticians. Her work showed how the structure of the mind (the patterning of information representation) may arise from the physical structure of neural networks.

Regardless of whether this specific account is correct, it seems clear from a variety of sources of evidence that the left and right hemispheres contribute differing forms of perceptual—and, we can infer, conceptual—structure. Whether tied directly to neural architecture as the Semmes hypothesis suggests or emergent from hemispheric cortical systems in ways yet to be revealed, human cognition appears to have an implicit structure that is assembled from alternate biological substrates.

Is this structure more than implicit? Can we experience holistic or analytic forms of information structure manifested as unique constraints within the fabric of consciousness?

Many times it seems that using the mind is automatic, so that we may be able to reflect on its products (the conclusions, images, or propositions that are delivered up) but not the process (the mechanisms of the delivery). We often know only through difficult inferences when we have failed to see the forest because of the trees. At other times, however, the conceptual structure itself seems to become a tangible presence in consciousness. For example, you may be trying to understand a financial transaction, such as buying a car. You think about the decision, yet you can't seem to get a general picture that implies whether to buy it or not. So you patiently analyze the information. You then sense that the application of your sequential, analytic reasoning is paying off— even before you know the answer. At that moment there is an awareness of the mental process, so that continuing to hold information in an analytic structure is a conscious choice made on the basis of monitoring your own mental process.

As a different example, consider those times when a frustrating problem has seemed insoluble despite repeated efforts to deal with it. Then suddenly, as if by accident, a novel understanding appears. Rather than the awareness of sequential, analytical reasoning described earlier, you seem to have no awareness of the underlying process. The familiar information just snaps into a new, vibrant relief, standing out against a broader context. It is as if you glimpse a holistic vista of the forest that immediately explains why the trees are in these particular positions.

We can consider an example in the life of Jared, whose brain we have examined in detail in the preceding figures. It is Friday morning before school, and his car refuses to start. Just last night he replaced the starter and put the battery on the charger, so everything should work. Jared now applies his analytic skills to the task. He breaks down the starter function into subcomponents, including the starter motor, the solenoid that engages the driving wheel, and the Bendix assembly that pivots the driving wheel into the teeth of the engine flywheel. The analytic skills of his left hemisphere allow him to order the causal sequence required to understand how each of the components must operate in sequential order to turn the engine over. Within this rational appreciation of the causal roles of each component, Jared can recall and test each element of his installation job.

When he convinces himself it is fully installed, but it still doesn't work, Jared maintains his analytic focus of attention. He again parses the problem into the sequential components. Could the starter motor be defective? Wouldn't the solenoid make noise if it worked at all? Didn't he make all of the electrical connections to the starter? In his effortful reasoning, Jared struggles with a methodical tenacity that usually allows him to complete a problem-solving task when others give up in frustration.

But not this time. Failing to find any explanation, Jared finally gives up. He knows he often gets fixated on a problem, exerting more and more focused effort but staying stuck on a rigid approach that has already proven ineffective. As he shifts his mental attitude, he steps back from the car and looks out the garage door. It's a beautiful summer morning and it's still cool out. Suddenly it occurs to him that he forgot to reconnect the ground cable to the battery. He feels stupid for this careless oversight.

More important, however, Jared gains insight into his mental process. He quickly recognizes that he did the right thing to give up and to shift his perspective away from struggling with the problem, thereby letting go of his strongly focused attention. He knows he always applies this focused attention with very good concentration, and it almost always produces results. However, sometimes this intense focus keeps him from taking in the big picture.

In this experience Jared gained a little more skill in auto mechanics. The more important lesson, though, may be what he learned from observing his own mental process. If we are honest, we must admit that insights into mental process are rare, fleeting, and not particularly reliable. It is undoubtedly hopeful to think that, without appropriate training and instruction, naïve consciousness is adequate to grasp the structural patterns in the ebb and flow of ideas. However, like Jared, we can, through occasional lucid reflection, find clues to the internal structure of intelligence. Even if we cannot apprehend it directly, we can sometimes see the effects of the mental organization taking form in the preconscious mind. We can then guess how this organization generates the latent structure of conscious experience.

Structure and Process in Experience

Insight into one aspect of the mind could open up possibilities for understanding others. When mental representations are ordered with a certain structure, we can reason systematically about the implications for cognitive process—how information is handled over time.

One suggestion for the way mental structure constrains mental process comes from theoretical physiology. Now, it may seem that physiology is so well understood in modern science and medicine that it needs no theory. Nevertheless, although many specific facts are known about physiology, it is still a challenge to understand how physiological processes are organized into subsystems, such as organ or endocrine systems, and how these contribute to the integrated functioning of the whole organism. Many theorists are now working with various approaches to dynamical systems theory, which models complex—and often seemingly unpredictable—interactions among system components that manage to achieve stability at the organismic (whole organism) level.

A foundation for today's dynamical systems theories was laid in the middle of the last century by Karl Ludwig von Bertalanffy with his general systems theory of biological development. Von Bertalanffy formulated basic principles for the way in which the structure of a system (a set of elements that interact) imposes constraints on the processes that occur within it. A closed system contains all of the interactions of the elements, while an open system includes interactions with things outside the system. In both cases, the system exists because the interactions are sufficiently packaged to separate it from the rest of the world. Von Bertalanffy observed, as had many biologists before, that biological systems begin their development with a diffuse and holistic structure. An example is the cellular organization of an embryo in the first hours after conception. Development then proceeds toward an increasingly differentiated structure. For example, the limbs, sensory systems, and organs of the embryo differentiate as separate elements of the organism.

This principle seems obvious enough that one might wonder whether it needs the special status of a formal theory. However, von Bertalanffy went on to observe that predictable modes of process follow from these forms of organizational structure. In the early holistic organization of systems, the elements exist in "dynamic interaction," with fluid, labile, and irregular transactions among the elements. In the later, more differentiated structures, the interactions become more systematic and routinized in repeatable and stable patterns, such as the well-known homeostatic mechanisms of organ systems that must become stable before the mature fetus is ready for life outside the womb.

Even before von Bertalanffy, theory-minded developmental psychologists of the early 20th century were impressed with the parallel between the developmental courses of biological and cognitive systems. Heinz Werner proposed that the mind emerges through an embryonic differentiation. The child's mind begins with a holistic, diffuse order and gradually differentiates into more articulated and systematized conceptual organization. But what about the holistic and analytic forms of psychological structure on the two sides of the brain? Do these engage differing modes of cognitive process?

Because the left and right hemispheres are operational at the same time, the holistic and analytic forms of structure might seem to coexist rather than representing successive stages of thought as implied by general systems theory. However, some observations on hemispheric contributions to learning by neuropsychologists have suggested the possibility of a developmental order to hemispheric influences on cognition that is reminiscent of von Bertalanffy's theory.

Around 1980, Elkhanon Goldberg and Louis Costa collaborated on theories of brain development in order to understand cognitive development in children. They reasoned from the left hemisphere's importance to well-rehearsed, organized, and routinized forms of thinking to theorize that the left hemisphere may be important to later stages of problem solving, once the issues and concepts have been well formulated. In contrast, they speculated that the right hemisphere may be important to early stages of learning and problem solving—before the structured patterns exist. It is as if each hemisphere captures a certain stage of development, so that each new process in the cognitive development of an idea or problem can be started fresh.

In formulating this theory, Goldberg and Costa drew from their neuropsychological studies of children's learning abilities and how these abilities are impaired by brain lesions. They considered the way in which learning proceeds over time, such as with a new skill like reading. However, if their reasoning is correct, we might expect a general trend toward reliance on right hemisphere learning mechanisms early in childhood, compared to increasing development of more organized left hemisphere operations with increasing maturity. Several observations of children's test performance have suggested that an increase in disciplined, formal thinking occurs in the period from 5 to 7 years of age, perhaps reflecting a maturation of the left hemisphere that prepares the child for the increasingly formal language and mathematical tasks of early schooling.

There is also some evidence of a strong reliance on the right hemisphere even in infancy, as the child is building early cognitive and interpersonal skills. Then, around the second year, a shift seems to occur toward greater reliance on the left hemisphere as the child achieves the articulatory movement and cognitive skills of walking and talking.

It is unlikely, of course, that this progression is absolute. From birth, both halves of the brain are functional and developing in parallel. But there may be a maturational progression that emphasizes right hemisphere control of emotional communication and perceptual development in the first year, which is replaced by the more organized and routinized cognition of the left hemisphere as the child develops the finely patterned motor skills of walking and talking in the second year. If this progression is accurate, it may represent the most extreme hemispheric asymmetry of life, in which there is a strong dominance of one half of the brain in ongoing intelligence. After these first two years, the integration may be more continual, with only transient episodes of dominance of one mode of structuring concepts over the other.

Although the developmental implications of hemispheric contributions are important for understanding childhood, they may also be significant for the short-term developmental processes engaged in learning a new skill or solving a problem even in adulthood. Conceptual structure may always constrain mental process. The holistic organizational skills of the right hemisphere may be essential to the operations of intelligence when a problem is new, confusing, or not well articulated. This may help explain the common assumption that the right hemisphere is integral to human creativity, a flexibility of mind that is particularly important when familiar and well-routinized ways of thinking are inadequate, such as when facing the challenge of new information.

It is easy (and it may be misleading) to idealize creativity as a human talent in the abstract. The productivity of effective people in any field depends on supplementing creativity with substantial training and discipline. In the long run, the popular generalizations about the right hemisphere and creativity may not say enough. The important realization offered by the research on hemispheric contributions to problem solving may be that certain qualities of the cognitive process, such as rapid and fluid interactions among ideas, may be inherent to certain qualities of cognitive structure, such as holistic and global forms of mental representation. The two hemispheres' differing contributions to cognitive structure are the best-documented characteristics of psychological function that are found to be lateralized in the brain. If we can develop valid theoretical models of the inherent relation of cognitive structure to cognitive process (i.e., the way information is organized over time), then we may find general principles of cognition that will help explain human intelligence in many specific domains.

Could these principles also inform subjective experience? Can we appreciate differing patterns of conceptual structure in the immediate operations of consciousness? Is it possible to adopt a holistic frame of conceptualization and then learn to use the process of intelligence that is inherent to that form of structure? Perhaps can we develop ideas and insights through understanding the general systems theory of biological development.

Even though holistic and analytic forms of cognitive structure may seem familiar and meaningful to issues of everyday cognition, they may not be under voluntary control in the simple way we often think about it. In fact, I am not certain that psychological structure, in the way we can infer it from evidence on hemispheric specialization, is readily apparent in the conscious experience of most people. After all, we have had many millennia of human thought and reflection. We have also had civilization and a written literature for the last several of these. Yet the holistic and analytic forms of cognitive structure were never identified in anything more than rough analogies. Only when evidence on the effects of damage to each hemisphere was examined scientifically did we discover these lateralized structures of human intelligence. Scientific analysis of evidence and rational inferences closely following from this evidence were necessary before we could infer, indirectly, the differing hemispheric representations of analytic and holistic mental structure. We found precious little of the mind's structure from introspection.

Over the last 50 years, scientific psychologists have become skeptical of the capacities of conscious introspection, with good reason. People act as if they have considerable self-knowledge and ready access to the contents and operations of the mind, but when tested for accuracy, introspection often proves an unreliable advisor.

Thus, it is at least possible that the alternate forms of psychological structure can be known abstractly and inferentially but not through direct conscious reflection. Certainly in his first years of school Jared rarely reflected on the increasingly analytic, modular, and automatic structure composing his cognition as he completed his reading and spelling assignments. He was certainly aware of the content of his concepts, the words he recognized, the correct patterns of letters that were increasingly familiar, and the meaning of the words. He experienced the conscious results of self-regulation: frustration when his skills were inadequate; satisfaction when his intelligence enabled him to accomplish the task at hand; and excitement when he sensed a new challenge was within his grasp. Still, for the most part, the mechanisms of the mind were taking form in a preconscious domain that was essential to Jared's experience but is known only by the impressions left in his awareness and the patterns that took shape in his behavior.

The interesting question is, what would happen if we gave a smart kid like Jared a meaningful education in neuropsychology (not as an academic topic but as a set of principles for understanding the workings of his mind)? Maybe then the structures taking on different forms in the two hemispheres of the brain could become understood as practical companions to experience.

For example, let us say you take a few moments to consider the events of the past week. At first you might simply try to remember some of the things that happened and to organize them in some way, as images of a calendar or scenes of where you have been. You find there is an important effect of applying analytic structure to your reflections. Things can be separated and delineated more clearly. You may analyze events in relation to what things were important to your family, what activities were more fun, or what events contributed the most to your career goals. Subjective qualities accrue to analytic thought, which tends to provide distance from events—objectivity that supports controlled, conscious manipulation of thoughts about the incidents.

In contrast, if you reflect on the past week in a holistic manner, the process may be quite different. First, since the structure of time is less definite than with an analytic approach, you might find it more difficult to determine exactly where the week ends in your imagination. However, you might find that once you achieve a holistic concept of the week, you can then keep track of various events better than if you analyzed only the specific constituent elements.

Furthermore, the emergent qualities of a holistic idea are sometimes unpredictable. The week may suddenly remind you of another one 10 years ago, or you may notice that you feel optimistic about this week's events for a reason that you can't trace to any single cause.

Typically, you will see that you naturally shift from the holistic mode to the analytic. Even if it were possible to emphasize one hemisphere's intelligence for a brief interval, it seems unlikely that we could stay fully asymmetric for long. As soon as you hold a holistic attitude, you find yourself recognizing something, such as the fact that last week seems to have gone by very quickly, and then you shift into an analytic cognitive operation, such as trying to understand what could have kept you so busy you didn't notice the time.

Although the two forms of mind trade and blend in ephemeral flashes, a few moments of study shows that each form of mental structure yields a unique order to subjective experience. For evolutionary reasons that remain unclear, whether by advantage or accident, the analytic and holistic structures of intelligence have taken up residence in different brain halves. This remarkable fact should invite not only scientific research but also exercises in efforts to train and engage one mode or the other. With training, we might expect more deliberate organizations of the mental process. With experience, we might expect more conscious realizations of both creativity and discipline.

Neural Modes of Perception and Action

Each cerebral hemisphere has all of the requisite neural subsystems of the whole brain, including the sensory, motor, emotional, and cognitive processes. Paradoxically, because of this completeness, the study of hemispheric differences gives only limited insight into the component mechanisms of the mind. We can see the specialized information-processing modes on the two sides of the brain. However, precisely because each hemisphere has all of the components of a brain (the thalamic, limbic, basal ganglia, and cortical networks), we find within each hemisphere a fairly complete brain. The component mechanisms—the parts that are hidden under the surface of consciousness—remain latent and obscured within these fully integrated cognitive units.

A more fundamental dimension of the mammalian brain does allow an effective decomposition of the whole function of the mind in order to gain insight into its workings. This is the input/output dimension, the separation of brain mechanisms into perceptual systems (those that take in information) and motor systems (those that create actions). One might think this would be the first of the brain's dimensions to be characterized by scientific research, yet even today this dimension remains poorly understood.

One reason may be that most psychologists think of sensory and motor mechanisms of the brain as "mere physiology." With the mentalism that has prevailed since well before Descartes, most psychologists assume, apparently without thinking clearly, that the mind and its cognitive mechanisms must be something more than body stuff. In fact, most psychologists have an immediate negative reaction to any proposal that a complex psychological operation (such as decision making) can be better understood by studying a sensory or motor mechanism of the brain (such as the networks that monitor actions). "That's just motor function," they say.

Considering the mental domain privileged over the bodily is a long-standing bias of psychologists and philosophers, but it is not inevitable. More than a hundred years ago, John Hughlings Jackson contemplated the relations of mental abilities to brain networks in patients with brain disorders and concluded that the separation of mental functions from sensorimotor processes made no sense. The brain evolved to regulate the motivational control of actions that are carried out by the motor system and guided by sensory evaluation of ongoing environmental events. There are no faculties of memory, conscious perception, or music appreciation that float in the mental ether, separate from the bodily functions. If we accept that the mind comes from brain, then our behavior and experience must be conceived of as elaborations of primordial systems for perceiving, evaluating, and acting. When we study the brain to look for the networks controlling cognition, we find that all of the networks that have been implicated in cognition are linked in one way or another to sensory, motor, or motivational systems. There are no brain parts for disembodied cognition.

Input and Output Architectures

As early brain researchers worked to understand the brain's control of behavior, they experimented with animals, mainly dogs, cats, and rats. Using a combination of lesion methods (causing damage) and stimulation methods (with electric current), they mapped the functions of the cortex. They found that areas in the posterior (back) regions of the brain were specialized for the senses: vision, hearing, and touch (smell and taste are deeper in the brain, near the motivational centers). When the auditory cortex was stimulated, for example, a dog might respond as if hearing something. If this area were lesioned, the dog would become impaired in auditory discriminations. The researchers also found areas toward the anterior (front) regions of the brain that are important to controlling actions. Stimulation of one part of the motor cortex would cause a leg to move, whereas stimulating another part might cause the dog to move its head and eyes abruptly to one side. This separation of motor function in the frontal cortex from sensory function in the posterior cortex is clearly delineated along the central sulcus in each hemisphere (between the pre- and postcentral gyri in Figure 2.8). The central sulcus divides the closely connected motor cortex from the sensory cortex. Both of these areas of cortex are organized in body maps, with the motor and somatosensory maps closely linked to each other through fibers that cross under the central sulcus (Figure 2.7).

With electrical stimulation as the primary guide in these early studies, researchers observed that some areas of the cortex were "silent," with no apparent effects from stimulation. Examining the anatomical connections among brain areas, they discovered that the silent areas of the posterior brain were often connected with more than one sensory cortex. This led to the concept of "association areas," regions of cortex that associated or integrated the primary sensory areas.

In the frontal lobe, the association areas were in front of the primary motor cortex or motor strip in the precentral gyrus (Figure 2.8). These frontal association areas were found to receive some input from posterior sensory association areas, but they are also heavily connected with the motivational centers at the core of the brain. To the early anatomists and neurologists, these connections suggested that the frontal association networks are able to integrate sensory data with motivational controls in order to plan and organize actions that meet the animal's needs for self-maintenance and reproduction.

Thus the input/output architecture of the mammalian cortex is organized with a fundamental division between the front (action) and the back (reception). For both the anterior and posterior cortical networks, the connections imply a kind of hierarchical pattern in which "higher" integration occurs first by combining input from the primary sensory cortices and then by organizing patterns that are fed to the primary motor cortex. Although it is difficult to separate the function of any one network from the others, each network must play a unique role in the operation of the brain and mind. In these patterns of anatomy we find intriguing clues to the structure of intelligence.

One clue is that damage to the parietal lobe in the posterior human brain (Figures 2.2 and 2.8) impairs the integration of visual information, particularly for visual attention to the opposite half of space. This integration may result from the convergence of multiple sensory modalities (channels) in the posterior brain, creating a holistic representation of space.

With a similar consistency of mental function with neural connectivity, damage to the frontal association areas often leads to deficits in planning and inhibitory control. Without inhibitory control, behavior becomes rash, crude, and impulsive. These behavioral deficits reflect the importance of the connectivity of the frontal lobe with both the memory functions and the motivational functions of the corebrain limbic regions. The massive frontal networks of the human brain normally allow for psychological (and computational) complexity in the motivational control of actions. Given the capacity of corticolimbic systems in memory representation, when this capacity is integrated with motor control we may find the development of the capacity for extended planning of actions before they are initiated.

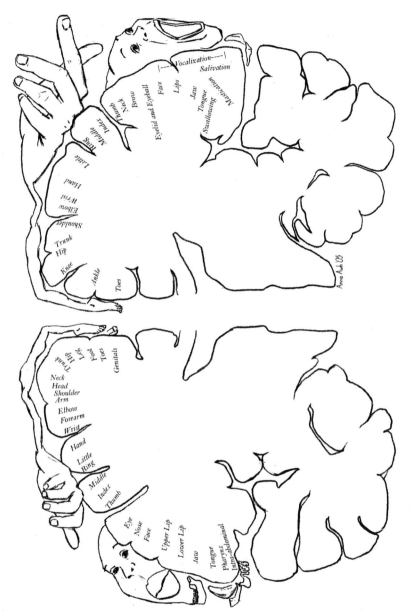

Figure 2.7. Homunculi of somatosensory (*left*) and somatomotor (*right*) maps.

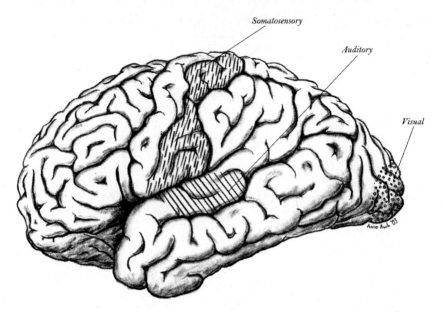

Figure 2.8. Visual, somatosensory, and auditory sensory cortices. The somatosensory cortex comprises the postcentral gyrus; anterior to this (toward the left in this figure) is the precentral gyrus, which contains the primary motor area.

Moreover, in no brain areas do we find areas of cortex dedicated to naked mental faculties. This realization might have seemed controversial to psychologists even a decade ago, but it is becoming increasingly apparent with results from neuroimaging, in which cognition activates sensory and motor regions even when no sensations or actions are required. Remarkably, this realization was already apparent to Hughlings Jackson more than a century ago when he observed that, when seen in the context of the brain from which it must emerge, the mind is but a great sensorimotor machine.

What led to this sensory and motor architecture of the cortex? Are there some requirements for information processing in the control of actions that are fundamentally different from those required for perception? A brief study of the transformations of input and output controls in the evolution of the vertebrate brain may give insight into these questions. The mind is achieved by a neural machine that must weave its most complex and abstract tapestries on the loom of sensorimotor networks. Yet it is also true that the evolution of mind has occurred by transcending the constraints of sensory and motor circuits and by interposing progressive degrees of complexity and delay between input and output. Along the way, it seems, human consciousness has emerged. Conscious experience appeared as a kind of self-conscious interloper, lingering in the indefinite delay between sensation and action.

Slow Reflexes

To understand this evolutionary progression, it is useful to consider the self-regulatory mechanisms of simpler brains. A general principle in the evolution of neural control systems is that simpler organisms respond with specific reflexes. A significant stimulus is linked by a reflex arc to an appropriate response. In contrast, more complex and highly evolved organisms show increasingly elaborate delays between stimulus and response. It is within the memory system interposed in this delay—and in the motivational controls that organize and direct the memory operations—that we find the increasing sophistication in the evolution of mammalian—and eventually human—intelligence.

The architecture of sensory and motor networks is critical to understanding this delay and the memory systems mediating it. The primordial architecture of the vertebrate nervous system can be seen in the vertebrae (the spinal cord), and the separation of sensory and motor circuits is already apparent in the form it will maintain throughout the neuraxis (the hierarchy of levels from spinal cord to forebrain). Figure 4.1 (in Chapter 4), which shows a cross-section of the human spinal cord, illustrates the sensory pathways in the dorsal (back) section and the motor pathways in the ventral (front) section.

Actually, even at the spinal level we must differentiate between the *somatic* sensory and motor pathways, which regulate the sensation and control of the body surface, and the *visceral* sensory and motor pathways (Figures 4.1 and 4.2), which regulate the sensation and control of the body's internal organs. The somatic pathways control not only the sense of the skin but also the special senses of vision and audition, as well as the striated skeletal muscles that allow both gross and fine movements of the body in relation to the external world. The visceral pathways not only sense the state of the internal organs but also act to regulate that state through the smooth visceral muscles of the internal organs. This separation of somatic and visceral functions is maintained to higher levels of the neuraxis, including the cortex. Figure 2.9 shows the representation of visceral functions in the limbic cortex, forming the core of the cerebral hemisphere.

Although we typically think of the somatic structures in relation to sensorimotor functions, it is easy to see that the visceral structures are equally important in the life of an organism. Through their interrelations with internal need states and their intrinsic homeostatic (internal control) mechanisms, the visceral sensorimotor systems provide motivational and emotional regulation of the somatic sensorimotor systems. Although the separation of dorsal sensory and ventral motor divisions becomes more complex at higher levels of the nervous system, it remains a fundamental organizing principle. Similarly, the delineation between visceral and somatic functions at the spinal level remains a key principle of nervous system function at the higher levels of the neuraxis.

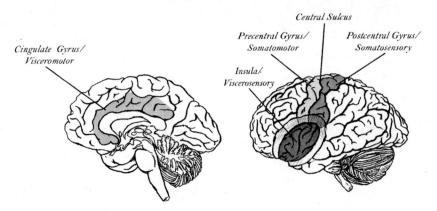

Figure 2.9. Medial view of the right hemisphere, showing the visceral cortex around the limbic core. Visceromotor functions are distributed along the dorsal (or superior) border (in the cingulate cortex), and viscerosensory functions are distributed along the ventral (or inferior) border (in the insular cortex).

Mind in the Gap

For simple vertebrates, not much intervenes in the reflex arc between stimulus and response. Even in humans, certain reflex arcs, such as that between touching a hot stove and pulling the hand away, are mediated in the spinal cord, linking with a few synapses (nerve connections) an effective connection between dorsal sensory input and ventral motor output. As each level of the nervous system evolved, it could not replace the previous level, which had to remain operational to maintain the survival and reproduction of each generation. Rather, each new level of the brain had to elaborate the themes of the old levels as it formed at the anterior (head) end of the neural tube. The neural tube is a major organizing form in embryogenesis, and it remains in the adult vertebrate brain in the form of the spinal cord. At the level of the brain stem, the basic organization of the spinal cord is maintained: The ventral circuits organize motor activity, and the dorsal circuits mediate sensory activity. Between these input and output functions are more complex integrating networks to mediate between them.

When this dorsal/ventral, input/output organization was elaborated into the cerebral hemispheres of the telencephalon, the plan became substantially more complex, as we saw earlier in this chapter. Furthermore, because our ancestors rose on their hind legs to get a better view of things, our brains are canted by 90 degrees, such that what was ventral is now anterior and what was dorsal is now posterior. However, the separation between sensory and motor networks was maintained. For the somatic system, this separation formed the major networks of the anterior (motor) and posterior (sensory) cortical systems (Figure 2.8). For the visceral system, the separation of sensory and motor functions was also maintained, but it remained within the dorsal (upper) and ventral (lower) divisions of the limbic system (Figure 2.9).

Within this input/output plan, the evolution of increasingly complex cortical networks allowed ever more complex and extended information structures to mediate the reflex arc. They are neural structures interloping between input and output. Rather than a fixed stimulus-response determination of the animal's behavior, the evolution of the mammalian forebrain resulted in memory systems that consolidate the information from sensory systems and organize it into patterns that capture its significance for the organism's survival. At the same time, these memory systems allow actions to be delayed when possible in order to formulate more effective—and often more effectively motivated—action plans.

In this process, both sensory and motor systems of the brain evolved complex and unique control systems that became integral to what we now call cognition. Some of these contributed to the formation of long-term memory. In this way they organized behavior in relation to motivational influences developed within the animal's historical experience. Other control systems were directed to the more immediate representation of information in what is often called working memory, carrying out the cybernetics of paying attention. Mediating these multiple layers of memory, the delay between stimulus and response opened up new horizons of behavioral and experiential complexity.

As they spanned immediate attention and long-term memory, the control processes shaping mental representations seem to have evolved differently within the sensory and motor networks of the mammalian neocortex. Understanding the differences may help us comprehend the biological subsystems of the mind. Although these information subsystems may not be obvious to casual examination of naïve experience, they nonetheless offer unique contributions to the structure of intelligence and to the subjective life of the mind.

Parallel Channels of Information Flow

In sensory networks, multiple streams of information may be represented in parallel. You can build a mental concept of your current surroundings that includes the sight of many things. This can be effective because, even though each visual fixation tends to focus on one object, you have a buffer of visual working memory that can organize the positions of many objects in a realistic, three-dimensional map of your surround.

Your current world map can readily incorporate the array of environmental sounds that arise from various vibrations within the local atmosphere. Even though you may need a focused auditory analysis—listening carefully—to identify an unusual sound, your auditory memory buffer naturally assembles the array of multiple current sources into a common soundscape. Although this "neuronal model," as Pavlov called it, quickly fades into the unconscious, it remains an implicit representation and an integral companion to present awareness.

We can discuss the organization of sensory experience through a thought experiment in which we imagine the experiences that a brain like Jared's might encounter on a typical afternoon. As Jared leaves his last morning class to have lunch at the diner on a busy street a few blocks away, his brain forms a neuronal model of his current surroundings. Even though he is not attentive right now to the sounds of the traffic, he would notice immediately if they stopped. His orienting response to an unexpected absence within the auditory scene would be triggered by a violation of the neuronal model that would arise from the discrepancy between his unconscious expectations and the world that actually appears to his senses.

Given the sensitivities of his primate brain, Jared is especially responsive to sight and sound. Yet his neuronal model also incorporates tactile sensations (the sense of things touching his body). Once he takes his place at the lunch counter, he quickly becomes unaware of the feel of the stool he is sitting on— that is, until his circulation causes trouble. Then, in a small episode of the homeostatic self-regulation of daily life, the developing pain of the impaired circulation becomes an organizing, motivating force. Jared is suddenly attentive to the feel of the stool, and he is briefly but effectively motivated to act to adjust his tactile sensation so that the pain is no longer a dominant element of his awareness.

In the assembly of the elements of concrete experience, another crucial experiential element is smell. Olfaction may be essential to the holistic experience of a place, even though it is usually a subtle component in the inventory of awareness. We may not notice how a certain smell colors the quality of a time or place until an unexpected encounter at some future time brings back a vivid memory of a unique episode of personal history. In the diner, the cook dumps raw onions on the hot grill. In a brief reflective moment, this smell takes Jared back to his childhood. He remembers a visit to his grandparents when he was twelve and his grandmother cooked something with onions, filling the house with a unique and defining odor. The nature of holistic memory is that it is organized through the self. So, as Jared mentally revisits his grandparents' home, he revisits not only the holistic experience of that time but also himself at the age of twelve. The nature of smell is that it pervades both the emotional responses and the holistic experience of a place and time. This kind of experience can be called *syncretic* because it fuses many elements into a single, undifferentiated concept. The result of accessing a syncretic memory is that one reminder brings back a complete episode of self and world. Jared remembers his grandmother's voice calling him to dinner and reflects on what it was like being a near-teen.

Although the sense of smell seems uniquely powerful in embedding both emotion and memory, the architecture of the sensory brain in general is well suited to creating larger wholes. Because information can be maintained in parallel, the concepts formed from sensory data tend to be configural, that is, shaped into patterns that integrate the entire sensory surround.

Within this embedding context, it is difficult to focus attention on an individual element. This is because normal perception naturally gravitates toward holistic integration, with all of the sensory streams able to receive some degree of mental representation at the same time. Holistic integration is a tangible quality of experience made possible by the parallel form of information structure emergent in the architecture of the brain's sensory processing systems.

A Serial Channel of Information Assembly

In motor processing, there are different constraints on information flow. The necessary process is not parallel but linear, such that actions can then be assembled effectively in a sequence in time. Even there, to be sure, a certain degree of parallel processing may be possible. Multiple input streams may be called on to influence a developing action. The act of shifting on the stool in the diner, for example, begins with Jared's sensation of discomfort. Although this sensation may not be conscious to any degree, it recruits an appropriate motivation for corrective action. The motivation organizes an implicit set of assessments from Jared's senses of touch, muscle placement, and balance as he prepares to shift. These can be activated more or less simultaneously and can converge in Jared's attention in parallel.

When, however, he carries out the action, it must be orchestrated linearly, as a sequence in time. Whatever parallel information channels are engaged in the preparation for the action, the action itself has a necessary serial order as Jared first tenses the muscles of his torso in anticipation of the new gravity configuration, then plants a foot for traction, puts weight on an elbow on the counter, then finally shifts laterally to bring the much-needed blood supply to his neglected and now painful left butt. Although some operations can be launched in parallel, it is the nature of motor programming that some actions can only follow other actions. The effect of this physical constraint is that most motor elements must be assembled in proper order to achieve a workable behavioral sequence. Because cognition is embodied and evolved from the sensorimotor substrate of mammalian intelligence, the subsystems of cognition that evolved from motor representations have retained many of the requirements for sequential assembly.

Nowhere is this more true than in language. Drawing on the well-developed sequencing skills of the motor system, human speech builds a serial, linear stream of meaning upon the scaffold of grammar. We can organize propositional logic, with meaning that is definitively specified, because we can sequence the ideas. Each idea builds in logical fashion on another, creating a cumulative structure of meaning out of a well-crafted serial order.

Cognitive Structure and Attention

The parallel versus serial forms of information flow within sensory and motor systems seem to have generated unique forms of attentional control in the evolution of cognition. As we saw earlier, an important clue to the parallel form of attention is that damage to the association cortex of the posterior brain (in the parietal lobe) may cause a "neglect" syndrome, in which the patient simply fails to notice things in the opposite half of space (see Figure 2.5). This seems to reflect damage to that mechanism of attention that normally updates the contents of conscious awareness when changes occur in the perceptual environment. An elementary form of this perceptual-based attention control is the orienting response, in which a change in the environment (like a noise outside your window) causes you to interrupt what you're doing, increase your arousal slightly, and orient your receptors (eyes and ears) toward the event.

Apparently, when the posterior cortex is damaged, this orienting mechanism fails to update consciousness for the associated (opposite) half of space, so that the corresponding awareness fades. I have watched a patient with a stroke in the right hemisphere eat all of the food on the right half of his plate, the half he could attend to effectively (with intact attentional control from the undamaged left hemisphere). He then looked up as if to ask for more. When I pointed out there was still plenty of food on the left side of the plate (the side that was neglected), he was surprised and glad to eat it, as if he simply had not noticed it. Through observing such deficits, we must infer that gathering experience from the world, even though it is not always conscious, is not entirely automatic. Rather, it requires physical mechanisms, neural systems for directing attention in space to integrate the information from the sensory channels. These are mechanisms of mind.

The orienting response seems integral to this gathering of information from parallel channels in the posterior brain. Through multiple brief orienting responses, various events in the perceptual environment may be attended briefly, building up a holistic internal model of the current environmental surround. In computational terms, we can describe this internal model of the environment as a *parallel distributed representation*. In experiential terms, this model is your ongoing, integrated awareness of the world of the moment.

Because each sensory system can call up orienting mechanisms in response to changes in its input stream, the posterior brain operates well in parallel. The net result is a holistic—and continually updated—model of the environmental context. Certainly there are conflicts, as when you become so captivated by a visual event (sparrows courageously mobbing a big crow outside your window) that you fail to attend to a significant auditory event (someone calling you who is getting irritated that you are not paying attention). But even with conflicts, the parallel architecture works well. Each sensory system makes its own bids for attention by creating a simple and effective way to regulate the traffic in the parallel cognitive architecture formed by the converging sensory channels of the posterior brain.

The control of action, however, requires a more focused, linear form of attentional control, one that suits the serial structure of action sequences. Various information streams must be coordinated to focus on the single sequence of activity required by an action plan. A central control process is therefore necessary.

Cognitive psychologist Alan Allport has pondered this problem and theorized that a common workspace may exist in the motor system of the anterior (frontal) brain that allows a single vector or direction of action to emerge from multiple inputs. Allport points out that a central workspace for motor plans and for the cognition depending on the frontal brain would be consistent with the finding from many experiments that cognition has "limited-capacity" resources. A limited capacity can be inferred when a person who is attending sufficiently to one task (such as trying to make a point in a conversation on a cell phone) has insufficient attention to give to another parallel task (such as changing lanes on the freeway). To the extent that the mind retains some residual executive decision-making capacity (which can never be taken for granted), the decision (avoiding a wreck versus making a point in the conversation) can be made to allocate the limited cognitive resources.

For most of us, the limited capacity of the human mind is not something that needs discovering. What may be a useful discovery, however, is the unique forms of attention that have evolved in concert with the information architecture of the brain. The arrangement of the sensory systems of the posterior brain, with specialized and relatively independent networks for sight, sound, and touch, leads to an inherently parallel information architecture in which each sensory channel feeds in its data, its chunk of experience, independently. The convergence of the sensory streams within this architecture leads to a dynamic and syncretic awareness of the current environmental context.

In contrast, the requirement for serial organization of the elements of a motor sequence leads to a more linear and restrictive form of attention in the anterior brain. In this more active form of attention, requiring psychological properties of conscious intentionality, the neural information control mechanisms must select only the dominant representation that emerges as the essential next step of the developing action plan.

Perception and Action in Experience

Thus, as human cognition evolved to extend the sensorimotor capacities of the mammalian brain, it emerged with an organizational form that continues to reflect the patterns of its ancient roots in the vertebrate neuraxis. As with hemispheric specialization, we can see how structure implies process. The organizational form of attention frames the ongoing information flow in thought and perhaps in subjective experience as well.

In philosophy, the study of the concrete appearance of experience is called *phenomenology*. Philosophers who have taken up this approach have argued that it may be instructive to examine the raw experience of reality without filtering it though preconceived interpretations. This approach is an important one. In fact I sometimes think the topic of this book could be called neurophenomenology.

Of course, we don't have conscious access to reality in a direct sense. We only have access to the mental (and neural) processes of our senses. The early German students of perception, the Gestalt psychologists, recognized this fact years ago with their principle of *isomorphism*. When they work properly, the mind's perceptual structures provide data that are parallel to (but not identical with) external reality. As we come to understand the forms of attention and cognition that arise from the sensory and motor systems, we can recognize that we are separated from reality not only by the channels of our sensory systems but also by the biases and inadequacies of our cognitive elaborations of those systems.

The senses are essential components of experience, of course, and there do seem to be differences among people in the extent to which they are sensitive to a given sensory modality. This sensitivity may be important to the way that some people can derive information particularly well from a sensory modality, such as sound. Musicians often seem to be especially sensitive to the qualities of sound. Whether learned or innate, this sensitivity may lead to the development of auditory intelligence and provide a capacity for fine, critical auditory discriminations. From this basis, motor patterns can be closely aligned with the sensory discriminations.

The implications for experience may not be limited to the first-order patterns in the sensory domain. There may be more complex qualities of consciousness that we can understand in relation to the parallel architecture of the perceptual brain. When each sense has its own integral attentional controls, there is no inherent need to restrict attention to one sense at a time. The contents of attention (what we can also call working memory or awareness and its immediate preconscious neighbors) can therefore be expanded to multiple sensory channels simultaneously. This allows a holistic, syncretic form of experience in which the fusion of multiple sensory features leads to synergistic qualities that are richer and more powerful than could be gained from each sensory element taken individually.

For example, as I walked along a path near my house one December afternoon, I was impressed at how dark it was that time of year in Oregon. Noticing this dim late afternoon daylight and the overcast sky, I was reminded of other December afternoons. The quality of the light was certainly important to this impression, but my experience was not just visual. Integral to the experience and the memory it called up were several sensory elements associated with walking through the outdoors: the feel of balance, muscle sense, and exertion as I moved over the climbs and descents of the path; the sounds of my steps on the path; the wind through the Douglas firs and madrone trees; the cool spray of the misty rain on my face; and the musty smell of the forest's decay that is always strong in the fall and winter rains here.

These elements are individuated, of course, only by focused attention, as I now reflect on the walk in order to write about it. At the time, they were of a piece, woven within one experiential whole. They were fully convergent because of the parallel architecture of the neural networks that funneled them into a single syncretic representation of this little episode of self-in-world. This is the "common sense" of the medieval philosophers, the central perceptual space.

For the motor system, the holistic nature of integrated mental representations is not an opportunity but a problem. Organizing coherent behavior requires taking the common, holistic representation of current awareness, integrating it with the appropriate motivational drive, and translating the result into a single course of action. In this process, multiple inputs must converge on a single output. The inputs include fairly direct ones from sensory association areas (for example, in visual guidance of a reaching movement), as well as more fully processed representations from limbic and association cortices (which can combine memory and motivational direction with the sensory guidance). These various control influences must be negotiated so that they guide a narrow motor integration process. By focusing and coordinating the multiple influences, the result is a single coherent course of action rather than a number of fragmentary impulses. The demands on organizational ability are considerable and require clearly focused cognitive and attentional capacity in order to achieve intelligent action.

For example, while on my walk I decided to leave the path in order to explore a stretch of hillside that looked as if it might lead to another ridge. The walking was fairly easy, and I ambled along through light undergrowth as I worried about how to write this section on the limited capacity of attentional resources in motor control. As I worried more and my ideas weren't getting any clearer, I found myself stopped in my tracks, blocked on three sides by thick bunches of alders.

With an aptness that I am afraid was lost on the alders, I had succeeded in demonstrating through my personal limitations what I could not explain in my writing. Traversing the hillside required only moderate attention, yet with my limited cognitive capacity fully occupied by the problems of writing, not enough mind was left for the serial, sequential task of finding my way. I became, this time literally as well as figuratively, lost in the woods.

Attention, Effort, and the Experience of Intelligence

The unique requirements for attention that are imposed by action planning of the frontal lobes has led to the proposal of an "anterior attention system" by one of my colleagues at the University of Oregon, Mike Posner. In studying brain activation measures (positron emission tomography, or PET) during a variety of cognitive tasks, Posner has observed that limbic networks of the frontal lobe, including the anterior cingulate cortex, are regularly engaged when subjects must exert effort to pay attention to certain elements of cognitive tasks.

These results seem to be consistent with the realization that cognition, even in complex forms, emerges from the sensory and motor systems of the brain. As it does, the cognition must elaborate on the control mechanisms that regulate those systems. Whereas the "posterior attention system" can operate in a reactive, responsive mode consistent with the orienting responses that can be triggered by each sensory system, the anterior attention system, which controls the motor mechanisms of the frontal lobes, has the task of selecting a coherent course of action. Because of this, the anterior attention system must represent values that prioritize choices. It must choose some actions while suppressing others. It must recognize and deal with conflicting choices. And all the while it must hold consciousness steady on the task at hand. This is hard work.

In fact, it seems that when the structure of intelligence requires linear and sequential information processing (i.e., when it must prepare for action), that is when maintaining attention requires the most effort. Active thought is difficult. Yet during certain phases of productive thought, little conscious mental result may show for the effort.

For an illustration, consider any task that requires planning. In business, for example, effective planning is a scarce resource. Most people can function once they know what they are supposed to do. Some people can even generate creative ideas for what could be done. However, very few can tolerate the extended struggle of planning a course of action: considering the conflicting options and opportunities, evaluating the likelihood of success in each possible outcome, and then charting a sequence of actions that will optimize both the chance of success and the gathering of information to guide future choices. This kind of planning is so hard it hurts. Maybe this is why many businesses look up from the distractions of daily events to find themselves lost in the alders.

Although there is of course great complexity to real-world problems of planning, the need for selecting a single action plan from an array of options engages an architecture of reasoning that retains its fundamental roots in the architecture of motor control. In the elaboration of decision making from action regulation, we may find consistent clues to how architectural structure implies an information process. Multiple information streams must be evaluated to allow a single behavioral choice. The structure of intelligent choice requires a restriction from many options to one decision, while paying appropriate attention to the value of each option. If the options are not evaluated completely and thoroughly, the decision will be impulsive and suboptimal. If the structure is not restricted and the choices are not narrowed, the decision may be delayed until it is ineffective. The motor attention system operates with limited capacity not simply because of too little memory but also because this cognitive system has evolved to converge on a single action.

Our awareness of the process of intelligence may itself be constrained by the structure of its physical, bodily implementation. Because the sensory systems are suited to funneling multiple information streams into a common experiential buffer, consciousness at any moment in time can span a broad perceptual array. Because the information flow must be restricted to select coherent serial actions, when consciousness is occupied by planning, it may require highly focused attention and experience.

Sometimes I wonder whether these structural realities have led to a tendency to define consciousness—and perhaps even cognition—in close relation to the sensory systems. At the same time this tendency may cause us to fail to recognize the intricacy and difficulty of action planning. Because sensory representations are salient and readily available to conscious inspection, it may be natural to think of the mind in relation to its sensory constructions. Certainly this was a dominant theme in British empiricist philosophy, which proposed that an understanding of experience should begin not with assumptions about idealized mental forms but with simple inspection of concrete experience. This approach has remained the dominant one as cognitive psychologists have applied experimental methods to their empirical analysis of mental processes, in which the mental operations of memory and attention are examined much more often in relation to sensory than to motor processes.

At the same time, our limited conscious access to motor planning has been an important realization of modern psychology. The belief that behavior is controlled by conscious intention has been a cornerstone of assumptions about human nature—from philosophy to religion to law. Yet an objective analysis of behavior, even a brief and informal one, shows that the belief in intentional control rests on shaky grounds.

William James once conducted an objective analysis of his own capacities in willpower. He did so while lying in bed one winter morning. Because it was the 19th century, James, who lay warm and comfortable under blankets in a room without central heating, faced early-morning demands on the volitional control of behavior that few of us now appreciate.

He formed a clear intention to get out of bed. Being a disciplined scientific observer of human behavior and understanding the philosophical importance of the doctrine of free will, James was sincerely impressed when nothing happened. Giving himself reasons, such as the lectures that needed preparing, and forming an even more vigorous and vivid intention to get up, James continued to observe his body lying passive, inert and warm under the covers, while his mind exerted its best effort of voluntary will.

Suddenly his ruminations on the mind were interrupted by the realization that he had promised to meet the dean at the university that morning. James's next conscious reflection found himself out of bed and fully engaged in the process of getting dressed and ready for the day.

The activity that had been so much the focus of his conscious intention was now liberated, not by an act of will but by the more effective intervention of a spontaneous, motivated, but effectively unconscious impulse.

James's scientific introspection provided a degree of objectivity on his own intentionality that seems not to occur to most of us. In many examples of modern psychological research, subjects assume they have conscious access to their intentions and motives, and, perhaps not unlike certain philosophers, they assume they consciously guide their choices and actions. Yet objective psychological analyses demonstrate that these assumptions are usually optimistic at best. With only the naïve view of the mind, people make up answers as if they should have privileged access to causal intentions when in fact the real actions of daily life can arise only from the preconscious void.

Although there are many factors to consider in understanding even a simple behavior, we may gain important insight into the limitations of human planning and decision making by studying the information flow of the motor system. With the restriction of the information structure required for action choices, mental representations must be restricted as well. These narrow representations then form the necessary substrate for awareness. The cybernetic result seems to be a high threshold in the motor systems of the frontal brain for mental representations to achieve consciousness. The experiential result is that many of our actions and ideas spring to life in a way that we can only admire—or regret—as observers of our own actions.

Lateralization of Sensory and Motor Systems

We can surmise, then, that sensory consciousness is easy to construct, responsive as it is to the information flux of the external surround. It is therefore easy to savor for both instruction and enjoyment. Motor consciousness, on the other hand, is inherently difficult to achieve, it must be assembled with effort so that it realizes motivational evaluations, and it is by its very nature limited when it does appear.

As I have developed the implications of the differing forms of representation inherent in sensory and motor systems, you may have noticed that a theme of parallel versus sequential control was also found for right versus left hemisphere specialization. This may reflect the fact that the dimensions of brain organization are not perfectly "crossed" or independent of each other. Rather, hemispheric specialization seems to have elaborated differently on the anterior and posterior attention systems. This differential elaboration might be a clue to how each hemisphere developed its unique cognitive mode. The right hemisphere's parallel and holistic cognitive skills and its ability to integrate spatial information may have developed by elaborating on the parallel attentional skills of the posterior brain. This possibility was in fact anticipated by John Hughlings Jackson, who observed that damage to the right hemisphere leads to a general deficit of "imperception."

On the other hand, the left hemisphere's capacity for sequential control and even its skills in analyzing events may stem from its elaboration of the attentional skills of the motor system of the anterior brain. Certainly clinical neuropsychological observation has shown that apraxia, the deficit of skilled action control, is more severe with left hemisphere lesions. In addition, it is a curious fact of human nature that most people show a dominance of the right hand for fine motor control. Left hemisphere specialization for praxis could explain this fact and would imply that the left hemisphere's cognitive specialization has elaborated on the uniquely focused and sequenced attention of the motor networks of the anterior brain.

Any differential specialization of this sort is naturally relative. There are certainly sensory systems in the posterior left hemisphere and motor systems in the anterior right hemisphere. In fact, proposing a theory of differential lateral specialization of this sort begs the question of how the left hemisphere, if it is specialized for sequential motor control, integrates the anterior and posterior brains differently from the right hemisphere. An important principle to keep in mind is that, whether it is in the evolution of a species or the development of a child's brain, specialization along one dimension (left/right) must be integrated with corresponding specialization in other dimensions (in this case, anterior/posterior).

Ethical and Aesthetic Modes of Experience

By the middle of the 20th century, neuropsychologists and psychologically trained neurologists had observed that damage to the frontal lobe could leave a person with intact sensory systems but with impaired higher control over motivational impulses. This seemed to suggest that a kind of pathological extraverted personality had been created by the unfortunate brain lesion. If so, then the theoretical implication was that the sensory systems of the posterior brain must normally tend toward an extraverted mode of attention.

Interestingly, frontal lesions often result in release of a grasp reflex, in which the patient will reach for and grab an object when presented with it.

To some observers, this idea of specialization of the posterior brain for external responsiveness was consistent with the lack of environmental responsiveness (the neglect syndrome) caused by damage to the posterior cortex. To balance this bias toward external responsiveness, the frontal cortical networks would have to apply an opposing bias toward introversion or the internal control of behavior. Perhaps relevant to such a general orientation is the release of the "avoidance reflex," which may occur with lesions of the posterior cortex (thus leaving the brain dominated by the frontal lobes). In this subcortical reflex, when the patient's fingers are touched, the patient withdraws the hand or stretches out the fingers as if pushing the stimulus away.

Although offering an important integrative framework for understanding the brain, these efforts to build a general theoretical characterization of anterior and posterior brain systems never became popular. As a result, they were never fully tested either by experiments or clinical study. The evidence of opposing anterior and posterior biases did, however, make a lasting impression on at least one neurosurgeon. Karl Pribram gave up his neurosurgical practice in the mid-20th century to study the psychology of the brain. To this day he continues to be one of the leading theoreticians of neuropsychology.

Pribram's studies of the brain led him to the anatomical finding that the frontal lobes are connected not only to the motor cortex but also to the limbic networks that regulate the motivational and emotional processes. This connectional evidence prompted him to reason that the executive functions of the frontal lobe—making plans, delaying gratification, inhibiting impulses—must rely on an elaboration of motivational input from the limbic networks. Taken together with the obvious role of the frontal networks in motor control, Pribram theorized, the frontal lobe's contribution to intelligence must be to bring values into decisions on behavioral choices. He speculated that this would lead to an ethical mode of intelligence, in which reasoning in relation to values and difficult choices must dominate experience.

A different bias takes form in the networks of the posterior brain, according to Pribram's analysis. Here again the connections of the sensory networks to the limbic regions of the brain formed an anatomical basis for Pribram's neuropsychological theory. With attention specialized for organizing converging inputs into a common world model, the posterior brain must be responsive to the motivational significance of the sensory data. Pribram suggested that this attentional orientation forms the basis for an aesthetic mode of intelligence, in which experience is guided by hedonic sensitivity to the sensory information flow.

For *aesthetic* experience, Pribram proposes that the motives are close to the sensory systems, taking in the hedonic qualities of the sensory process. The values of this mode are closely attuned to the receptive and holistic attention of the sensory brain.

For *ethical* experience, Pribram suggests that motives operate to guide the decisions on action choices. The values of this mode must be constrained by their functions, so that, rather than a receptive openness to sensory experience operating in the posterior brain, the values of the anterior brain must compete to motivate a limited set of action choices.

One neuroscientist derogatorily described Pribram's ideas as "neurophilosophy," meaning that his notions are too speculative and "squishy" to qualify as real science. But for me, this would be a compliment. Pribram's ideas do qualify as neurophilosophy, and I find them deeply thought provoking. Pribram speculates about the brain's mechanisms in a way that can inform us about the meaning of experience.

Often I have found that there is an opportunity for appreciating a simple pleasure of life, such as walking through the university campus on a spring day, but my mind is so engaged with anxiety over some current problem that I am operating on less than 10% of my hedonic bandwidth. Truly beautiful events and qualities of the world are readily accessible in the dynamic flux of reality, but they simply do not register in my awareness.

It may be useful to know about the nature of attention in the posterior brain and the responsiveness of orienting mechanisms in order to open sensory channels to savor their qualities. Interestingly, it often seems that a reflective mental process requires a certain degree of relaxation; the anxiety of worrying about things engages not only a focused mode of attention but the motor system and muscle tension as spontaneous side effects as well. Apparently because of the mechanistic link of anxiety with the motor system of the anterior brain, relaxing muscle tension serves to decrease anxiety and allow a broader, more aesthetic mode of awareness.

In a similar fashion, anxiety and tension may be integral to the effortful attention that emerges from the motor system. An anxious attitude, in fact, may often be essential for getting things done. An effortful task (such as when I was writing a difficult section of this book) frequently seems to bring out anxiety and frustration. For me and maybe for others in similar situations, working effectively requires tolerating uncomfortable and even escalating levels of anxiety and frustration. Often I find that I seek out distraction under those circumstances in order to avoid the pain of the anxiety brought on by the struggle. The result is that I do not get the job done.

In this section we have seen how Pribram's aesthetic versus ethical modes may help explain the different forms of experience that emerge from the sensory versus motor systems of the brain. Pribram's theorizing may offer both philosophical and practical insights. By understanding the sensory and motor mechanisms of mental processes, we can see their effects in subjective experience. We can even imagine taking hold of them more effectively.

Structures Around the Limbic Core

The third major dimension of cortical structure we will consider in this chapter might be called a *radial* dimension of the brain because it goes from the center out to the surface of each hemisphere. It could also be called a *core-shell* dimension in that it shows the organization that extends from the inner core of each hemisphere out to the sensory and motor modules on the hemisphere's outer surface (the shell). Each cerebral hemisphere can be thought of as half of a cantaloupe, shrunken until it is heavily wrinkled. The opening of the half-sphere is drawn in at the midline of the head and wrapped around the inner core of the brain stem. This opening of the hemisphere forms a border or limbus around the brain stem within the medial (middle) wall of the hemisphere (Figure 2.4). This inner core of the hemisphere is connected to the brain stem by circuits that make up the *limbic system.*

Levels of Corticolimbic Representation

In the back of the brain, the sensory modules for vision, hearing, and touch are connected to the association areas of the cortex. These allow the association or integration of the senses. These regions are then connected to the limbic networks at the core of each hemisphere (Figure 2.10). In the front of the brain, the pathways are reversed, with most of the connections beginning in the limbic cortex, projecting to the frontal association areas, and then continuing on to first the premotor and then the motor areas (Figure 2.11).

The general outline of this scheme was known by the middle of the 20th century, due to methods of tracing anatomical connectivity such as strychnine neuronography. The poison strychnine was placed on a location on an animal's cortex, causing abrupt seizures in that tissue, and electrical recordings were then made from various cortical areas to determine where the localized seizures spread and thus where the target tissue location sent axonal (nerve fiber) connections.

In the later decades of the 20th century, advances in tracing fiber connections allowed major insights into cortical connectivity, thanks in great part to quantitative studies of connectivity conducted at Boston and Harvard universities. Quantitative studies are those in which the question is not just what areas are connected to other areas but also how many fibers are sent from one area to the next. With this precise new evidence, a picture of the functional tapestry of the cortex has emerged. We are still only beginning to appreciate the significance of this picture.

At a general level, four levels of the cortex were found. At the limbic core (the medial opening of the hemisphere) is a level that can be called the *paralimbic* cortex. Connected to that is a level of cortex that can be called the *higher-order* or *heteromodal* association cortex. These terms imply that, for the back

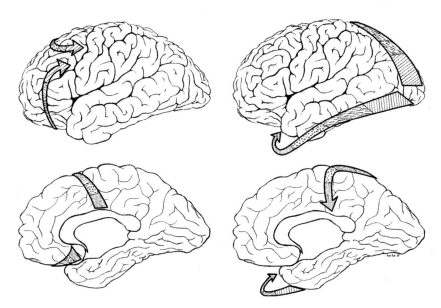

Figure 2.10. *Left*: primary direction of corticolimbic traffic for organizing output from limbic integration toward specific action modules in the motor cortex. *Right*: primary direction of corticolimbic traffic for integrating perception from specific modules in the sensory cortex (in this case the arrows start from the visual area) toward the limbic cortex. For both input and output, the arrows show separate network pathways for traffic for both the dorsal (*top*) and ventral (*bottom*) surfaces of the hemisphere (after Pandya and Mesulam).

of the brain, two or more sensory modalities (e.g., touch and hearing) are integrated within that cortical network. The next level of cortex is called the *primary* or *unimodal* association cortex because only one sensory modality is integrated. The next level (and the farthest away from the limbic territories) is the *primary* sensory cortex, such as for vision or hearing.

The same four levels of cortical networks are found in the front of the brain, starting with the paralimbic, then a higher-order association cortex. However, then the parallel to the primary association cortex in the back of the brain is called the *premotor* cortex; that in turn is connected to the *primary motor* cortex, mirroring the architecture of the primary sensory cortex in the posterior brain. The general appearance of these regions in the normal brain is shown in Figure 2.11, from the interior or medial view of the right hemisphere.

A schematic for how these might appear if the cortex were unwrapped and extended like petals on a flower, stretching out the adjacent cortical levels, is shown in Figure 2.12. This is a highly schematic view because, if unwrapped accurately, each petal would take on a quite different shape from the others. Nevertheless, the overall architecture is captured in this schematic.

Structures of Intelligence 79

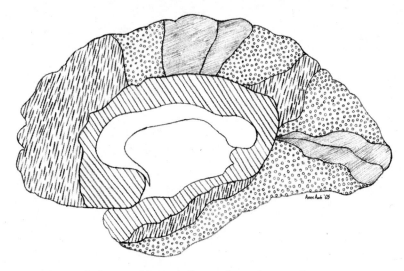

Figure 2.11. Medial view of the right hemisphere showing the major cytoarchitectonic divisions. Shaded = primary sensory or motor neocortex. Stippled = unimodal sensory or motor cortex. Dashes = heteromodal association cortex. Diagonal stripes = limbic cortex (after Mesulam).

The quantitative studies of cortical connections begun in the 1970s, particularly those by Deepak Pandya and his associates at Boston University, showed a remarkable order to the pattern of connectivity of these regions of the cortex. Many neuroscientists have emphasized the imponderable complexity of the cortical connections and areas. Some have even concluded that each mammalian species has a unique and fully idiosyncratic pattern of cortical areas that cannot be compared with that of other mammals. However, when Pandya and his associates conducted their quantitative studies, counting the actual density of the connections among the cortical areas, they found a common and well-ordered architecture in the cortex of primates. In fact, other studies suggest that this pattern in its general form is characteristic of higher mammals, including humans. Two major principles were clear.

First, *connections stay at their own level.* With the exception of "adjacent" connections (paralimbic connects to higher-order association, higher association connects to primary association, etc.), connections from one level go primarily to other brain areas of that same level. Thus, limbic areas project primarily to other limbic areas and not, for example, to association or primary sensory cortices. Unimodal association cortices project primarily to other unimodal association cortices, and so on. In this way, each level seems to constitute a functional unit of the cortex, cutting across both sensory and motor domains. The idea of connections staying within their own levels is illustrated by the concentric lines between the petals of the cortical flower in Figure 2.12.

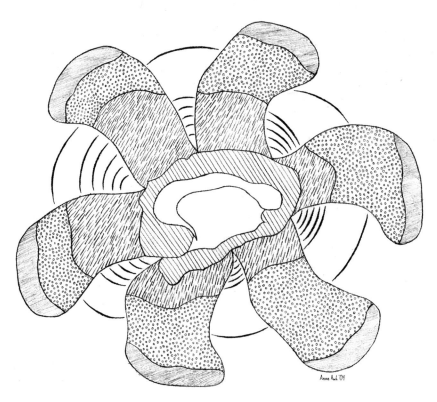

Figure 2.12. Schematic of the medial view of the right hemisphere "unfolded," like the petals of a flower, to separate sensory and motor pathways. Each pathway connects the primary sensory or motor cortex (shaded) to the limbic cortex (striped) with two intermediate regions of association cortex, one for unimodal (stippled) and one for heteromodal (dashed) representations. The lines between the pathways represent the density of interpathway (and within-level) connections. This density is greatest across the regions of the limbic cortex (here shown as continuous). The within-level connectivity decreases across the heteromodal regions of the pathways, becomes sparse across the unimodal regions of the pathways, and is nonexistent across the primary sensory and motor regions (except for the interconnectivity of the primary motor and the primary somatosensory areas).

Second, *the greatest density of connectivity within a level is found at the limbic core.* There is then a progressive decrease in within-level connectivity moving out toward the primary sensory and motor modules. This is also illustrated in Figure 2.12, with many lines between the levels of a petal near the core and fewer out toward the tip of the petal (the primary sensory or motor cortex of the shell). In fact, the primary sensory and motor cortices can be accurately described as *modules* because each is an isolated island. Each island is connected with the diencephalic thalamus but with no other cortical areas except the adjacent unimodal association cortex of that sensory modality or motor area.

The exception to this pattern is that the primary motor cortex has point-to-point connections with the primary somatosensory cortex, linking the two in what some researchers call the *sensorimotor cortex*.

Between the densely interconnected limbic core and the isolated primary sensory and motor modules, the two association levels show *intermediate* densities of interconnectivity: The higher-order heteromodal association cortex is more densely interconnected across broad cortical areas than the unimodal (i.e., modality-specific) association cortex.

It is well known that each primary sensory cortex has its major input from the thalamus of the higher brain stem. Visual input, for example, comes to the primary visual cortex from the thalamus. However, the connections of the primary visual cortex (beyond the local ones intrinsic to itself) are then restricted to just its adjacent neighbor, the unimodal visual association cortex.

In the frontal brain, the architecture is a kind of mirror image of the posterior sensory networks. The primary motor cortices also have their output connections through the subcortex. Thus, both the primary sensory and primary motor cortices are "islands" within the cortex, yet they form the final links with the outside world (in sensation or action). In this way, the primary sensory and motor cortices could be considered the interface with the outside world, forming the "outer shell" of the brain and mind.

Bidirectional Traffic and the Connectional Order of Perception

As a student of the brain for many years, I have been fascinated by this evidence of the organization of cortical connectivity. It is part of the fine order of nature, not unlike the circulatory veins of a leaf and the crystalline structure of a snowflake. Just visualizing this structure offers a glimpse into the beauty of nature. For scientifically minded people, a simple understanding like this may be an adequate reward for the intellectual work it requires.

However, does this structure imply a psychological as well as a neurological order? As we study this tapestry of cortical connections, is there some sense in which we are looking at the fabric of experience?

One problem is understanding how the representation of mental functions could emerge directly from the pattern of neural connectivity. If it does, then many of the traditional assumptions about the brain must be reconsidered. We knew that the sensory and motor areas of the cortex were dedicated to those specific functions, but almost everyone agrees that these are just sensory and motor functions, not the mental qualities of cognition. We have traditionally recognized that the limbic networks at the core of the brain are important for visceral processes—controlling the body's internal workings. Still, we always assumed such things would be bodily stuff, not integral to the mind. Where, then, could the mind be found? All that is left is association cortices in the parietal and temporal areas of the posterior brain and especially in the large prefrontal cortices in the anterior brain.

If connectivity implies function in a direct way, then the association cortices are intermediate in the level of integration they perform. The greatest specification—and therefore local, modular action—must be in the primary sensory and motor cortices. Through similar reasoning, the greatest integration must be in the paralimbic cortices at the core of the brain, with intermediate levels of integration in the heteromodal and unimodal association cortices. Could this be a clue to the order of experience?

An important fact is that the connections go both ways. In the perceptual pathways, not only do "forward" or "inward" pathways run from primary sensory cortices down to the limbic core of the hemisphere, but other pathways, sometimes called *back projections*, also run from the core back out toward the neocortical shell. In both directions, the connections making up the pathway proceed from one adjacent network to the next, so that there are four basic network links (primary sensory, unimodal, heteromodal, and limbic going in the neocortical-to-limbic direction) in each pathway (Figure 2.12). Remarkably, the density of the connections is as great for the outward projections (e.g., from the center of Figure 2.12 outward) as they are for the inward projections.

We have always assumed that the inward projections are the basis of perception because they carry the data from the sensory cortex to be interpreted somehow in the association cortex. But if the mind is embodied in its connectional architecture, then each direction of traffic must be equally important. In some way that we do not understand, the memory consolidation process must reach out from corebrain limbic networks to shape the unfolding perception, forming in the process four interdigitated representational levels.

When I look out my window this dark, moonless night, I try to see the outline of a butte that I know rises on the horizon across the valley. When the sensory data provide no structure—because it is dark—could there be a structure of mind in this blank perception, one that emerges from the patterns of cortical connectivity? I can see mainly darkness, but could there be four levels of experiential order to my perception?

Cortical Interactions in the Consolidation of Experience

Try as I might, I look, but I can apprehend no perceptual structure. I know from my studies that there must be structure within this operation of consciousness, but it remains largely implicit. I know the look of the butte very well, so I can expect the visual qualities, and there is enough light from the nearby city that as my eyes adapt to the dark I can see the outline of the butte on the horizon. I can even imagine the texture of the trees with enough vividness that the imagination is almost visual. So perhaps what is happening is that the structure of my perception is emerging not from explicit consciousness but from the implicit organization of memory.

Memory also imposes itself involuntarily. In my idle reflection as I try to appreciate the darkness, I am reminded of another dark night years ago, when my wife, her sister, and I hiked up the butte on a late September afternoon. We had no lights, and darkness fell before we could return to the trailhead. With the appearance of this memory, I remember trying to look into that darkness as well, seeing no form consciously but somehow managing to stay on the trail.

If there is a phenomenological lesson from this brief exercise in mind wandering, it is that memory is the constant frame of awareness. I can relate to perception objectively and isolate the sensory qualities that would occur to anyone trying to see in the dark. I can then discriminate what my knowledge of this visual scene allows me to expect, as memory projects an almost hallucinatory perceptual expectancy. However, when I reflect even briefly on the subjective quality of this exercise, I find the process is shaped by my own personal experience. Certainly in this moment of mind wandering, it is not particularly significant that I recalled this specific incident of my life, but this personal operation of memory is not an unusual occurrence. For everyone, personal history is an implicit companion to the subjective quality of life.

The English language, in fact, recognizes the integral role of memory in consciousness. The word "experience" describes both present awareness and past history. It may be that this usage reflects the recognition that each instance of awareness leads to the accumulation of effects that make one experienced. Yet both psychological and neurological analyses show that awareness without memory is thin indeed. If pushed to the limit of immediacy, consciousness can allow only the most limited sensory registration. It is the capacity for some extension over time that allows consciousness to integrate the meaning of sensation, actions, feelings, and ideas. Within the brain, the evidence has shown that this incorporation of memory with experience engages specific patterns of cortical connectivity. In fact, this integration of consciousness runs across the four levels from the sensorimotor modules to the limbic core.

As with most of our knowledge of the brain, the early evidence came from clinical observations of the effects of brain lesions. Damage to the temporal lobes of the brain, particularly if it involves the limbic cortex, leads to amnesia, the inability to form new memories. The amnesia is most obvious when the damage is bilateral, implying that even one hemisphere's limbic circuits can maintain a degree of memory capacity. Although there is usually some retrograde amnesia, meaning a loss of memory dating to hours or days before the lesion, most of the loss is anterograde, meaning that new memories cannot be formed after the lesion.

One of the first amnesic patients I examined in my work as a clinical psychologist had attempted suicide by stopping along the road and routing the exhaust of his pickup truck through a hose and into the window of the cab. Someone passing by saw the hose attached to the exhaust. They looked into the truck, saw him passed out, and opened the door, saving his life.

However, his memory was already damaged. Perhaps because they were the first regions of the cortex to appear in mammalian evolution, the limbic regions of the human temporal lobe (including the hippocampus in particular) have a fragile blood supply. As the oxygen capacity of the blood was impaired by the buildup of carbon monoxide, this man's limbic regions were extensively damaged, even as other brain areas were spared.

The effect was a dense and permanent amnesia. Amnesia is simple to test, but it is often missed by physicians or psychologists because they fail to test memory capacity specifically. If I had not known what had happened to this man and had simply conducted a brief interview, I might have missed his severe memory deficit. He did not seem particularly depressed, was socially appropriate, and demonstrated good language skills. He answered my questions reasonably and showed insight into the factors leading to his depression and suicidal thoughts. However, starting with a few hours before his suicide attempt, he could not remember anything that happened a few minutes before the present moment. Remarkably, he was unconcerned with this deficit and just tried to make up stories that would seem plausible to explain the massive gaps in his knowledge.

For many years clinical cases such as this have shown that the limbic areas of the temporal lobes are essential to memory. However, systematic animal research was necessary to clarify the role of the limbic areas in relation to other areas of the cortex. If the limbic nuclei or cortices are damaged in an experimental animal, the animal cannot form new memories. In fact, if the connections between the limbic areas and the other cortical areas (neocortex) are severed, new memories cannot be formed. Remarkably, however, old memories remain more or less intact unless there is very extensive damage to the neocortical areas.

These studies lead to a clear picture of the anatomy of memory. The final storage of a memory must be widespread within the neocortex. However, the storage process (the creation of lasting experience) requires an intact limbic system and intact connections between the limbic cortex and neocortex. The implication is that some interaction occurs between the limbic areas and the other three levels of the cortex (which we can group together as the neocortex) to allow memories to be formed. This interaction achieves the consolidation of memory.

From several fascinating sources of evidence, experimental and clinical, we have learned that the consolidation of memory takes time. At first, the evidence from animal studies showed that after a learning experience (such as associating a stimulus with a response to get a reward), the brain must be left alone for a few minutes for the memory to be consolidated. If, for example, the animal's brain is given an electric shock, the consolidation process can immediately be disrupted. Whatever the limbic system is doing to the neocortex, this process is interrupted, and the memory is lost. Some truly amazing results were obtained by studying the memory loss of patients that received electroconvulsive shock therapy, which caused a gradient of impairment, with recent memories impaired more severely, as expected. What was amazing, however, was the evidence of measurable impairment of memories from many years before the treatment. Such an effect implies that that the consolidation of memory may not be complete in a few minutes or even hours or days but may extend over years.

Consolidating Experience

Thus there is a process to forming memory. But is it relevant to subjective experience? Is the two-way traffic (from the limbic core of the hemisphere out to the lateral neocortex and back again) an integral process underlying conscious experience? Or is it just a neural mechanism that forms a biological substrate that is psychologically irrelevant?

Certainly it makes a kind of biological, adaptive sense that two-way, corticolimbic traffic occurs in the consolidation process. However, does this connectivity have psychological meaning, implying that we reach out to the sensory data to process it somehow expectantly from representations within memory that are held in the heteromodal association and limbic cortices? Could this be a kind of anticipation that engages perception, where we do not simply take in the data of reality and interpret it, but begin with our interpretations based on previous experiences as we engage the sensory data in the perceptual process?

We can consider these issues with another story about Jared's brain, which we last left at the lunch counter. Jared's reflection on the time he spent at his grandparents is interrupted when his friend Andrew comes into the diner and joins him at the counter. Jared has known Andrew since grade school, and they often hang out with the same friends, guys who love cars. As he often does, Andrew starts in with a story of how he was racing someone, this time a guy in an old Toyota. Even though Andrew's adventures are usually at least half imagined, he is still one of Jared's best friends, and Jared lets him tell his story. Jared is only half listening until Andrew asks if him whether he wants a new transmission for his Civic.

Jared's old Civic has a bad fifth gear, and he has to clip a bungee cord to it to keep it in gear on the highway. Andrew says he can have the new transmission for $100. Jared would love to have a new transmission, and he knows they can go for ten times the price Andrew has named, so his suspicions are immediately aroused. When he asks Andrew where he got the transmission, Andrew says he picked it up on the street.

Jared is now attentive not only to Andrew but also to the other people in the diner. He knows immediately that Andrew is talking about a transmission from a stolen car. He becomes concerned that someone else will hear. At first he is tempted to tell Andrew that he wants the transmission. Not only does he need it, but he also knows he will never find one for this price.

Yet as Jared looks into the mirror on the wall facing the counter, checking out the customers behind him, he also looks at himself and Andrew at the counter. It takes him only a minute to decide he doesn't want the stolen transmission because he has always been opposed to lifting cars. Even though it would be easy to take this transmission, he knows he would regret it. When Jared turns it down, Andrew gets angry. He tells Jared he'll never get a chance like this again and then walks out.

After Andrew is gone, Jared pays the bill, goes outside, and sits on the curb. He adopts a cool posture (his typical attitude to present to the public) and watches the traffic go by, just as he and his friends often do. As he looks around idly, he thinks about what just happened.

Andrew never has any cash, so he wouldn't have a transmission to sell unless he had helped steal the car. Jared had heard guys bragging about the thefts, which often take place in a mall parking lot. Usually, there would be several kids in on the heist, one or more to serve as lookouts and one to shim open the door and run through the pass keys that usually turn over the Japanese ignition locks worn from a few years' use. Andrew must be hanging with hoodlums, and that probably means other thefts. It also means that Andrew is running the risk of getting locked up. Jared is worried about his friend but he doesn't know what he should do about it.

In this story, Jared's intelligence has been challenged by another problem. This time it requires a psychological rather than a mechanical solution. We can again analyze the structure of Jared's understanding of the problem in relation to the structure of his brain. This time we will consider the architecture of Jared's corticolimbic connections (between the limbic cortex and the heteromodal, unimodal, and primary levels of the neocortex) as the traffic across these connections arbitrates the patterns integrating memory, expectations, and the sensory data.

In Jared's experience, the problem started as he realized the implication of what Andrew had said. In a neuropsychological analysis, we need to start with perceptual and cognitive processes that are so elementary that people take them for granted. Hearing Andrew's words began with the neural processing of the speech sounds in Jared's primary auditory cortex. This is represented by shaded section of one of the sensory pathways (petals) in Figure 2.12. Almost certainly, the interpretation of each word engaged the unimodal auditory association cortex as well (stippled area). Furthermore, we can assume that there was important feedback from the unimodal association cortex back to the primary auditory cortex. This counterflow from Jared's knowledge of English must have shaped his comprehension of each word as he listened to the rapid flow of sounds in Andrew's speech. The primary auditory cortex is mapped by frequency (high or low pitch of the sounds), not by higher perceptual objects such as words. Therefore, processing at deeper or "higher" network levels is required to form the perception of a word that derives meaning from the sounds. Considering the connectivity of the networks as shown in Figure 2.12, the sound data arriving in Jared's primary auditory cortex can be formed into comprehensible words because the word representations in the deeper association cortex (striped and dashed) "reach out" to the incoming sounds and give them form.

All of this is automatic, with no registration in consciousness, because Jared is an experienced interpreter of English speech. If Andrew were to speak in Spanish, then Jared would have a more conscious knowledge of the discrete speech sounds because they would be isolated from the embedding word meanings. As it is, Jared's consciousness has no residue of the formation of word meanings from speech sounds. He simply hears the words. Even then he remembers only their meanings.

In studies of the brain regions that are essential for speech, both clinical cases of brain damage and neuroimaging research make it clear that the interpretation of speech requires heteromodal association networks. Particularly important to the comprehension process are the heteromodal regions of the temporal-parietal cortex in the left hemisphere. In the architecture of the corticolimbic networks, this would activate not only several levels of the auditory pathway but also cross-pathway integrations of the temporal-parietal lobe (e.g., between the visual and somatosensory heteromodal areas), as can be visualized in Figure 2.12. In addition, even though we are emphasizing speech comprehension, it is clear that the motor areas of the frontal lobes are involved, as if we understand speech by covertly producing it.

Even with this complexity of the neural systems, what we are describing so far is the generic comprehension of spoken language. Jared's interpretation is more complex still. First of all, he was not only processing Andrew's words (primarily in his left hemisphere) but also Andrew's tone of voice, which was somewhat stressed when he began talking, in that his humor was more forced than usual. With his well-developed auditory intelligence, Jared sensed this. The interpretation of prosody or tonal qualities of voice is carried out within the right hemisphere of most right-handed people. This nonverbal interpretation seems to parallel the verbal processing of the left hemisphere. Furthermore, Jared was not only listening to Andrew but watching him as well. He could see

Andrew's nervousness in the tension of his body posture and in the tight little smile that Andrew flashes when he's anxious. The interpretation was then quickly supramodal, engaging visual qualities as well as auditory ones, to create concepts of nonverbal gestures that integrated Jared's sensory space.

We can begin to see how the full architecture of the corticolimbic networks (within-pathway, as well as cross-pathway) becomes engaged in the complex patterns of real experience. The visual and auditory information must be processed in its own right in order for us to comprehend words or to get the gist of a tone of voice or facial expression. Then it must be combined to create a larger whole, such as an integrated impression of the meaning of Andrew's communication.

Nevertheless, the input direction of processing is only part of the story. Jared quickly infers the implication of Andrew's story—that he's stolen a car—because Jared has domain knowledge that can create the context for Andrew's communication. This is knowledge that is specific to the domain (in this case the domain of the stylized culture of young car enthusiasts). Most of the other people in the diner could hear everything Andrew said but not understand that it implied theft. In Jared's interpretation, we see a complex base of knowledge held in memory, and we see it reaching out to constrain and enrich the processing of a communication.

The result is that a complex interpretation is apprehended quickly and holistically. The memory-based knowledge quickly forms an implicit expectancy, and this then frames Jared's interpretation of Andrew's next statement—that he got the transmission on the street. For many people this answer would not make sense, but Jared's knowledge had already primed him to expect an answer of this sort.

The reverberation between memory and perceptual representations continues as Jared goes outside and sits on the curb. He imagines what it would have been like for Andrew to steal the car, and his visualization is strongly engaged by this process. This extended rumination over meaningful information means that Jared will remember it. The corticolimbic processing, back and forth along the pathways in Figure 2.12, must be what consolidates the new information within Jared's long-term memory. Somehow the consolidation is driven by the knowledge that this information is significant.

In contrast, Jared will soon forget the details of Andrew's story about racing another car, even though this took as much time in their interaction as the mention of the transmission. This is because the racing incident was not emotionally significant to Jared, particularly in the context of Andrew's previous stories about similar incidents. Knowledge places constraints on the process of perception. The significance of the perception, perhaps gauged by the reverberation caused in limbic networks, directs the incorporation of the new data into the knowledge base of personal experience.

The Mnemonic Structure of the Unconscious

We have now opened up two critical questions in understanding the neural roots of intelligence. One deals with representation—how the deeper patterns of connections in brain networks can explain the pattern of knowledge. This is the first question addressed in Chapter 3. The other deals with control—how the mind's capacity for memory is regulated by the significance of the information. This is the second question taken up in Chapter 3.

At this point, without insight into these questions, our analysis of the physical basis of a mind like Jared's must remain rather vague. We do know that cognition, such as Jared engaged in over lunch today, must involve the sensory and motor networks. This in itself is an important fact. It provides a basis for understanding the embodiment of the mind in biological processes. We know that the sensory and motor networks—the representational mechanisms that interface our internal experience with the outside world—are relatively isolated modules. We know that they interact through a structured, layered set of connections with deeper, association networks of the cortex. We know that the association networks are then connected at the core of each hemisphere with the limbic networks, where they are linked to the emotional and motivational circuits of the subcortex and brain stem.

Finally, we know that some arbitration must cross these multiple network levels from the core of each hemisphere to the sensory and motor modules of the interface shell. In some way, that is how we consolidate information within long-term storage. Even at this point, with this outline of cortical connectivity, we have a basis for reasoning in a systematic way about Jared's experience.

Let us consider how Jared's processing of Andrew's communication will interact with his ongoing experience as it unfolds over the rest of the afternoon. After sitting quietly for a few minutes at the curb, Jared gets up and walks a few blocks back to high school so he can attend his afternoon classes. In history class, however, he does not pay close attention because all he can think about is the trouble that Andrew is courting if he is stealing cars.

However, Jared's thoughts of Andrew are themselves distracted after class when he talks with a girl, Kim, that he has admired for some time. Kim is good looking, and she shows an enthusiasm for the ideas presented in history class that Jared finds impressive. Still, Jared feels disappointed whenever he thinks about Kim because he's always believed she was too smart for him. She argues so well in class—better than the teacher—while Jared can hardly string two words together.

As they are leaving class this afternoon, he tells her he liked what she said today about the Civil War, specifically that the textbook failed to explain why people would go to war with other citizens in their own country. Kim does not put him down but seems genuinely interested. His spirits buoyed by Kim's enthusiastic response to this conversation, Jared follows an impulse and asks her to a party tonight night that his friend Tim is having. When Kim agrees, Jared is elated. He has to work to appear cool and collected as he starts toward his next class. He bounces along and can hardly find his way to biology lab.

As is the way with young minds, Jared's experience is captured by powerful responses to events. As a result, his concerns over Andrew must take their place within the dynamic processes bidding for his limited memory capacity. He simply doesn't think of Andrew again the rest of the day—until, that is, he is back in his room after school. He is trying to catch up on his history reading before getting ready for the party. Even though he wants to understand it so he can talk intelligently about it to Kim, the material on the Civil War gets only minimal stage time in the theater of Jared's consciousness. Instead, the primary actor on this stage is Kim. Having been curious about her all year, Jared recalls several fantasies he entertained about her as he sat in the back of history class.

As his mind wanders, Jared's thoughts become unfocused. Unexpectedly, Andrew comes to mind. Jared remembers that Andrew got angry when he refused the transmission because Andrew offered Jared a really great price. Suddenly Jared realizes that, at that price, Andrew was basically offering him a gift of the transmission. Andrew obviously valued him as a friend and wanted to help him. He could have gotten a lot more for that transmission anywhere in town.

Jared immediately gets warm and feels embarrassed and uncomfortable. He is ashamed that he did not realize what Andrew meant. He is also curious to understand his own mind, how he failed to notice something so important. He realizes that it was because he was so preoccupied with his own ethical dilemma—and impressed with his own righteousness—that he was blind to Andrew's gesture of friendship. No wonder Andrew got mad.

Just as we took apart Jared's initial experience of his conversation with Andrew and analyzed its components in relation to the architecture of his cortex, we can do the same with this example of delayed cognition. This cognition is recalled from memory and involves no immediate sensation or action. Therefore, we may not think to consider whether it would engage the sensory and motor regions of the cortex. Yet this is what neuroscience research implies, that when we engage in cognition related to a sensory or motor process—even from memory—we do so by reengaging sensory or motor networks. Jared can almost hear Andrew's voice. This is not a hallucination, but it is clear that he can now sense, within his memory of Andrew's voice, that Andrew was hurt at his refusal.

Of course, sensory representation is only part of Jared's skill at empathy. Other brain networks must be engaged as well. How did Jared understand the emotion in Andrew's voice? He probably used his auditory cortex and even his motor cortex as he recalled the memory of Andrew's voice. Research suggests that we perceive the actions of others not just by using our own sensory cortex but also by mirroring those actions with our own implicit action plans in the motor cortex. These representations that formed in Jared's neocortex (toward the sensorimotor shell of each hemisphere) were essential to allow him to recall and reflect upon the full meaning of the interaction.

Recalling this event was of course necessary before Jared could reflect on it. We must therefore ask what contributions were made by Jared's deeper cortical networks, those in the association and limbic cortices, to allow him to recall the memory of Andrew's anger so it could be reinterpreted in sensorimotor terms. This brings us back to the key questions of representation in distributed networks and control from motivational processes. Where was this memory all day, while Jared was distracted with Kim? How could it present itself in a new form, more fully understood and processed, when it had been completely absent from Jared's consciousness? Must we assume that integral mechanisms of Jared's experience were self-organizing, out of awareness, in his unconscious mind? If so, must we assume these mechanisms were carrying out not only implicit actions and sensations but also implicit emotions, such as an empathic resonance to Andrew's hurt feelings? These are complex questions, even when posed in the form of a simple story about Jared's afternoon. Even as we struggle to work out a meaningful understanding, a few principles can be stated with confidence.

First, the intelligence that Jared brings to understanding his interaction with Andrew is as rich and complex as any example of human intelligence. The human brain is fully challenged by the problems of social interaction. These problems are as difficult and as important as any of those presented by technical, artistic, or intellectual work. This principle should be easy to understand. It is only because we sometimes confine our concepts of intelligence to skills in school or at work that we may fail to recognize it.

Second, certain physical mechanisms (processes of sensory and motor activity of the body) are integral to the structures of intelligence. In ways we are just beginning to understand, the mind is embodied. This principle is somewhat more difficult to understand because, within the naïve view, the mind seems separate from the body. In the chapters ahead, we will see additional evidence that supports this principle of the embodiment of mind, and I will illustrate it with concrete examples.

Third, the structures of intelligence are achieved through neural connections, such that the pattern of those connections implies the pattern of mind. Although this third principle is critical for understanding the scientific evidence of the brain's neural structures, it is perhaps the most difficult principle to understand, especially when stated in abstract terms. In the next chapter we will turn to studies in computational science that offer somewhat more concrete, practical insights into the functional properties of connections and distributed cognitive representations. By studying computational models of distributed information representation, we will find powerful metaphors for how the mind's knowledge can be represented in the brain's structures. This analysis of the representation of information will show today's remarkable convergence of neuroscience and computational science.

3
Principles of Representation and Control

Representation refers to the way information is held in the brain and mind, how it is stored or encoded. Chapter 2 outlined several ways in which understanding the connectional structure of the human cortex implies its representational capacity. But how exactly does this work? How can memories be coded within the pattern of neural connections?

These questions are important because it is difficult to be effective users of brains if we don't know how they work. At a physical level, there must be mechanisms that encode the data of experience within neural networks. These are bodily mechanisms of the vertebrate brain that have evolved over several hundred million years to create the intelligence that underlies our daily experience. Each of us senses our dependence on this intelligence in one way or another, yet we have precious little understanding of its workings. We are more or less aware and more or less self-aware but only vaguely in control.

In the example of Jared's afternoon, we have seen how the normal experience of an adolescent often involves formative episodes of self-awareness. When Andrew offered him the stolen transmission, Jared considered his own values as he looked beyond the counter and saw himself in the mirror. His decision not to take the transmission was then deliberate, based squarely on his concept of himself as an honest person. Later, when Jared realized that Andrew was making him a gift of the transmission, he again integrated this experience with his understanding of himself and tried to understand how one set of values (being honest) had blinded him to other important values (being sensitive to a friend).

Although these are formative experiences for his sense of self, in both cases Jared has only a loose degree of control over his own experience and behavior. Like most of us, his day is unfolding, and he is trying to keep up. He maintains a sense of agency—of making his own choices—but his insight into his own mental processes and his real control over them is limited. Perhaps if he studied the brain's functional structures, such as we just did in Chapter 2, Jared could find a roadmap to his neural networks, a reference guide to the interwoven wetware of his mental resources. Certainly our current knowledge provides only an outline at best. However, what if science continues to make progress in computational analysis of the brain for another few decades? Will we be able to teach a kid like Jared how to program the code of his own corticolimbic networks?

Maybe the limited and fleeting consciousness of our mental lives is a simple result of our ignorance of the brain's mechanisms. Now there are contextual constraints forming across the multiple levels of the corticolimbic hierarchy, but, though we are largely unaware of it, we remain naïve interlopers in mind space, constrained passively and unconsciously. What if we could reach down and take hold of the actual, physical mechanisms of arousal, attention, and memory? Would we then achieve more conscious self-direction?

In this chapter we will study principles of information science that may help explain the brain's operation. These principles show how artificial neural networks can achieve distributed representation—the spread of information across widespread artificial neural networks. These principles must teach us about the nature of distributed representation in the brain and mind as well. At the same time, we will take up the question of *control:* how neural processes can excite, inhibit, and somehow thereby self-regulate.

To be sure, we are a long way from understanding the neural code. Nonetheless, by studying the workings of distributed representation and control, we may learn how to reason from the patterns of brain anatomy and physiology to understand the structure of experience.

Information in Machines

From an analysis of the neuropsychological evidence, as reviewed in the preceding chapter, modern research and clinical observation are showing us the outlines of the brain's connectional structure, but what exactly does this teach us about the qualities of mind? How can its functional process—the one that constitutes subjective experience—be captured by neural connections? When we commit an experience to memory, it becomes part of the self and coded in some way within the brain's physical connections. How?

The most important scientific foundation for this question comes from computer science. Computational models are now showing us the principles of parallel, *analogical* forms of information representation. Certainly logical operations, as difficult as they are for most people, have been the easiest to simulate in computers. In contrast, it is the formation of coherent *analog* conceptual gestalt patterns that has proven difficult for machines. In recent years, however, modern computational theory has been teaching us about analogical concepts. These are packets of meaningful information about the world that are analogs of reality. They are iconic mirrors of its qualities rather than logical propositions about it. The theory and research on distributed representation *(connectionism)* is showing us how analog concepts could be represented in the pattern of connections among a dense web of neurons.

When machine intelligence was announced as the next step in computer science in the 1950s, philosophers were quick to argue that subjective experience has mental qualities that a machine program could never capture. On its surface, this argument seems to be commonsensical. However, modern parallel-distributed computational models have shown properties that are remarkably similar to human mental representations, at least when compared to the traditional digital, symbolic models of computer processing. Examining the nature of these distributed computational models provides an important scientific basis for reasoning about the brain's connections.

The Life of Nerve Nets

Parallel-distributed models of information representation are often called *connectionist* models. They represent information not within the neurons themselves—the nodes of the network—but in the strength of the connections between them. For someone who is used to modern digital computers, a connectionist form of representation is foreign and unintuitive. We think of information as being represented in the words in the computer's memory, the individual units of bit patterns, with these computer words then processed by instruction operations (for example, calculation or logical comparisons) performed on those words. This is a *symbolic* form of representation: Each word in computer memory stands for a specific thing and holds the information that the computer program deals with. It is also a *logical* form of information control in that each of the instruction operations is consistent with the logic of the computer language (e.g., Basic or C^{++}).

A connectionist program is neither symbolic nor logical. The individual units are "dumb" neurons in that they do not hold any symbolic information but simply "fire" or discharge, depending on the summed weight of the connections they receive at any given time. Of course, these are just computational routines and bear little resemblance to brain neurons. However, the connection patterns formed across the network capture the essential representations that allow complex, analogical perceptions or decisions. These connection patterns are not discrete logical computations but complex and dynamic ones that quickly become difficult to predict (or even analyze). By organizing enough of these dumb neurons and adjusting the connection weights adaptively (getting closer and closer to a desired outcome), remarkable structures of analogical information can be created.

Neurophysiologists have long believed that the brain's information-processing functions must be achieved somehow by altering the patterns of connections among its neurons. Studying in Vienna toward the end of the 19th century, Sigmund Freud conducted neurophysiological research as part of his medical training. He tried to develop a theory of nerve nets—the network patterns among neurons—that could carry out mental functions. He speculated, for example, that neurons that were active together would strengthen the connections (now known to be synapses) that mediate this joint activity. In today's computational neuroscience jargon, this is the principle of "fire together, wire together."

Freud's theory of neural connectivity and memory was thoroughly ignored by his contemporaries. The difficulty of bringing neurological concepts to important problems of the human mind led Freud to abandon the study of the brain and to focus exclusively on psychological theory. It would be more than 50 years later, in the mid-20th century, before Freud's doctrine of functional connections would be reinvented, apparently independently, by a young Canadian psychologist, Donald Hebb. For his part, Hebb's theory of neural connections was also discounted by his contemporaries in neurophysiology and psychology, that is, until 30 years later, when computer scientists rediscovered Hebb's principles. As they built their connectionist models, the computer scientists found that "Hebbian" connections—those that increased in strength when the connected neurons are coactive—could form the basis for self-organizing (pattern-generating) parallel-distributed networks. These were nerve nets with a life of their own.

The work on nerve nets has progressed slowly for many years, yet it is beginning to provide mechanisms for the way in which the interconnections among neurons could explain the functioning of not only artificial neural networks (computer models) but also brains. This work provides a basis for the principle that motivated the study of brain anatomy in Chapter 2. This principle could be stated as "connectivity implies function."

Connected Representations

The early searches for mechanisms of mental representation by Freud and Hebb were primitive. Very little that looked like mind was explained by these theories, but they were looking in the right place. After more complex connectionist networks had been constructed in the 1980s, the representations started to look not only mindlike but also more naturally mindlike than any of the digital computer models of information processing that had been developed in cognitive psychology. A brief review of the learning shown by these computational models is highly instructive. We will see that, even if these models are still only metaphors for the mind—and even if the brain somehow represents information in a fundamentally different and still unknown way—the connectionist models are much more powerful metaphors than any we have ever had before.

An important advance was the recognition that the connections of artificial neural networks function best when they are sorted into "layers" or groups of neurons on different levels. These levels can be densely interconnected within each layer, but they are segregated so that connections between the levels do not dominate the within-level connections. Interestingly, as we saw earlier, the mammalian cortex evolved with adjacent levels (progressing from limbic to neocortical areas) that are themselves interconnected in a similar fashion.

Another important advance in distributed representations was the development of a method for "training" the weights or strengths of the connections of the model. The object was to be able to adjust these weights in ways that progressively construct a representation that achieves the desired goal. For example, let's say we want to construct a connectionist network that discriminates men from women, using only black-and-white photographs of their faces. The input layer would be neurons or nodes, each of which reflects the grayscale value of one pixel of the photograph, where the 2D photograph is represented, for example, as 640 by 480 pixels. That would be 307,200 pixels, and our computational model would require as many neurons in the input layer. Because the output layer would have only two neurons (activate the M neuron for males and the F neuron for females), a large reduction is obviously required. We would construct a hidden layer to mediate between the input and output layers, and then we must find the pattern of weights across all three layers that will allow a photograph to be presented to the input, classified correctly by the network, and result in the M or F neuron being activated appropriately.

Exactly how we adjust these weights turns out to be a sticky problem, as we will see. However, the theoretically important result was that, once the weights could be adjusted, the artificial neural networks did represent complex concepts, such as the difference between male and female faces. Conventional digital computer simulations had never been able to form this kind of representation. These were pattern representations, configural concepts that psychologists had long described as gestalts. The connectionist model clearly stored the pattern representations in the distributed network (when a face was presented, the simulation could correctly say M or F). And yet these were qualitative, configural concepts (the femininity of a face) that had never before been achieved by computers.

The biologically plausible behavior of connectionist networks extended to a number of fascinating features. For example, they showed "pattern completion." You could blank out many of the pixels of a photograph, present it to the network, and if it was well trained on many photographs (and had an output layer appropriate to report its internal representations), it would fill in the missing pixels with reasonable approximations, just like the human mind does.

These networks also showed "graceful degradation." If you damaged many of the neurons (the damage is in this case computational, i.e., by setting their weights to zero), the output would become less accurate—but only proportionately so. In contrast, damage to *any* of the elements of a digital computer (or even a digital computer model of mental operations) may cause a catastrophic failure since the causal chain of the instruction processing is broken by the failure of that one link.

Finally, connectionist network models showed "increasing differentiation" of their representational capacities as their training progressed. In the example of male and female faces, the initial training may lead the network to recognize only the obvious photos (for example, where hair is a major clue). However, after more training, even subtle gender discriminations become possible. The representations thus become more differentiated with experience. As both parents and psychologists have observed, this progressive differentiation of concepts is the canonical route toward complex intelligence taken by the child's developing mind.

A number of principles could be derived from the observations on connectionist networks. Perhaps the most fundamental one is that *information is relational.* The information that achieves such interesting functions in the parallel-distributed networks is formed through the *patterns* of relations among elements, not the content of elements themselves.

Distributed representation may be simply a property of these computing models, but it may also be a property of brains. What we see in connectionist models may be a clue to the mechanism for the relational nature of information: Information is formed by connecting elements. If so, an interesting problem is where the connections stop. Does information become increasingly rich as its extended interconnections become increasingly developed?

The Stability-Plasticity Dilemma

As with any new theoretical advance, connectionist theory is not without its problems. Early in the study of connectionist networks, the network models showed a problem at the core of the representational process. Although many people thought it would be soon solved, it remains a core problem to this day. Call it the Old Dog Problem. Once a network is trained, it is difficult to teach it new tricks. If you do manage to do so, it forgets its old tricks. This is an important point.

Take, for example, a network that recognizes the gender of faces. If we teach it to discriminate between adults and children, it will learn this new discrimination but immediately start to forget how to discriminate male from female faces.

In technical terms, this problem is called the *stability-plasticity dilemma.* We could state it as a principle for computing—and maybe life: *You get stability (keeping what you know) or plasticity (learning something new). Take your pick.*

From the perspective of computational mechanisms, this makes sense: The same weights (connection strengths among the nodes or "neurons") are used for every representational task. If we take the weights that are well tuned to make the gender choice (male versus female faces) and readjust them to represent an age choice (old versus young faces), then the connection strengths that were good for the gender choice are now changed, and that concept is thereby degraded.

Maybe this is why old dogs—and, come to think of it, old scientists—don't learn new tricks. It's not just that they have the connectional inertia of their extensive old knowledge (compared to the puppies and graduate students, who are both eager and clueless), and it's not just that they have to put out the effort to learn the new tricks. Instead, it is that the very learning process itself causes them to forget those old tricks that they worked so hard to master.

What if this is a fundamental problem of intelligence in distributed systems? The point is important to grasp in its naked simplicity. If you have only one brain and it functions as a distributed intelligence system, then there is basically one representational network—one concept—that you must use all the days of your life. It's your self.

To the extent that various purposes and situations are completely independent, then there is little interference, and new learning proceeds with minimal disruption of your old knowledge. But to the extent that there is overlap in information content between current events and your historical experience base (i.e., the historical self), then learning has a cost. The cost of new learning may be the abandoning of your old self.

The Advantages of Architecture

Once we have a theory of how representations are formed in distributed networks, we can begin to understand the implications of differing patterns of network architecture. These have been tested by computational simulations that construct differing forms of connectional architecture and then examine the results in the patterns of representation that can be formed. Through this pioneering modeling work, we are learning how to construct complex and interesting patterns of information representation.

One key advantage was mentioned earlier. It turns out to be very useful to separate the nodes of the network and the patterns of connection among them into different layers. In fact, an important advance in parallel-distributed or connectionist models was the creation of hidden layers of nodes that were directly connected neither to the input (data to be processed) nor the output (decisions to be made) but instead were interposed between these levels. The improved representations of these networks (shown by their more effective decisions) suggested the advantages of architecture. It is useful to have certain domains of the connectional architecture constrained by neither the input nor the output, at least not directly. Rather, these connectional domains are allowed to form patterns that are somehow intermediate.

The advantages of differing architectures may be most important in relation to the stability-plasticity dilemma (SPD). The destructive interference of new learning on old patterns is greatest within a single, fully interconnected network. On the other hand, if a particular network can be separated from the systemwide changes caused by new learning (such as in a specialized module), the old learning of that network module can be protected.

Of course, as a solution to the SPD, this is cheating; the specialized module would be degraded just like any other one if it were equally open to the new learning. However, if we cannot have an elegant solution to the SPD, maybe we can still learn from methods of coping with it. Research on different representational architectures has shown us how to achieve practical solutions to artificial intelligence. Perhaps more important (as we saw in Chapter 2 and will study further later on), the computational research is showing how differing conceptual structures might be achieved by the architectural features of complex connection machines, such as brains.

The specific patterns of cortical architecture may imply specific advantages of mental representation. Although each of these patterns and advantages must be examined in its own right, we can look for general principles that describe what happens as we move from a diffuse and dynamic system—wherein all of the elements are interconnected—toward a more differentiated system—wherein more specific and constrained information relations are formed. The key point is that *differentiation of function within a network may allow the stability of certain functions, whereas others are modified by new learning.*

We saw the natural progression toward increasing differentiation in the example of computational networks that discriminate between male and female faces. First they separate on the basis of crude features, such as hair. Then they learn to discriminate progressively more differentiated qualities, such as the fineness of facial features and the general outlines of bone structure. To describe this increasing complexity, we can steal a principle from the theory of psychological development, one that suggests how structure may provide organization to information systems: *Development tends toward increasing differentiation.*

I have no problem stealing this principle from the psychological theorists because they stole it from the biologists who were studying embryological development. The biologists formulated the principle of differentiation as they observed the embryo develop from a single cell into organized clusters of cells, then into primitive wormlike shapes of the embryo. Finally the fetus sprouts the increasingly fine features that are unmistakably characteristic of its species. This development recapitulates the broad outlines of the phylogenetic progression.

In the application to mental development, psychologists observed that children begin with holistic, undifferentiated concepts. For example, a child learns about a dog and then a short while later is excited to point out another dog— only this one is a cow. The attentive parent helps with the differentiation process by explaining the difference between dogs and cows. The increasing complexity in the child's intelligence is achieved through increasing structural differentiation, that is, the separation of finer distinctions among the increasing variety of conceptual elements.

In the application to computational models, differentiation may be a way to understand the semantic advantage of a specific connectional architecture. The meaning, or semantic function, of a network may be allowed greater complexity as its architecture becomes more differentiated. The mass action that is dominated by the stability-plasticity dilemma is most characteristic of fully interconnected (i.e., dynamic-holistic or architecturally undifferentiated) networks. Differentiation may be the most critical structural process in a distributed representation system because it carves the defining boundaries in relational space.

As we saw in the last chapter, a growing body of evidence is charting the network architecture of the human brain. Each of the architectural dimensions of the brain, whether anterior-posterior, left-right, or core-shell, suggests a kind of differentiation, a parcellation of relations that moves the brain's networks from an undifferentiated mass toward a more specified functional architecture. The principles being developed in the computational science of distributed systems may offer tools for building a theory to explain what these dimensions of the brain mean for the structure of the mind.

Recall the structural difference between frontal and posterior cortical networks reviewed in the last chapter. The anterior or frontal brain must organize information in the service of actions. With this task, the frontal networks face the problem of serial order—how to assemble a sequence of actions in linear order from the range of possible movements that may be activated in the current relational matrix. This task requires a form of differentiation, inhibiting many actions while selecting only one.

A different advantage of architecture may obtain for the posterior brain. In Chapter 2 we considered Allport's idea that, whereas actions must be selected singly, sensory systems can be combined in parallel, providing a comprehensive map of experiential space. From the computational principles in the present chapter we can infer that the more holistic connectional architecture in the posterior association cortex may be more susceptible to the stability-plasticity dilemma. As new elements are incorporated into the current percept of the sensory surround, the percept changes.

The perceptual representation of this present episode in time, this episodic memory, is thus ephemeral, changing holistically as each new event is incorporated within it. This contrasts with the separation of influences within the elements of motor sequences. With greater differentiation applied by the motor networks of the frontal brain, each action must be integrated only with its immediately adjacent actions, thereby avoiding the fusion of relational influence that would paralyze us if we tried to carry out—all at once—every motor element of an action sequence.

The anterior-posterior dimension of brain network architecture thus seems to exemplify forms of neural structure that confer unique capacities in the relational structure of information. We can take another clue from biological and psychological theories and recognize that holism and differentiation are not the only structural mechanisms describing complexity in either organisms or minds. After differentiation has increased the complexity of conceptual elements by articulating each element in terms of its unique features, it is necessary to organize the information set into integrated patterns. These integrated patterns cannot simply fuse meaning again; they must retain the differentiation of elements.

This process is described as *hierarchic integration* because it is not a regression to the holism of the primitive stage of development but rather an integration added on top of the differentiated complexity. Even as the differentiated parts are retained as separate, they are joined in higher-level wholes that achieve system-level functions. In the example of learning to recognize faces, the perceptual knowledge represented within a computer model—or a human brain—gains increasing differentiation of features that are recognized as separate elements. And yet the elements are also integrated—hierarchically—into whole patterns, such as males, females, young males with fair complexions, or adult females with angular features.

In Heinz Werner's theory of psychological development (borrowing liberally from concepts in embryology and evolution), the progression toward increasing complexity is described by the *orthogenetic principle.* "Orthogenetic" here means "general development." The orthogenetic principle states that *wherever development occurs, it proceeds from globality and lack of differentiation toward increasing differentiation, articulation, and hierarchic integration.*

Are there structural features of the brain's anatomy that could suggest computational implementations of hierarchic integration? Could these then allow the integration of experience without succumbing to the stability-plasticity dilemma? These questions will guide our research into the brain's mechanisms in the next several chapters.

Certainly this research must be tempered by the realization that we are just beginning to integrate connectional principles with neuroanatomy. Furthermore, many scientists consider the very general concepts of differentiation and integration as too vague to provide strong direction to scientific theory. Nonetheless, I am convinced that the principles of computation in distributed-information systems are providing explicit guides for interpreting the advantages of each dimension of the brain's architecture. These principles may also explain—at least in their broad outlines—the processes of learning and development that shape increasingly complex information systems in each child's mind.

Learning From Errors

Even before they face the problem of maintaining the balance of plasticity, connectionist computer network models must deal with the task of finding the right weights that learn the concept. For the gender discrimination task, for example, you present the network model with a face, and it computes its best guess (male/female) based on the weights it has among its nodes (neurons). Initially the weights are random, so the discrimination (male/female) is also random. So how do you tune the weights to give a better answer?

The answer is training. As we saw earlier, computer scientists devised a way to "train" their connectionist models to give the right answer. They created algorithms (numerical methods) to "propagate" the error from the decision (saying female if the face was a male) back to correct all of the connection weights of the network. That way, any weight that had changed in a certain direction when the output was wrong was switched in the opposite direction. Similarly, if the decision was a correct one, then all of the weight changes were kept because they were going in the right direction. This simple "learning" algorithm (called "back propagation") allowed connectionist networks to be "trained" to achieve remarkable feats of concept formation.

The detractors of connectionism, of course, cried out that this was blatant cheating. The networks didn't really *learn* on their own but had to be "supervised" and trained weight-by-weight to give the right answer.

Now, everyone had to admit that the networks *did* learn discriminations—and, by implication, concepts—that were never before captured in computers. However, the process of training in the early connectionist simulations was indeed artificial. For each error, you had to take the degree of error and go back and poke it into each connection weight throughout the whole network. This was inelegant. Moreover, even in computer simulations, this was tedious enough that critics complained that such simulations were irrelevant to the real-world learning of the human mind.

I sometimes wonder whether these critics of connectionist models ever had children, or, if they did, whether they remember what it is like to train a human child to make elementary discriminations. It's not like the little creatures simply resonate to the sensitivities of the culture and spontaneously acquire the social graces. Without extensive and patient training, they are glad to relieve themselves on the floor. Getting children to assume even a modicum of civility requires training with a level of detail and patience that makes the back-propagation training of a connectionist network seem quite reasonable.

Kids do, of course, find ways to learn. They learn to self-regulate their actions to get what they want. If we could understand how they do this, maybe we would find clues to the evolution of control in large-scale computational networks. If so, maybe we could take this understanding—now in explicit scientific terms—and give it back to children. We could teach them how to understand their own mental process as they gain the self-awareness and abstract intelligence of adolescence.

We can derive a new principle of representation from the research on training artificial neural networks: *Concepts—like kids—adapt. Or not.* In connectionist modeling, getting the networks to form concepts requires making them adjust themselves to decrease the error between what they imply (the actual output strengths) and what they should imply (the goal output strengths). Here we have an important fusion of the two properties of intelligence that we are searching for in this chapter: representation and control. The concept that is achieved by the network is its representation, regardless of whether it is a concept of a face or some other complex pattern. Furthermore, the network achieves this representation only by integrating control—the adaptation to the goal—within its very structure. If minds handle information as machines do, then adaptation must be integral to the representations of intelligence. When they stop adapting, the networks stop learning.

Understanding a Friend

To exercise these computational principles, we can return to the thought experiment on Jared's interaction with Andrew. The question is how distributed representation may figure in Jared's understanding of this interaction. We can begin by considering his immediate understanding that the transmission was from a stolen car. First, we will think about the formation of this concept in the general terms of parallel-distributed representations. Then we will apply the computational principles to Jared's brain in order to study one of the dimensions of the structure of intelligence that we examined in Chapter 2, specifically the differing architectures of the left and right hemispheres. We can then reason through some of the neural structure that must be integral to a concept like the one Jared is struggling with.

When Jared heard Andrew offer him the transmission for $100, he became uncertain and even confused because this price was too low. This fact became potential information because it was not congruent with Jared's knowledge of his world. In addition to the uncertainty, it was potential information because it was valued (because Jared needed a transmission). Jared's neural arousal increased with his interest and uncertainty. As his attention became focused, he took in Andrew's communication more intently than before. When Andrew said he had gotten the transmission on the street, Jared quickly surmised it was from a stolen car.

When I initially presented this interaction to you, I naturally moved on to more complex issues, such as Jared's desire for the transmission and the conflict this created with his concept of himself as an honest person. However, if we stay focused on the concept that the transmission was stolen, even this elementary process of concept formation presents an important illustration of the structure of a mental representation.

The Spread of Meaning

First of all, Jared formed the concept quickly but not immediately. If we could examine his mental process in the 1 or 2 seconds during the formation of this concept, we might find that it required connections between broadly distributed mental elements. Furthermore, these connections were made in a coherent, orderly fashion. To understand how they might have been formed, it would be helpful to review the evidence on cognitive structure from research on spreading semantic activation.

These experiments make use of the fact that cognitive processes take time. The mind is often fast, but it is always finite. Therefore, pioneers in cognitive psychology such as Michael Posner have realized that they can use the physical time of mental operations—the study of mental chronometry—to analyze important features of mental structure.

In this line of experiments on spreading activation, the question is, in conscious experience, how long does it take to access the meaning of a word? Researchers present a word (called a target) to a subject and then ask the subject simply to press a button if the target is a real word or another button if it is a pseudoword. For example, "flarb" is a pseudoword, so you would not press the "real word" button, but "tire" is a word, so then you would.

Once your speed of activating the word meaning is established, we move on to another experiment. A split second before the target word is presented, another word is presented, called a "prime." If the prime is related to the target word (for example, if the real-word target is "tire" and the prime is "wheel"), then you would be faster to press the real-word button to "tire" than if the prime is unrelated (for example, "basket").

This result shows that, within that split second between the presentation of the prime and the presentation of the target (for example, 100 milliseconds or one-tenth of a second), a mental activation of some sort must have spread between the prime and the target, facilitating mental access to the target.

This is interesting enough, but it gets better. If you make the split second a little longer (e.g., 200 milliseconds), then your recognition of the target will be influenced by words that are not only closely related (like "wheel" is to "tire") but also more remotely related (like "engine" is to "tire"). It is as if some gradient of "relatedness" in the mind requires a physical operation to be traversed. We know it is a physical operation because it takes a predictable amount of time.

Through many such tests, experimental psychologists have determined that the spread of meaning—from an idea to its close associates and then to more remote ones—happens over time in a regular and predictable gradient of relatedness. They have called this process "spreading activation." The implication of this term is that concepts are related in memory in coherent networks of meaning and that regions of these networks can be activated through use.

When we apply this evidence of mental structure to complex, real-world concepts, we see that the spread required to form a complex concept like Jared's must be extensive, and it must be fully distributed across diverse informational elements. There is no logical rule that stated that cheap parts are stolen. You would infer this if you were a kid in this particular social circle but only because of an extensive context of knowledge of life on the street. It could have required 50 or 60 social interactions with his friends before Jared built the knowledge base that allowed him to take in Andrew's statements, then form the concept that the transmission was stolen. Jared was not conscious of all of those interactions when he formed the idea; he did not experience a spreading of this new information across the memories of the last 3 years of his life. However, a kind of parallel-distributed activation must have occurred to relate Andrew's words to Jared's extensive knowledge base. As this occurred over the few seconds that Jared listened to Andrew, a new concept of the transmission as stolen goods took form within Jared's awareness.

This is an implicit form of knowledge. If you asked Jared how he made this inference, he might say something about what he knows about parts, cars, and his friends, but this would be a rationalization, not a coherent introspection. In actuality, Jared's concept was formed quickly but not instantly through a parallel-distributed process in his preconscious mind. The effect is similar to the pattern completion that has been demonstrated for connectionist computer simulations, for example in recognizing a face from a few fragments of a picture. As a result of this process of semantic activation, what was delivered into Jared's awareness was a suspicion, a tentative conclusion, with little knowledge of the extensive—and distributed—mental development that brought it forth.

Many forms of real-world knowledge are like this. Rather than being explicit, denotative, or logical, they are implicit, connotative, and analogical.

Neural Structures of Meaning

Although we have considered Jared's concept of the stolen transmission in psychological terms, as if it were a single coherent mental thing, it may have a more complex existence within the physical architecture of Jared's brain. When we consider its structure in neuropsychological terms, we encounter evidence that even an everyday concept like this may have a complex componential structure. Components of this concept may take form differently on the two sides of Jared's brain.

In Chapter 2 we saw that the left and right hemispheres seem to support differing forms of network architecture. These different physical structures may be relevant to their dissimilar psychological structures: the left hemisphere's analytic skills and the right hemisphere's holistic skills. Some psychologists studying spreading activation questioned whether the left and right hemispheres contribute differently to conceptual structure, even in the highly organized domain of language.

Even though the left hemisphere is most important to language, the right hemisphere also seems to play a role that is consistent with the right hemisphere's holistic conceptual skill with nonverbal information. Patients with right hemisphere damage often fail to get a joke or the gist of a story even though they understand each word and sentence. The implication seems to be that the right hemisphere forms a holistic conceptual framework that then serves as an embedding context for the left hemisphere's more analytic parsing of specific verbal propositions.

Considering this theoretical framework, several psychologists adapted the spreading activation experiment to test the left and right hemispheres separately. One way to do this in an intact person is to present the prime and target words to one visual half-field (such as to the right of the fixation point of vision). Because the visual fields are crossed (Figure 2.5), the initial visual input is then to the hemisphere opposite that visual field (e.g., the right visual field projects to the left hemisphere).

What happens with this experiment is that the spreading of meaning across the elements of memory seems to occur differently on the two sides of the brain. Furthermore, there is a suggestion that qualitatively different mechanisms may be involved, with a form of inhibitory control observed in the left hemisphere that is not seen in the right.

Consistent with what would be expected from the brain lesion evidence, the left hemisphere shows a tight pattern of semantic activation in which close associates are primed ("tire" primes "wheel") but remote associates are not primed ("tire" does not prime "engine"). In the right hemisphere, on the other hand, the remote associate *is* primed, consistent with a more holistic, expansive pattern of semantic (meaning) memory activation on the right side.

This would make sense with the language problems from left and right hemisphere brain damage and with the general theory of analytic (tight) concepts on the left and holistic (loose) concepts on the right. However, what was unexpected was the time course of these effects when they were examined more closely. The left hemisphere *does* show activation of remote associates if priming is tested in the first 100 or 200 milliseconds. Then, after another 100 milliseconds, it shows activation only of the close associates. This finding implies that, in that fraction of time, the remote associates have been inhibited. In contrast, the right hemisphere does not show this inhibitory effect, so that the remote associates remain activated.

The integration of the neuropsychological evidence with the cognitive psychological evidence thus suggests a unique insight. It points us to ways of thinking about the differences in control process, as well as the resulting representational structure on the two sides of the brain. The right hemisphere seems to show the more basic or primitive mode of spreading the representation of meaning across cognitive elements. The left hemisphere also shows this mode—at least at first—but then it evidences a secondary control process that somehow acts on the primitive spread to restrict it to the closely connected meanings.

Further research on these hemispheric differences may lead to an improved theoretical understanding of the mechanisms underlying the left hemisphere's language skills. Tight control of the meaning in the relations among words may be important to the orderly sequencing of words in sentences and to the cognitive skills that follow in the left hemisphere. These include not only the advantages of grammar but also the rational control of logical inference. When psychologists diagnose the severe mental illness of psychosis, one of the first clues is the loosening of associations, in which the patient's thought processes bounce from one meaning to the next without maintaining logical order. This may reflect the loss of rationality in a specific formal sense as the left hemisphere fails to provide its monitoring and sculpting of semantic associations into tight, controlled, sequential, and thus verifiable meanings.

Yet understanding life's mysteries is not always possible through logical inference alone. Since we know that right hemisphere damage causes deficits in understanding stories and jokes, there is a place for the right hemisphere's simpler algorithm of spreading meaning across associations, however remote. This broader engagement of meaning and prior knowledge may be as important to the intelligence of everyday life as the tight linguistic structures of propositional thought.

Physical Limits of Mind

Research on hemispheric contributions to intelligence has many questions to address, but it may be useful to reflect on the simple fact that we are conducting a scientific analysis of the mind's operation in relation to brain tissue. Cognitive psychologists such as Posner have taught us how to study the time course of mental operations with great precision. Not only is this temporal analysis a powerful experimental tool that allows us to look into the unfolding of meaning over each brief interval of time, but it also gives us insight into the physical constraints of the mind. The mind extends in time. There is no magic here that escapes the physical constraints. If there is not enough time, the meaning does not spread.

Moreover, the mind extends in space. This is the physical space of neural tissue. Although the evidence is still preliminary, we have clues to the structure of this physical extent. In fact, if we remember the anatomy lessons in Chapter 1, we will recall that the lateralized forms of network structure implied by Josephine Semmes's studies may be responsible for the hemispheres' differing forms of semantic structure.

The left hemisphere's restriction of spreading semantic activation (from a diffuse spread back to a tight center) may reflect the same mechanism that allows its specialization for like elements in any given cortical location in Semmes's analysis (Figure 2.6). The inhibitory control confines the activation of meaning to the close associates. In contrast, the right hemisphere's more primitive mode of diffusely spreading meaning may be consistent with its more widely interconnected pattern of cortical organization, with any given cortical location integrating unlike elements.

Of course, Semmes's best evidence showed differing representations of sensory and motor maps within the two hemispheres. In contrast, the studies on semantic structure deal with more abstract, cognitive representations. However, if a common theoretical approach proves workable for both sensorimotor and verbal semantic domains, this would be further evidence of the common basis of embodied cognition.

As we apply the reasoning from computational science's parallel-distributed representation, we can see that both hemispheres of the cortex appear to operate through some form of parallel-distributed representation. The evidence of left hemisphere semantic inhibition may be an important clue to how it achieves the more symbolic, logical operation of language and mathematics. Just as it was for many years a quandary how a computer operating on digital logic (addresses and instructions) could achieve mindlike functions such as recognizing a face, it is now a fundamental problem to determine how to get a parallel-distributed neural network to achieve the organized, sequential control of propositional logic. The left hemisphere of the human brain clearly does this, and yet it is undoubtedly a massively parallel computational network. By carrying out careful experimental studies of its cognitive process, we may learn from the left hemisphere its secrets of computational order that allow symbolic operation.

The importance of today's scientific analysis of the mind is not just the direct conclusions it provides but also the integration of the differing theoretical perspectives on the mind. Even though the computational models are still "toys" that serve primarily to illustrate the principles, these principles are illuminating. We can now review the evidence on the connectional structure of the cortex in a new light, with a more explicit interpretation of the implications of anatomy with regard to the mind's function. Even though the cognitive research on hemispheric semantics is still preliminary, it links powerful laboratory methods on language analysis to the classical neuropsychological evidence on functional asymmetries between the hemispheres. We can analyze both brain and mind in both space and time, within a single theoretical model.

Bringing Neural and Computational Models to Mind

Can this analysis be extended to the mental experience of everyday life? The answer depends on the breadth of theorizing we can tolerate. In an interesting twist of perspective, the degree to which we generalize the ideas of left and right hemispheric computational mechanics may depend on those mechanics themselves. If we adopt the tight semantics of the left hemisphere's fully developed associations, then remote associates are suppressed. Only exact implications are allowed to enter consciousness. This is the mode preferred by many experimental scientists, in which the conclusions can be linked through direct implication to discrete observations of the experimental data. A theoretical model must be developed to make the work scientifically interesting, of course. However, in psychology and cognitive neuroscience, this model typically stays very close to the laboratory and makes predictions about similar experiments rather than the psychology of real life.

To apply neuropsychological reasoning to everyday mental function requires dealing with substantially greater complexity than the cognition that is captured in laboratory experiments. This complexity can be approached with more inclusive concepts that capture the gist of the meaning even if the specifics are fuzzy. And of course these are exactly the ideas that draw on the right hemisphere's more primitive, loose, and less differentiated semantics. To extend a scientific analysis to everyday experience, we must accept a looser level of description and a greater degree of ambiguity.

The challenge of bringing a theory of mental function to everyday cognition is a daunting one. Complex processes of mind are engaged by even the most routine activity of social interactions. To continue to study this complexity, we can again consider Jared's understanding of what Andrew said but this time in relation to the contributions from both cerebral hemispheres.

After Andrew leaves, Jared goes outside the lunchroom, sits on the curb, and watches the traffic. His experience is coherent, with a more or less linear train of thought running through his consciousness. Nonetheless, even though Jared has no conscious knowledge of the contributions from the two hemispheres, we can infer that each of these contributions provides a different pattern of elaborating the meaning of the ideas that Jared engages. Each of these hemispheric ways of thinking has its advantages.

Many times compelling ideas (even if they are disturbing or repulsive) capture attention only because of their attractiveness and not because of their logical sufficiency. This may be happening with Jared's idea that Andrew stole a car. If Jared applies a tight semantic analysis to what Andrew said, there is no direct implication of a theft. The critical thinking that Jared must do now extends beyond the literal interpretation of Andrew's statements. He must reason about things such as the likelihood of the theft occurring generally, whether Andrew has been hanging with new guys who may be suspect, whether there is some other explanation for the low price, and, finally, how to interpret Andrew's statement that he got the transmission on the street.

The critical thinking Jared applies in this process depends in no small part on his left hemisphere's capacity to structure each semantic relation with a high degree of specificity and precision. We can even draw from the computational and neural evidence to speculate on the physical pattern of meaning that is formed in Jared's left hemisphere in each instance of his reasoning. What are initially broad meaning patterns become sculpted into tight webs of verbal representation by the left hemisphere's unique inhibitory bias. Because of this restriction to close associates, it may be that the tight spheres of verbal meaning in the left hemisphere have greater stability and salience in conscious experience when contrasted with the more diffuse, primitive, and intuitive spheres of meaning in the right hemisphere.

However, if Jared has access only to narrow, literal ways of thinking, this will not achieve a particularly rich understanding of the situation. The idea of a theft emerged as a spontaneous insight, a gestalt that was formed from the relational pattern of information that linked the relevant facts in Jared's understanding. This pattern could form only within the broader context of Jared's domain knowledge, which includes a rich array of experiences with the guys he knows and the stories he has heard. The broad, hierarchic, conceptual integration of this contextual knowledge requires the unique semantic and neural network properties of the right hemisphere.

Imagine that waves of activation spread broadly on the right side of Jared's brain, allowing inclusive and holistic patterns of thought. Yet, with this breadth comes correspondingly less instantiation of the pattern in conscious experience, and the meaning consequently tends to be distributed at the fringes of Jared's awareness. This may give the right hemisphere's semantic contribution the quality of an impression or intuition rather than a clear and distinct realization.

These patterns of thought must be dynamic and ephemeral, shifting with each new association that is uncovered and each thought that competes for Jared's attention. The selection of the thoughts that gain attention must be made on multiple grounds, including coherence, salience, and most of all significance. It was the significance of the idea of the stolen car that galvanized Jared's attention to this idea in the first place.

How is significance attached to a cognitive representation? This must be the most important of control processes, ensuring that the scarce resources of the mind are allocated to motivational priorities. Yet in many studies in experimental psychology and cognitive neuroscience, the motivational significance of cognition is poorly understood. It sometimes seems as if the person in today's research studies is a cognizing machine, bound to process information regardless of its motivational or emotional value.

Yet human cognition is always motivated. In reality, an important value of testing scientific theories against the cognition of daily life is that we cannot ignore the obvious facts, such as the importance of motivation to the thinking of real people. Exactly how do we bring motives, such as the fear of danger to a friend, to bear on our scientific analysis of the neurocomputational basis of concepts? How do certain representations gain motivational significance that energizes their informatic tenure in the life of the mind?

Important clues come from studies of animal learning. Although interest is now increasing in researching human emotional and motivational processes, the studies of animal learning have provided the most fundamental scientific basis for understanding animal information, in which motives are intertwined with cognitive representations.

To be sure, this is laboratory research, so it examines the cognition of small animals (mostly rodents) because they are easily kept in the laboratory. As a result, some readers might question the relevance of this work for their own questions about subjectivity. It may not be immediately obvious what they have to learn, personally, from rodent consciousness.

Many readers, even if they have had a psychology course, may be surprised by the developments in animal learning theory that typically fail to make it to the textbooks. Animal learning theory has taught us not only about motivation but also about complex cognition that we must infer goes on in the brain of rats. The highly strategic choices that rats show in structured learning experiments are impressive. Of course, there are important aspects of the human brain that are simply undeveloped in the diminutive rodent cerebrum. Yet the research is showing that the basics—the fundamental questions we must address to understand the mechanisms of experience—are remarkably similar for all of the unique variants of mammalian neural architecture.

Information in Animals

Representations in connectionist networks might not be associated with value. Because of this, according to our analysis of Shannon's definition of information in the Introduction, these representations may remain patterns of data, never becoming real information. How do we know what representation should be formed? How can goals or motives be monitored in a way that directs the formation of the defining weights in the network? How can such concepts of goals or motives be created in relation to the organism's fundamental values—its survival needs?

We have learned from artificial neural networks that external feedback from errors is difficult to incorporate within the connectional patterns of the network. Could values somehow be integrated within the network from the outset? In this regard, research on rat choice offers insights into the nature of human values and self-control. Problems of learning that seem to be intractable for machines are solved quite readily by these modest animals.

To be sure, the behaviorists who studied rat learning did not expect insights into information. Yet this body of evidence has shown definitively that, for mammals at least, learning is not a passive association of stimulus and response. Rather, it proceeds from the animal's cognitive representations of future events. These representations include integral representations of values, and the animal's elementary expectancies consequently serve as reference frames for evaluating events and regulating the consolidation of experiences in memory. Even simple animals routinely carry out complex processes of motivated intelligence that we cannot begin to approximate in our machine models.

The Remarkable Discovery of Rat Plans

The goal of learning theory, of course, was not to attribute mental activity to animals. Quite the contrary, the goal for behaviorists was to explain behavior—habits—as passive effects of training, with no reference to the animal's internal mental processes. Based on Pavlov's pioneering observations at the end of the 19th century, behaviorists attempted to derive the simple rules of behavior from reflex mechanisms, without intervening internal (cognitive) mechanisms.

In Pavlov's original studies of digestion and salivation, a reflex is observed when a biologically significant stimulus (such as the appearance of food) elicits a reflex or unconditioned response (such as salivation). From this primitive control of behavior, the behaviorists attempted to examine the principles of association (conditioning) through which new, learned stimuli gain the ability to elicit the behavioral response.

Studying the biological roots of animal behavior naturally begins with reflexes. The simplest creatures, such as worms, respond reflexively to stimulation such as pain, for example by twisting away from the painful stimulus. Furthermore, the elaboration of more complex behavior in evolution has involved not only more complex instinctual patterns but also the capacity for learning.

If there is a general principle of behavioral evolution in higher animals, it is that an increasing delay is interposed between stimulus and response. This delay allows the greater flexibility in behavior that has allowed complex animals their success in adapting to changing environments. This delay is made possible by the mechanisms of memory. A workable theory of memory, therefore, becomes the key to unlocking the more complex problems of experience and behavior.

How does an animal learn? How does it remember the important facts of its situation, so that it can form the new habits required in a novel environment? One of the first formulations of a principle of learning was Thorndike's law of effect: If a behavior results in satisfaction, it will become more strongly associated with the situation in which it occurs. In the early decades of the 20th century, psychologist John B. Watson popularized Thorndike's law as a fully sufficient explanation for not only animal but also human behavior. Watson termed this approach *behaviorism*.

Maybe it was an inferiority complex in relation to the scientific rigor of their physical science colleagues. Whatever these realities of academic culture, it was behaviorism—and not cognitivism—that attracted the staunch adherents in the early days of psychology in the United States. In fact, behaviorists became not only staunch but also positively fundamentalist. Learning, and by extension all of human behavior that extends beyond our base instincts, was seen to proceed through a process of reinforcement, a strengthening of habits by rewards and a weakening of them by punishments. Because the process was rigidly mechanistic, even the powerful theoretical developments in control theory in the middle of the century did not seem relevant.

Control theory may have been one of the more significant intellectual advances of the 20th century. It presented a formal theory for the way in which processes of feedback and feedforward could be used in controlling ongoing processes with great precision. This approach, termed *cybernetics* by its leading proponent, Norbert Wiener, found immediate applications in electronics. Concepts of cybernetics were then seminal in the foundation of computational science. From the outset, however, cybernetics was recognized as integral to animal and human behavior. The subtitle to Wiener's 1948 book was *Control and Communication in the Animal and the Machine*.

However, the behaviorists wanted simple, not complex, notions of control. And, of course, they were positively repulsed by any suggestion that behavior was determined by an animal's capacities of cognitive representation. So control theory did not revolutionize the study of behavior the way it should have. Eventually it was not just the experimental psychologists who signed on to behaviorism. By the 1960s, when cognitive psychology was on the rise, the largest U.S. university departments that were training clinical psychologists had taken up behaviorism. These psychologists proposed that stimulus-response reinforcement principles are adequate not only to explain human behavior but also to treat the varieties of human mental illness and emotional disorder.

Even during what now seems a dark age of psychological science (the Age of Behaviorism), certain brave intellects never faltered in their display of conceptual learning. These were the rats. In thousands of learning studies, these sturdy creatures lived out their lives in small chambers, with electrical switches measuring out small increments of food or water only when their vigorous actions met the stringent study criteria. The rats had their behaviors reinforced, to be sure. They were reinforced up, reinforced down, and, undoubtedly, in unpublished conference proceedings at least, reinforced sideways as well. But throughout these trials, the rats held out hope. They showed that, in their own small ways, they had plans.

Hope, of course, is not something that behaviorists expected to find in the rat mind. In fact, the whole idea of behaviorism was that mental processes could not be studied with an objective, materialistic science. Therefore any mention of cognitive representation was impolite, and the idea that there could be cognition going on inside a rat was indeed offensive to the behaviorist mentality.

I admit that "hope" may be a somewhat optimistic term for rodent cognition. You will certainly not find any mention of rat plans in the psychology journal articles of the times. However, in their responses to the behaviorists' experiments, generation after generation of rats showed a steadfast capacity to expect good outcomes and to adjust their behavior accordingly. The facts of rats' learning abilities have important theoretical implications, so it is worthwhile to consider the accomplishments of these admirable creatures. Furthermore, this is not just for the philosophical insight or general inspirational value we may gain (although to me both are noteworthy). The evidence on the mechanisms of rat learning provides instruction in the motives and values—the adaptive controls—that appear essential to the evolution of mammalian intelligence.

The Partial Reinforcement Extinction Effect

One finding had to do with the consistency of reward. Rewards that were more consistent were found to be *less* effective in controlling behavior in the long run, when compared to inconsistent rewards. Said differently, rats that had been given variable schedules of rewards (sometimes getting the reward but sometimes not) were *more* persistent in their habitual actions when the reward was removed, compared with rats that always received the rewards.

Learning theorists were quick to accept the reality of variable reinforcement effects and to cite them in the textbooks as "laws of behavior." However, behaviorists such as B. F. Skinner ignored the logical problem that these effects invalidated the mechanisms underlying stimulus-response learning. According to Thorndike's law of effect and similar statements of reinforcement theory by Hull and others, the rats with the steady rewards should have had their responses reinforced more and therefore should have shown stronger habits.

Although the facts of learning experiments presented a logical problem for behaviorists, learning was always second nature for the rats. In fact, if you imagine that you are the rat, the partial reinforcement extinction effect is not surprising at all. If you were sometimes deprived of the reward, you would be used to the fact that it might not appear. Furthermore, you would naturally expect it to come back eventually, so you would keep trying. In contrast, for consistent rewards, you would be likely to give up when they stop, concluding they would never come again.

Of course, rats don't think like us, and it is undoubtedly unscientific to assume they do. Yet, because the animals respond differently to variable versus consistent rewards, we must infer (even using the most critical logic) that the rats develop an internal representation (a concept of some sort) of the reward probability. This representation of reward probability is scientifically necessary to explain their response to the removal of the reward (i.e., their unlearning of the associated behavior, also called the *extinction rate*). Something more than a mechanical reinforcement of habit strength is going on here, and it has to do with the cognitive representation of anticipated events that must be formed in the rat's brain.

Maybe we shouldn't call it hope. After all, this is just a rat in a cage. And maybe we shouldn't call it a plan. The poor rat probably has only the vaguest degree of consciousness and maybe none at all. Yet no matter how you look at it, this is an *expectancy*—a representation of future outcomes—that must be recognized as determining how the rat responds to the withdrawal of rewards. In addition, it is a *valued* expectancy, one that anticipates the hedonic quality of forthcoming events. Whatever it is for the rat, it sounds like hope to me.

A theory of reinforcement of habit strength, in which the residual strength of the behavior is a passive effect of training, simply does not explain the facts of these experiments. This is the reality of modern animal learning research, and it should be more widely appreciated. Rather, the findings imply that the rat has an internal representation of the probability of reward, and this representation— this concept or expectancy—interacts with the changing events (the loss of reward) to determine the behavior that follows (the rate of extinction or loss of the learned association). We can formulate a principle of animal learning that captures these effects: *Rats know what's happening.* Even these modest mammals have accurate expectancies about future events. They then use these expectancies as guides to evaluate discrepancies when actual events fail to meet the rat's contextual model for what's happening. Rats learn by metering events against their expectations.

The Magnitude of Reward Extinction Effect

A closely related paradoxical reward effect occurs for big rewards. If a rat is trained with large rewards for its actions, it shows faster extinction (unlearning of the response when rewards are stopped) than if it was trained with small rewards. Again, such a principle would not be surprising neither to dog trainers nor to corporate management. However, according to the law of effect and to classical behaviorism, big rewards should lead to strong habits, not weaker ones. Apparently, rats form an expectancy of the magnitude or value of the reward (a big hope), and their behavioral response to nonreward seems to be gauged in relation to the difference between this nonreward and that expectation. In handbooks of dog training or personnel management, this might be called disappointment.

The Successive Negative Contrast Effect

Rats are especially quick to show their awareness of the value of events when they are cheated out of the rare food treats that become the highlights of their impoverished laboratory lifestyles. Although it may not sound like a delicacy to us, for a laboratory rat it is a treat to get bran mash, especially in contrast to sunflower seeds. So if a rat is given bran mash after it presses a bar in its cage, it learns to work the bar for the mash quite sincerely. If it is hungry, the rat will also work for sunflower seeds but never, apparently, with as much enthusiasm.

The interesting result for learning theory happens when the rat is trained with bran mash but then switched to sunflower seeds. Reliably, its work output drops below what it had been when given sunflower seeds all along, as if the rat felt it had been cheated. "Well, of course," you might say. This sounds like what happened when the people in engineering were given bonuses last year. When they didn't get bonuses this year, they were all whining in no time, whereas the poor souls in customer support were happy to work for their paltry wages all along.

Again, we must be careful not to draw too close a parallel with human experience. Managing personnel certainly requires more skill than training rats. What is important for a scientific psychology is the minimum inference that is necessary to explain the behavior. Under classical reinforcement theory, habit strength is directly proportional to the magnitude of the reward. In this analysis, sunflower seeds are sunflower seeds, and they should be no less rewarding just because you've been spoiled with a bran mash diet.

Yet the rats act as though they know the difference, and they're disappointed when they don't get what they expect. This implies that their representations of the learning context and the behavior tracking those representations involve an inherent expectancy of the value of the reward they should receive. Rat learning is not responsive to the value of the reward in an absolute sense, as a homeostatic reflex might be, but rather to the expectancy the rats develop for what they should get in that situation.

Another principle thus suggests itself: *Hope is the measure of the present lot.* Perhaps, not unlike rats, we ascribe value to events, not in relation to an absolute appraisal but in relation to an expectation of what should be happening.

Cybernetics of Perception and Action in Time

Modern studies of animal learning have thus led us to conclude that rats and other mammals of similar intellectual complexity act upon concepts. This is a logical necessity implied by observing the conditions of their behavior. But why do these animals form expectancies that then guide their behavior? We can find interesting answers to these questions by applying the control theory that the behaviorists could not seem to understand.

We have seen that the evolution of complex brains involves an increasing delay between stimulus and response. Simple creatures have reflexes that link the sensation of a stimulus to an action. More complex creatures interpose memory, a delay, and the action may become a more complex response as a result. It becomes a response not just to a stimulus but also to the confluence of the stimulus and that particular context.

The complexity of response organized within this delay requires intelligence. Yet this capacity to represent a situation flexibly is still organized within the bodily mechanisms that guide sensation and action. The rat's behavior requires highly structured motor patterns. In controlling those patterns, the rat needs working memory to organize the sequences of actions. Before you get to the good-smelling thing, you have to get through this wall. In this way, the motor pattern needs a plan to guide it, an expectancy for action and its rewards.

Furthermore, the sensory system is clearly integrated within the plan. The rat forms sensory targets that are criteria for the actions. Getting closer to the good-smelling thing is a major criterion, but in the real world of rat problem solving, there must be intermediate criteria (getting around the wall) that can be held as effective goals in the short term. The sensory representations become guides for action, integrated within the motor plans as essential self-regulatory set points. The successful rat needs vision.

Thus, a concrete analysis of sensory and motor controls may be useful to keep the rat's cognition within the necessary temporal context. Expectancies are formed in mammalian brains because the cybernetics of sensory goals and motor plans require not only integrating over recent history but also projecting this history to anticipate the future. When the control systems are only reactive, they are slow. Slow is not good when you're a rat or any other creature that needs more than its reflexes and primitive instincts. When control systems are predictive, they are fast, particularly when the future is like the past. Rats have evolved to a level of complexity that allows their sensory and motor patterns to reach from the past into the (probable and hoped for) future.

Human brains, of course, have evolved to elaborate their memory structures to allow much greater spans of personal history to be brought to bear on present plans. This allows elaborate planning for future hopes and schemes. Nevertheless, as we studied the sensory and motor structures of human intelligence in Chapter 2, although they are perhaps less concrete for you and me than for a rat, we must recognize that these bodily structures (within their visceral and somatic compartments) still form the foundation for human experience and behavior. The lessons we learn from rats about expectancies guiding behavior may be extended similarly. These lessons suggest insight into the spans of memory and attention that allow us humans increasing flexibility and choice in organizing behavior in time.

Learning and Being Informed

Another lesson that rats taught us is the "blocking effect." This lesson turns out to be fundamental in bringing information theory to mammalian biology. It shows that learning is dependent not on simply pairing stimuli but rather on the information value of events. To understand information value, even for rats, we have to consider the cognitive and motivational questions of uncertainty and need.

A rat is first presented with a tone at one end of its cage. At the same time it is subjected to a painful electric shock through the floor bars, zapping its feet. After a few repetitions, the rat learns that, when that tone comes on, a shock is the next item on the agenda. It is then time to relocate. This is called conditioning of a place avoidance response to the tone cue, beginning with the shock as the unconditioned stimulus.

Although avoidance conditioning is readily established, something different happens if the rat first learns that a light signals the shock. Then if the tone is presented together with the light, the rat fails to learn that the tone is *also* a shock warning. This occurs even if the pairing and repetitions of the tone and shock are identical to the no-light training.

This is another problem for passive learning theory, which predicts that the repeated association of the tone with the shock should give the tone punishment value through a passive conditioning, regardless of whether there was a light present or not, but that does not happen. The rats show a blocking effect, through which prior training with the light seems to block the aversive conditioning (association) of the tone cue. How can we explain this?

One might say that, once the rat learned that the light was the cue to retreat, then the tone was not informative. The tone carried no new information, and therefore the rat ignored it. No wonder it wasn't learned. In fact, the scientist who first brought these blocking effects to the attention of the animal learning community in the 1960s, L. J. Kamin, drew similar conclusions about his rats. At a national meeting, Kamin told his colleagues that his rats learned new associations *only* when the cues were *surprising* to them.

Now, proposals about the subjective lives of rats—such as what they find surprising—are what drove behaviorists to their limited worldviews in the first place. How are we to understand surprise in the rat mind? Behavioral scientists are not even trained in human phenomenology, so how could they be trusted to interpret the subjective qualities of rodent consciousness?

Nonetheless, other experts in learning theory quickly recognized the empirical, predictive accuracy of Kamin's observations. When they did experiments to test Kamin's assertions, other researchers found that the rats *did* seem to learn only when they were surprised. In an influential series of papers, R. A. Rescorla and A. R. Wagner described the experiments that allowed them to build on Kamin's observations to formulate an explicit theory for how the animal's expectancy regulates learning. Specifically, they showed that the best predictor of the rate of learning of a new stimulus is the degree to which that stimulus is discrepant with the animal's expectancy for what should happen in that situation.

Remarkably, formulated in this way, even elementary animal learning turns out to be controlled by the information value of the signal. Blocking occurs because, if a stimulus is fully expected, it carries no information. It provides no reduction of uncertainty.

This line of evidence, when followed to its logical conclusion, suggests a fundamental relation between learning and information. It's not just that learning requires an informative (i.e., an unexpected) stimulus. The process of being informed is equivalent to learning. *To be informed is to have learned.* By the time information has been assimilated—by the time it is comprehended and your uncertainty is reduced—you are already transformed. You have learned.

Furthermore, consistent with the stability-plasticity dilemma, by this point learning has already exacted its price. Once you have learned, your old self is not to be seen again.

The Rodent Economist

Because I am not a behaviorist, I can allow myself to infer what must go on in the confines of the rat's modest cranium. The logical inferences from the rat's behavior are easy to draw and yet somewhat surprising. Rats approach the learning context with an expectancy of reward. Because this representation is modified by experience, the expectancy includes not only the probability of reward (as evidenced by the rate of extinction or decline of the response habit) but its magnitude as well (also evidenced by the rate of extinction to nonreward).

As any freshman business student can tell you, these properties of rat expectancy contain the essential elements of economic forecasting. The expected value of a future event (EV) is computed by multiplying the magnitude of its value (V) by the probability it will happen (p). Thus $EV = V \times p$. As evidenced by the rat's response when rewards fail to appear, the rat uses its concept of expected value as a continuous criterion for evaluating current events. Specifically, the discrepancy between the representation of expected value and the current outcome seems to be computed in some implicit fashion because *that* is what determines the change in the rat's behavior. If the rat's actions are not rewarded, its persistence depends on its computation of expected value, a computation that is based on experience with the probability of the reward.

The computation is actually more dynamic than textbook economic forecasting. The rat brain computes the discrepancy between the expected value and the current outcome, and it uses this to adjust behavior. If a reward fails to appear, this failure is evaluated in relation to the previous probability of the reward. If that probability was low, the rat is *more* persistent, a complex but effective cybernetic bias to match behavior to the likely future probability of success. If the reward is noticeably smaller than expected, the rat is *less* persistent, again because this discrepancy signals a change in the environment that warrants adaptive regulation. If a signal occurs that is not expected and yet this signal is significant for rewards or punishments, the rat learns that signal quickly because the discrepancy signals that the current expected value representation—the concept of the adaptive context—is defective and must be updated.

The rat's self-regulation is thus cybernetic, operating according to the control theory principles of feedback and adaptation. It is also representational, in that certain concepts of events and their values are held in memory in some form and then guide its behavior. Remarkably, the rat's cognitive evaluation is also highly dynamic, rather than static, in that it continually computes the deviation of ongoing events in relation to the concept of expected value. It is this deviation or discrepancy (such as in relation to large rewards) that leads to changes in behavior.

We might say that, as they studied rat learning and memory, the behaviorists expected that rats would be bookkeepers. They would passively record the events of their environment and dole out the passively associated actions accordingly. Instead, the rats turned out to be entrepreneurs, forecasting both the likelihood and value of future events, looking out for downturns, and capitalizing on opportunities. When the expected events did—or did not—come to pass, the rats were prepared. Like good economists, their concepts of expected value were ready to serve as set points for evaluation and self-regulation. The outcome of the evaluation, based on the discrepancies from the set points, was then the control mechanism for learning. Even the rat is an information machine, turning events into information in the context of present values.

The extension to human economics seems straightforward. *Economics is what remains when hope is done.* Hope is the guiding motive engine, and what we capture with our economic analyses are the concrete residuals of its effects on our choices. The motive force for business is the expectation for good outcomes. This expectation is a subjective process—and a valued one. Whether it is considered as consumer confidence, business management confidence, or worker confidence, this fundamental psychology of business expectation is as dynamic and fragile as any form of hope.

There are certainly effects of aggregating expectancy across the business elements of a country or a global economy. Yet, as evidenced by business cycles and market crashes, when it is frustrated, the human expectation for good outcomes responds—in aggregate form—with the same deflation as the loss of hope in any rodent brain.

Losing Hope

And when the market fades, hope is lost.

As sturdy as they are in facing the unmentionable deprivations of laboratory captivity, even rats can be frustrated to the point that they lose hope.

Specifically, behaviorists sometimes rigged the experiments so that the rat's coping efforts repeatedly failed. The animal receives a noxious punishment (a shock, bright light, or loud noise) regardless of its behavioral effort. Then hope fades.

Without its motivating, valued expectancy, the animal soon manifests a condition of *learned helplessness*. Once in this condition, the rat becomes withdrawn and listless, and—importantly for learning theory—it fails to respond to learning opportunities that would lead to successful outcomes for a euthymic (normally happy) rat. From this evidence, we see that positive expectations are essential for normal learning.

In fact, the condition of learned helplessness is a good laboratory model for the anhedonia and loss of motivation seen in clinical depression in humans. Several features of the impaired neurophysiology in helpless rats parallel those observed in clinically depressed people.

These findings are important to understanding and treating depression. They are also important for a theoretical understanding of motivated learning in general. The valued expectancy shown by normal rats is not a fixed cognitive capacity of the animal's brain. Rather, it is a capacity that requires an adequate motivational (perhaps we could say emotional) basis. When frustrated to an extreme degree, the animal becomes depressed and withdraws. This appears to be an evolved capacity, apparently as a coping mechanism to conserve resources in a severely punishing environment.

Whatever the evolutionary cause of depression, the effect is theoretically significant. Without the normal motivational bias to expect rewards in each environmental encounter, the animal or the person becomes incapable of effective learning.

Human clinical depression is a phenomenon I have studied for many years, both through clinical work with depressed people and in my laboratory studies of how they think. It is easy to consider depression as a disease that is somehow distinct from normal emotion and self-regulation. Although it *is* a disease and in fact may be life threatening, there is extensive evidence that the effects of depression in pathological states are the same ones that operate in normal people. Normal people work toward goals, whether projects at work or in their backyards. In reaction to failure, the human brain's response will likely include depression, not unlike the rats' brain response to continued frustration. Depending on the seriousness of the failure and the reactivity of a person's mood systems, the depression may be transient and hardly noticed, or it may be extended and subjectively vivid.

To illustrate that emotions exert dynamic influences on the brain and mind, my associates and I have titled our research project "Depression and Anxiety as Neural Control Processes." The hypothesis is that the emotional responses that guide expectancies and react when expectancies fail are integral to normal as well as pathological learning. Although this hypothesis can certainly be debated, it is clear that the pathological emotion of clinical depression applies a strong control bias on a person's representational process.

A depressed person sees the world and the self through a markedly negative bias. This is not just a subjective feeling: The depressive bias causes the person to lose hope that things could be better, even to the point that suicide seems like a reasonable choice. This is not just an effect on immediate consciousness: The depression impairs learning ability just like learned helplessness does in rats. The result is that the depressed person feels unable to struggle with life challenges and therefore invites a spiral of further failures and greater depression. Although hope may often be an illusion, clinical work with human depression shows it may be a necessary one. Without it we flounder.

Clues to Evolutionary Structure

Although the rat's expectancy is automatic, primitive, and (we must assume) unconscious, it provides a remarkable device for behavioral self-regulation. This ability seems to reflect the memory capacity that appeared for the first time with the evolution of the extensive corticolimbic networks in the mammalian brain. As some of the more sophisticated learning theorists came to understand the integral role of expectancy in rat behavior, they questioned whether simpler vertebrates, such as fishes or turtles, would also show such effects. With limited corticolimbic networks, these creatures may not have the full complement of cognitive ability that we find in mammals. M. E. Bitterman, in particular, provided evidence that reptiles do not evidence expectancy-based learning, even though these animals do show the passive association learning implied by the law of effect.

A remarkable feature of the vertebrate brain is that it evolved through the progressive elaboration of certain basic structures. The early evolutionary biologists called this elaboration process one of forming "terminal additions." The newly evolved structures are added on top of the old ones at the end (termination) of the embryological process. It is as if each new addition is an afterthought.

Certainly there are important exceptions to this rule, enough that most modern biologists question whether it is reliable and are glad to throw it out as a historical error. However, it remains a fundamental organizing principle for the evolution of the vertebrate neuraxis. The principle of evolution through terminal additions explains why, in every mammal brain, we find the major circuits of a fish brain within the brain stem (including the paleostriatum of the basal ganglia) and then the reptilian circuits on top of that (including the neostriatum of the basal ganglia); only then do we find the mammalian circuits on top of those (including the highly developed limbic system and cerebral hemispheres).

Impressed by Bitterman's findings on the lack of expectancy-based learning in reptiles, as well as by the implications of these findings for the evolutionary structure of learning, Abram Amsel questioned whether rats might progress through a developmental stage in early life at a time when the fish learning systems would be operational and yet the mammalian learning systems would still be immature.

Amsel took young rat pups out of their nest at a point when they are little more than fetuses and cannot stray far from the mother's nipple. In order to keep the mothers from attending to the pups and confusing the experiment, Amsel and his students sedated the mothers. They then placed the pups at a distance from the nest and examined the learning abilities of the pups as they tried to reach the mother.

In several experiments, the Amsel laboratory found that the pups did not evidence the partial reinforcement extinction effect. Specifically, they did not extinguish their approach to the mother even when they were blocked from suckling. Yet just a few days later in their development, as additional brain systems became mature, the pups *did* show the partial reinforcement extinction effect, as if their criteria for learning had suddenly shifted to a new level.

The implication was that there are multiple learning systems in the mammalian brain, and, as they become mature, they are stacked vertically, one on top of the other.

Some readers may notice a queasy feeling as they read of Amsel's experiments and conclusions. Does this mean that the human brain exhibits an evolutionary order as well and that, without sufficient vigilance, some of us could be in danger of slipping into attitudes or patterns of behavior that are, well, reptilian? Amsel points to earlier experiments with normal college students that might give us pause on this issue.

These studies examined eyeblink conditioning, in which a cue such as a light or tone becomes the conditioned stimulus for the eyeblink caused by a puff of air. Fortunately, the college students who were subjects in this experiment showed expectant learning, as is befitting our revered human position in the phyletic order.

However, if during the conditioning procedure the students were given a separate cognitive task that drew their cognitive representational resources away from the eyeblink conditioning procedure, they showed only the limited, fishlike form of passive association learning (and extinction) for the eyeblink response.

For many psychologists at American universities, these conditioning results were a big relief, finally explaining those outbreaks of unusual undergraduate behavior during final exam week. With their limited cognitive resources overloaded by cramming for exams, the students' behavior naturally regresses to premammalian forms.

In theoretical terms, these results imply a hierarchic structure in mammalian learning and memory systems. Within it, a primitive learning mechanism may be engaged if the more complex expectancy system is otherwise occupied.

Expectant Action and Perception

The research on animal learning shows that a complex expectancy system is integral to mammalian behavior. Although human learning must be more advanced than the more elementary forms of simpler mammals, the mammalian adaptive controls that lead to expectant engagement of the world may be integral to each of the human structures of intelligence that we examined in Chapter 2.

Considering the differentiation of action networks in the anterior cortex and the perceptual networks in the posterior cortex, for example, the integration of these networks with the elaborated memory systems of the human brain allows extended cognitive representations to mediate between the perception of events and the actions taken. This key aspect of mammalian learning appears to become most sophisticated in humans: An increasing representational capacity acts to separate stimulus and response. Just as stimulus-response associations cannot explain mammalian learning, expectancy may be both a necessary explanation and a requisite development in human intelligence. It is not enough to delay responses based on an extended memory capacity and greater recruitment of networks of relevant meaning. People, like other mammals, need to act in the world rapidly, using their powerful memory capacities as effective guides, not as drags on response speed. The evolutionary solution seems to have been expectancy, which creates memories of the future that reach out to oncoming events based on predictions built up from previous learning.

Consider Jared as he drives home after school, taking the long way through the countryside (Figure 2.5). Many of his driving skills are simple sensory-motor patterns that have been developed through practice and automatized; consequently, they no longer appear in his conscious experience. As Jared negotiates a turn, for example, it is not just a feedforward operation that is directed by his initial guess as to how much to turn the wheel to make this turn. If so, he would be wrong many times and quickly end up off the road. Rather, he has continual visual input of the changing scene during the turn, providing him dynamic feedback on the progress of the turn and allowing him to adjust the wheel accordingly. In addition, he senses the feel of the drag of the wheels from the centrifugal force of the turn, both the rear driving wheels and especially the front steering wheels, letting him know how the "lateral g" of the centrifugal force acts against the car's suspension. Jared has learned to integrate these features seamlessly and unconsciously as he sees the turn, judges his speed, and evaluates the turning process. He has practiced these skills sufficiently that he is now aware only of the high-level operations because the more elementary sensory and motor cybernetics are subordinated to his automatized neural operations.

Even these elementary cybernetics engage expectancy, as Jared's knowledge of making previous curves with this specific car sets up a neural model of how the curve should go. This expectant model becomes the template for experience, and Jared then fine-tunes this template on the basis of the ongoing feedback. Even the visual perception he uses for the feedback must be expectant to a degree. He adjusts his motor action (turning of the wheel) as he anticipates the projection of his current speed and position to guide the turn. If he used only his current speed and position, he would be correcting too late and very likely already going out of control.

Just as the expectancies that guide animal learning are valued and motivated, so are Jared's expectancies that guide his driving. We hope that there are abstract expectancies that represent the potential dangers of fast driving and of failing to anticipate the mistake of another driver. At the same time, Jared manifests the expectancies of a healthy young man who is confident in the skills he has developed and optimistic that he can handle whatever challenge comes along. As he drives, the critical structures of Jared's intelligence are fine-tuned by motivational influences that are as important to his driving as the vision of oncoming objects or the feel of the wheels on the road.

Transparency of Mood Control

Is Jared aware of the motivational influence that shapes his expectancy? In clinical psychology we see the interdependence of emotional controls and cognitive representation in the negative expectancies of depressed persons. Perhaps more impressively, in patients with bipolar disorder, whose mood may swing from depression to mania, we see the grandiosity and exaggerated positive expectancies in the same person once they shift to a manic mood. But can people like Jared understand the dynamics of representation and control consciously in the flow of subjective experience?

Judging from the extreme influences of moods, such as those described by the subjective reports of people with manic depression, we cannot. The evaluative bias of moods appears to be an intrinsic, unseen influence that shapes the consciousness of the naïve mind. When depression sets in, your experience is *not* that you are the same person and that the world is the same, except that now you feel bad. Rather, you feel as if you are worthless and your future in the world is hopeless.

Even this is too tentative to describe the experience of the depressed person, who feels no uncertainty. You *know* you are worthless, and you *know* the world is hopeless. The fact that you felt fine or even manic the previous day seems irrelevant. Mood seems to act as a motivation in the immediate unconscious, and the motive-memory you should have—of how you felt yesterday—seems unable to provide any perspective whatsoever.

Although there may be factors that lead some people to be especially susceptible to this depressive bias, the basic mechanism of mood-dependent cognition (control-biased representation) seems to be integral to the normal brain. If a person who has never been clinically depressed abuses stimulant drugs, then when the drug-induced euphoria fades and the person "crashes," a negative bias in cognition occurs during the drug-induced depression that appears identical to that in clinical depression.

Given considerable evidence that normal mood biases are different from pathological mood biases only in degree, we must conclude that Jared is unable to understand the influence of his adolescent confidence and optimism on the risks he takes while driving. Operating from the fog of the naïve mind, as Jared experiences the invincibility of youth, he simply believes that he can continue to push the limits of his car and his driving skills and that he will succeed.

Elation of a Crush Encouraged

To frame the causal roles of mood-based controls and cognitive expectancy within a thought experiment, consider Jared's experience and behavior when Kim spoke pleasantly with him after class. Harboring a crush on Kim all year, Jared had often become slightly depressed walking out of history class. That was when he imagined the failure he would have trying to relate to Kim. She was really smart, and he never knew what to say. When she talked with him today and even seemed interested in him, Jared's mood began to alternate between a growing panic and a mix of (hopeful) anxiety and elation.

The effects of the increased elation were apparent in many aspects of his behavior, such as his elevated tone of voice, rate of speech, and the extra bounce in his step even as he consciously tried to maintain a cool and relaxed look. A positive expectancy was integral to this good mood. As the mood developed, it applied its inherent positive control bias to Jared's cognition and to his sensation and behavior. As the two students stood outside and talked, Kim's face was so beautifully animated and her taut young body was so mesmerizing that Jared wanted to reach out and grab her.

Fortunately, Jared's anxiety was sufficient to belay that action, but he *was* impulsive enough to ask her for a date. Later, driving home, he would cringe at even this action and flush with embarrassment at the memory, knowing that coming on to someone he had hardly spoken to before was a rash impulse.

Nevertheless, Kim agreed to go on the date. In the natural ecology of adolescent relations, Jared's hopeful impulse was actualized by a good outcome. As he tries to find his way to his next class, the realization of this fact causes Jared's elation to surge. Certainly he knows he is happy and that this is because of the interaction with Kim. But, for him, the world itself has changed. With the limited perspective of the naïve mind, Jared does not fully grasp how his elation shapes his interpretation. Rather, for some reason, the old school looks better than it ever did, even if he can barely concentrate enough to find biology lab.

Jared's worry about Andrew is now a vague remnant of the past, supplanted in his consciousness by this new world, one fully transformed by his elated infatuation.

Conscious Learning

In this chapter we have examined the representation and control of information in abstract, scientific terms. However, as I have illustrated with the examples from Jared's activities, the principles of information in both machines and animals may be meaningful for human psychology as well. It is not clear that the principles I have chosen are the best ones or even completely accurate. But, when it is eventually formulated correctly, the scientific study of information promises a more explicit understanding of the human mind, not just in the abstract but also in the concrete appearance of personal experience.

Given the current progress in computational and brain research, this understanding may be close at hand, and yet it will be difficult to bring to consciousness. When examined critically, the naïve view of the mind is surprisingly limited. We can grasp the contents of experience, the *results* of information processing, but we have only the most limited access to the processing that generates them. As we speculate on the science of learning and what it means for understanding personal experience, we must rely largely on rational inference. Consciousness itself provides only the weakest of guides.

We saw the limitations of conscious access to the mind's workings in Chapter 2 as we examined the structures of intelligence provided by the brain's anatomical connectivity. We are beginning to build an effective science of psychological neuroanatomy as we study the functional significance of neural connectivity. We can see the significance of architecture in the left and right hemispheres, in the motor and sensory systems, and in the core limbic versus exterior sensorimotor interface networks. In the computational science approach to cognitive representation, the connectionist models are teaching us how structure implies function and how representational capabilities emerge from the architecture of network connectivity. The combination of cortical anatomy and connectionist theory is powerful. Even if we do not yet understand it in any detail, we nevertheless know that we can now look into the architecture of the mind.

In this final section of the chapter I explicitly reason from the principles from information theory and animal learning to consider the question of human experience. Although this analysis is a preliminary, cursory one, it will become clear that the principles of information representation and control in brains and machines are relevant to the basic questions of human knowledge and experience. When we apply these principles to the human brain, it should also be clear that the processes of mental representation and control are only partially voluntary and only somewhat available to conscious inspection. The more we understand scientifically about the fundamental mechanisms of experience, the more we find they extend beyond the narrow perspective of the naïve mind.

Information and Change

Perhaps the most important realization from studying elementary forms of representation and control is that information requires change. To be informed is to learn. Because of this, representation and control are inseparable in a dynamic, adapting cognitive system.

We saw this when we studied the way in which complex representations—concepts—could be formed in artificial neural networks. Even simple models of distributed representation show us the importance of connectivity. These models then teach us how to understand the brain's connectivity (Chapter 2). In each case, the representation of information in artificial networks required some mechanism for changing the connection weights adaptively (i.e., for learning).

In most cases, the mechanisms used for "training" the networks in machine learning are inelegant and not a very good approximation of the learning that goes on in biological systems. However, within each simulation, some method of learning is required. We need some adaptive adjustment of the representation in order to model the target output. If we analyze the element of information in the artificial network, the essential atom of information, it would be each adaptive adjustment. Information is change, constrained within context.

A different perspective on the complementarity of representation and control was given by the review of animal learning principles. There we found that, at least in mammalian learning, control cannot be described without representation. Mammals represent their worlds. In fact, they continually represent their desired future worlds, and these expectant cognitions are the guides for learning. They also determine what constitutes information. Although the simple association between stimulus and response may shape the behavior of fish and reptiles, learning in mammals is contextual and inextricably bound with an animal's expectancy for what should happen in a given situation. Change requires the information of the context.

Thus, in simple cognitive systems, we can see what must shape the primitive elements of the information structure of the human mind. Can we find these fusions of representation and control within the structure of conscious experience?

The Reach and Span of Consciousness

We have seen that information can be defined by the reduction of uncertainty. In the distributed representations of machine simulations, uncertainty can be defined by the mismatch between the system's output (its current representation) and reality (the target output). In mammals, we find that learning within brains is also guided by uncertainty, in this case by the discrepancy between a valued expectancy and what actually happens. Considering these principles of information control together, perhaps we can better understand the relationship between uncertainty and information. Cognitive systems at a particular level of complexity may form expectancies as a natural imperative. When these systems gain a sufficiently dynamic memory capacity and when they must use this capacity to organize actions, cognition becomes expectancy.

If consciousness is not an epiphenomenon but rather is integral to human intelligence, then it may be linked in a fundamental way to this bias toward expectancy as a guide for processing information. This would imply a forward bias to consciousness, an anticipatory quality that is seldom recognized. Certainly the mind is considered to have voluntary control, and the associated sense of agency implies the capacity to decide actions in the future. Yet when considering consciousness itself, philosophers have often emphasized the present contents of consciousness, one's private subjectivity, as the defining feature. For their part, when psychologists are asked about consciousness, they typically show their functional bias and emphasize the memory capacity required for conscious experience. We need at least some memory capacity to hold the elements of thought to allow momentary reflection.

Yet if the simple representation of content were adequate to define consciousness, then we would find it in a video camera. Maybe there is a new perspective that can be gained from an information-theoretic (machine learning) analysis of animal (mammalian) learning. Processing information in a dynamic (time-varying) cognitive system requires expectancies. If so, then conscious information processing—conscious learning—may involve encountering the data of experience expectantly. To be conscious is to lean into the future. We meter our awareness of the world by our expectations of it.

Interestingly, because learning involves change, any experience that involves learning is dynamic; as a result, the contents of experience are continually engaged in becoming something different. Of course, most of experience—and learning—is unconscious. As a result, subjective participation in most of experience is as simple as vaguely recognizing the significance of events and then paying attention at some future point and finding ourselves transformed. However, to the extent that we gain awareness of the transformation inherent in the learning process itself, then consciousness is expectant, as the potential of the self and the uncertainty of the present combine to spill into the coming time.

Reaching for Information With the Self as the Only Handle

The forward-looking aspect of consciousness may be a novel implication of an information-theoretic analysis. Yet another aspect of expectancies, their interpretive nature, has long been recognized in studies of perception. The constructive approach in psychology argues that perception, at least in any complex form, requires interpretation.

The notion that we can perceive *without* applying interpretive biases is sometimes called "the dogma of immaculate perception." The implication seems to be that, if unencumbered phenomenology really existed, it would represent a miracle of biblical proportions. More accurately, both cognition and the perception of events requires the interpretive structures that we bring to the occasion. To the extent that these interpretive structures are active and motivated, they form expectancies.

To the extent that they incorporate the residuals of experience that are encoded within our representational networks (our concepts), the interpretive structures of consciousness must be historical, the developmental products of personal life histories. Reasoning with structural principles, we can see that only the least-integrated concepts—those that we fail to integrate within the larger knowledge-relational context of personal intelligence—escape the informational inertia of personal history.

The temporal span of consciousness thus increases in both directions—future and past—with the increasing complexity of experience. The evolution of mind between stimulus and response can perhaps be apprehended in the twilight between the asserted history of self-knowledge and the tentative projection of the future. With increasing evolution of the mammalian brain, not only do our expectancies reach farther into the future, but they also integrate more of our personal history. For people even more than rats, expectancies must be forged out of personal experience.

Yet even for people, these patterns of expectancy are seldom accessible to conscious inspection. We are biased to expect regularities of the world, such as the interaction patterns of our childhood social and emotional relationships. We expect these implicitly and unconsciously. When people act otherwise, we are confused. Experience, in its conscious as well as unconscious forms, is a function of the historical self.

Learning or Being

Thus, as we consider the temporal integration required for a complex, adaptive information system, we find constraints on the span of experience that impinge on an increasingly complex awareness. With the historical self as the embedding context for consciousness, we are aware not so much of the world as it is but rather of what we want it to be—or, depending on the extent of personal trauma, of what we fear it will be.

Coupled with this expectancy bias as the base of consciousness, the interesting dialectic implied by information theory is that we become informed only as our expectancies are violated—as events are not what we want. Therefore, when we are fully certain and when expectancies are met by ongoing events, information (even when its data are present in the flux of reality) remains invisible. It is only when we are uncertain, when expectancies fail, that we can be informed.

If these principles of expectancy and information are correct, then perhaps the stability-plasticity dilemma is a reality for human learning, as well as for machine learning. Stability is the steady state, when expectancies match reality. Of course, the more complex your expectancies are, the more they are prepared for disconfirmation by some element of ongoing reality. When a discrepancy is encountered, then you face the stability-plasticity dilemma. You can stay the same (choose stability), in which case you are uninformed, but at least you preserve the historical self. You can change (choose plasticity) and become informed, but in the process you have to give up the old self and confront the painful novelty of a new identity.

Is this process of negotiating expectancy and learning a conscious one? Probably not. People often report that learning is difficult, but they are talking about a pragmatic, almost mechanical difficulty. You never hear people say that information is too difficult to assimilate because it means giving up their old self.

Rather, learning engages an unconscious struggle with the inertia of the self, the relational mass of personal information. And when it is done, we are transformed and cannot remember being otherwise. When we are children, the self is light, and we are fully transformed by each instance of reality. When you are a child, a summer seems to last forever, and at the end you are a new person. When we grow old, the self becomes increasingly crystallized and ever more massive with accumulated wisdom. As a result, with increasing age we incorporate progressively less information—unless, that is, we are able to tolerate what must become the price of the unrestricted learning that is now available in the Information Age: a continually labile and childlike identity.

Knowledge, Growth, and Digestion

Does becoming informed actually require a transformation of the self? Can't we just add a little information to memory without having to face an identity crisis? Is the structure of the self really the only vehicle that can incorporate significant new learning? This makes learning sound like eating.

In fact, one of the most influential scientific theories of learning was based explicitly on an analogy with digestion. This approach was developed early in the 20th century by a young Swiss biologist, Jean Piaget.

As a child, Piaget had become interested in evolution. He became fascinated with mollusks—shellfish—in the lakes near where he lived and wondered about how they evolved. He recognized that a key issue for the survival and adaptation of any species is the process of digestion. In this process, external elements (nutrients) are taken into the body, and the body's structure incorporates these elements. Piaget described this mode as *assimilation* because the body's structure dominates that of the external elements. We could describe this as a mode of internal stability.

However, there is also an opposite process, in which the body must change in response to the new elements. Piaget called this process *accommodation*. We could describe this as plasticity, in that incorporation forces a change in the internal structure.

It is through assimilation and accommodation that an organism interacts physically with external reality. These become critical processes for Piaget's model of evolutionary adaptation, an impressive insight for this young student of evolutionary biology. Based on his initial scientific publications, Piaget was offered the position as head of a natural history museum, a prestigious scientific post at the time, but understandably he had to decline the job offer because he was only 12 years old.

As a curious scientist, Piaget continued to apply his analytic reasoning to other interesting events in his life, such as the birth of his first child. Like all new parents, Piaget was profoundly impressed by this experience, so, like a good scientist, he embarked on an ambitious program of intellectual inquiry with his children as the experimental subjects. He reasoned that if we could understand how intelligence develops from an infant's vague awareness into the fully instantiated adult form, we could construct a *developmental epistemology*, a scientific understanding of how knowledge comes about through processes of biological change. Piaget's model for the growth of intelligence was taken from his earlier biological studies: the digestion of material from the outside world. This time, rather than nutrients, the "digested material" was information.

Piaget formulated principles of cognitive development that remain important research topics today. Although many developmental psychologists have taken issue with the specifics of Piaget's observations, his general theory continues to provide a powerful model for understanding the integral relations between structure and information in cognitive development. Learning is not just the linear accretion of data. It also requires an accommodation of fundamental changes to the structure of the mind itself. In this way, Piaget's ideas were able to provide an organizing framework for a critical task of developmental neuroscience: understanding how the child's learning continues the task of morphogenesis that was begun in embryonic development.

Dialectics of Information Accommodation

Because they lead to the transformation of the old self, the changes required for accommodating new information cannot be accepted without a struggle. Both this struggle and the dialectical nature of learning it causes are foreign to most of modern psychology. However, this dialectical nature of learning has been recognized by many classical analyses of how ideas grow over time. In Socrates' teachings, progress in ideas is achieved through a kind of structured argument, in which initial assumptions are tested by critical reasoning; as a result, truth emerges from the challenge of the dialog. Socrates' student Plato generalized the dialectics of this process—the confrontation between an initial idea and its critical analysis.

Building on the Greek fundamentals from Socrates to Heraclitis, German idealist philosophers such as Hegel continued to emphasize the developmental nature of knowledge, its transformation and progression through time. Hegel formulated the dialectical process in the opposition between ideas that could lead to a new, more complex structure of a more developed idea.

The Hegelian analysis was brought to our modern era through Marx's application of the dialectical model to the economics of the industrial revolution. The hegemony of the capitalist industrialists would be broken by the collective, antithetical will of the working class. Although the struggles of the modern global economy suggest these issues were not resolved as easily as Marx envisioned, dialectical processes clearly remain active in the struggle of economics and ideologies in modern societies.

Thus, if we consider the theory of how ideas progress in a political and philosophical context, there is good reason to think that the process is one in which transformation of the existing order is required for progress. Although Piaget's approach is consistent with this dialectical analysis and developmental reasoning, it is rather more dynamic than the linear accretion of information that is implied by many of our modern assumptions about how people learn.

Nonetheless, when we search for first principles, such as in the realities of machine learning or the courageous confidence of the rodent economists, we may come upon a similar model of learning, of increasing information complexity through self-transformation. When rats expect and receive their familiar good outcomes, the events of the environment are assimilated into their expectancies. There is minimal learning. But when expectancy is violated by incongruent events (i.e., when uncertainty is high), then opportunity exists for dialectical progress and accommodation.

Change engages the poles of the dialectic and is effected by the opposing mechanisms of the digestive process. When we choose stability and the continuity of personal identity, we can assimilate only events that the current representational model expects. When we choose change, we accommodate. Identity is sacrificed, and we are informed.

The Negative Sculpting of Network Architecture

Can we integrate machine learning and animal learning within the same theoretical model? Could we then use this model to interpret the brain's connectional structure? To explore this reasoning, we can relate the ideas of this chapter to the questions about the connectional structure of the brain in Chapter 2.

In many ways our reasoning has become inverted. In Chapter 2 we studied the network architecture of the human brain with the rationale that connectivity implies function. If we know what is connected to what, then our connectionist theory can suggest how mental function is therefore achieved. Admittedly, this is a somewhat vague hope. In fact, as we have studied connectionist machine learning more carefully in this chapter, we have found that connections lead to a primitive dynamism of distributed mental representation, in which stability can be maintained only at the expense of the plasticity of new learning.

As a result of this analysis, the differentiated patterns of network architecture that we studied for the human cortex in Chapter 2 can be seen to serve a *negative* function, preventing connectionist representations from interfering with each other. For example, the separation of individual sensory cortices may have been necessary to allow each sense, such as vision, to have its own network architecture in order to fully elaborate the sensory data without being constrained by the general connectional inertia of other sensory or motor processes. Similarly, the specialized architectures of the left and right hemispheres may allow domains of intelligence that, although they communicate through the hemispheric commissures, are afforded a degree of isolation and thus freedom from interference.

Even in the patterns of network specialization within each hemisphere, such as suggested by Semmes's research and theory, we may find that connectional architecture is determined by the reality of the stability-plasticity dilemma. The right hemisphere's holistic operation may reflect a neural structure that allows the more holistic, dynamic, and primitive connectional operations of transformation and change. The left hemisphere may have evolved a more discrete and modular organization for the very advantage of avoiding this dynamism and thereby achieving greater local, modular stability.

Network architecture thus implies cognitive function not only in the positive sense of allowing the distribution of connectional relations but also in the negative sense of protecting certain differentiated representations from degradation through diffuse interactions. It is quite interesting to consider both of these perspectives on architecture in relation to the morphogenesis of the cortex in the developing brain. In the embryonic brain, the connections are massive, and the process of functional differentiation of the cortical architecture—both in the womb and in the early months and years of life—is one of subtractive elimination. Only the used connections are retained. The learning process is thus one of negative sculpting; the sculpting is the mechanism of default, and only functional activity allows connections to be retained.

The brain's process of architectural differentiation—both in cortical anatomy and in cognitive individuation—thus appears to be sculpted negatively rather than positively. The connectional differentiation is a way not only to embody the functional patterns of the individual's experience but also to protect those patterns from the haphazard plasticity of a holistic and fully interconnected distributed representational system.

Strategic Learning in a Distributed Cortex

Does animal learning contain lessons for the evolution of distributed representational systems? It is in the evolution of mammals that we find the first appearance of the massive ordered reticulum—the networks of billions of neatly structured neurons—in the cortex of the telencephalon. This radical architecture seems to have brought the capacity for increasing delays between stimulus and response. Perhaps it also brought the inherent representational limits of a distributed connectional system. The plasticity that would support self-organizing structures within a developing cortex may also support the destructive interference with existing representations that goes with this plastic capacity.

One developmental mechanism for coping with this problem is term limits. Restrict the period of plasticity in the morphogenetic process. Let the juvenile animal remain plastic (during the process of subtractive sculpting of the network architecture), then at the point of sexual maturation, turn off the plastic adaptive mechanisms. The mature animal can then quit learning and get down to the business of performing, with its memory protected from interference. This strategy seems to be universal among mammals, including us, with some degrees of infantile plasticity remaining in the slower-developing species.

The more complex mammals, such as rodents and people, also seem to have evolved highly strategic learning mechanisms that are perhaps tuned to cope with the distributed representational capacity of the mammalian cortex. Whereas the primitive vertebrate (fish) learning strategy is a passive association, we have seen that mammals hold a contextual expectancy, a concept of what should happen next in the present world. Because this concept is positively motivated, the organism evaluates events in relation to how they fit this valued world model.

We can speculate that this form of motivated expectancy has evolved to be aligned with the computational realities of the massive mammalian reticulum. With this context model as the mental map of ongoing experience, the events of the world are assimilated into the model. It may adapt slightly, but only slightly, because events are largely congruent with it. For the most part, certainty holds.

As shown by the animal experiments we have reviewed, significant new learning occurs only when events are discrepant. This new, accommodating learning process seems to require a different mechanism that must disrupt and change the previous contextual model. There is evidence that the recognition of discrepancy and the focusing of attention that are required to support new learning are carried out specifically in the frontal networks of the mammalian brain. This shift toward accommodation, of changing the internal representation to incorporate the necessary new learning, seems to reflect a separate strategic learning mechanism, one that is geared to maintain the focus of attention sufficiently to overcome the connectional inertia of the mammalian cortex. This is an explicit neurocybernetic strategy for encountering and integrating new information.

Dialectical Science

In this chapter we have seen the problem of learning in distributed machines: the choice between learning and being the same. We have seen the unique strategies of learning in mammalian brains: Hope looks to the future, and it must be dashed when events prove it wrong and demand radical new learning. I have argued that the mammalian learning strategies may have evolved in concert with the massive distributed cortex of this lineage, gating the assimilation or accommodation required to manage and effect representational change.

However, even if this argument would qualify as neuroscience, where is the philosophy? Do people think like this? When you must face a situation in which new learning is required, are you aware of the philosophical dilemma of choosing between this new learning and the old self you know and love?

Certainly not. People don't think of learning as a dialectical dilemma, whether for rats or themselves. As a result, we can reasonably ask whether this line of reasoning has any merit for personal philosophical instruction. It is one thing to speculate about the computational basis for the neuroanatomy and future outlook of struggling rodents, but it is another thing (most likely straining credulity) to argue that the same process of information accommodation shapes the daily exigencies of conscious experience. Nevertheless, there is direct evidence of the dialectical nature of learning, discovered in a careful study of information and change. This was change neither in machines nor animals nor even politics but in the theoretical progress of science.

Historian Thomas Kuhn studied examples of progress in scientific theory. These examples are times when research has led to new insights into the nature of reality and caused us to give up previous concepts because they no longer fit with the research data. The classic example is the Copernican revolution, the recognition that the earth is not the center of the heavens. This particular theoretical change required a major shift in perspective, and it was bitterly opposed. Kuhn studied many examples of progress in scientific thinking—new learning about nature. These were often smaller advances than the big events in popular scientific history and were recognized primarily by scientists, but they took on a similar form.

Kuhn observed that textbooks emphasize the cumulative nature of scientific progress, as if research simply adds more facts to the body of knowledge until a new view of things, a new theory, becomes obvious. What he found instead was more complicated and more interesting than the textbook account. Kuhn learned that a major change in the ideas of a field is typically resisted by most of the people in that field. This resistance remains strong even when the evidence for the change has been gathered and widely disseminated yet logically conflicts with the established theory.

Maybe the function of information in scientific theory is the same as in machine and animal learning. Information is created only as expectancies are violated. When facts are expected by the theory, incremental progress occurs (what Kuhn calls "normal science") but no real change to the theory itself. This is like the contextual model of the rat, which is able to assimilate small events while avoiding real change in the distributed representational structure.

It is when discrepancies arise that the old theory fails. However, in the interesting paradox of uncertainty and information, those who remain certain (the stalwarts of the Old Guard) are blind to the new information. Only those scientists who are prepared to disbelieve the established theory are capable of engaging the uncertainty that allows new facts to become information.

What happens next is what Kuhn called a scientific revolution. The old theory is not just modified. Rather, it is destroyed by those who recognize the incompatible evidence and who then have the opportunity to see the world in the new light of a novel theoretical perspective. Assimilation has failed, and accommodation has brought about a new order. In the dialectics of intelligence, it may be that information invariably presents the stability-plasticity dilemma. In the process, we face the ontological ambivalence of information, which offers us the choice between identity and learning.

Principles of Representation, Control, and Experience

One principle is that *concepts are distributed*. This principle comes both from observing the brain's densely interconnected network architecture and from observing in computational science that complex, analogical concepts are formed through distributed representations, which causes meaning to be contextual. The basic form of meaning that we perceive in experience is gained neither from logical operations nor from discrete motivational links but rather from an embedded network of associations that make up the background network of whatever apparent elements appear in consciousness.

Certainly meaning is conditioned by personal significance. Some of the connections that are engaged must have motivational value; otherwise the data of experience fail to gain the motive value to become informative. However, within the rich matrix of associations that is inherent to distributed representations, meaning may be gained by multiple points of personal significance, only some of which may be apparent within the focus of attention. Only the surface of meaning is conscious.

These qualities of meaning lead to a related principle: *Just as meaning is contextual within a distributed representational mind, knowledge is relational.* Knowledge is not the possession of isolated facts but the capacity for manifesting the relations among multiple factual elements. Expert knowledge is created by efficiency in this process of relational organization, with the result that multiple relations are implicit within the mental operations and do not need to be articulated individually in order to be influential. The structural mechanisms of relational knowledge are differentiation and integration. First there must be differentiation of conceptual elements (probably with inhibitory delineation, as we have seen) and then hierarchic integration of sets of these within more abstract categories.

The implications for experience are interesting and may be relevant to the boundary between conscious and unconscious processes. Explicit semantics, in which meaning is restricted to one or a few highly probable associations, is relatively easy to hold within conscious experience. The limited capacity of working memory can then be allocated to holding the limited relations that are sculpted by the inhibitory control that defines direct linguistic implications and discrete perceptual objects.

Another general principle comes from animal learning studies: *Complex brains form expectancies.* They then evaluate events in relation to them. Learning, the process of accommodating information, only occurs when you are surprised—when there is uncertainty and it is reduced by the information.

This principle has started out as an empirical fact or at least a generalization of the facts of animal learning capabilities. The only way to explain how mammals learn is to infer that they form expectancies. This has not become a recognized theoretical principle yet, but it should. The explanation for this must be that expectancies are the natural extension of the increasing memory capacity of mammalian brains. If a brain gains an extended capacity for representing knowledge of reality and acts in time, then forming expectancies—and evaluating events in relation to them—may be the inevitable outcome. When events are unexpected, uncertainty arises. We may then be informed and changed.

The principle that complex brains form expectancies leads to a corollary (a consistent subordinate principle)—that complex brains must have mechanisms for fine-tuning the certainty of current expectancies. These mechanisms would set the gain, as it were, for recognizing discrepancies and processing information. In some modes of experience (for example, in the elation of success), expectancies may be biased toward certainty, thus blinding us to discrepancies. In other modes (for example, in anxiety in the face of a threat), we become uncertain. Then we are capable of becoming informed.

Thus the motive control of uncertainty is a theoretical question that follows closely behind the recognition that complex brains form expectancies that must be negotiated with the process of being informed. When this reality is considered together with the principle that concepts are distributed, we have a fundamental set of ideas that prepare us for the analysis of the unconscious mechanisms of adaptive memory consolidation in the next chapter.

One more principle completes the set: *The mechanisms of experience are largely unconscious.* This is what we see when we bring principles of machine and animal learning to human experience. These fundamental mechanisms of information processing must be integral to human learning, yet they appear to operate largely out of awareness.

We are conscious of engaging events, to be sure. As we attempt to predict and control those events, we maintain the illusion of agency, that is, of being in control. However, in the critical process of learning, we negotiate the struggle between identity and information with only a minimal capacity for self-monitoring. The dialectical process of learning has been recognized since the time of the early Greeks, the time of some of the earliest ideas present in the historical record. And a careful historian such as Kuhn finds that this struggle of change remains the basic process for learning even among professional learners today, the scientists. Nonetheless, as evidenced by the surprise and consternation engendered by Kuhn's revelations, even scientists, who are supposedly trained in deliberate reasoning, are scarcely aware of the process.

If we have such limited insight into the mechanisms of knowledge development in one of the more advanced of our intellectual endeavors, science, then we can hardly assume much conscious insight into the workings underlying everyday experience. Rather, the assumption of conscious control of the mind may be an illusion that we adopt to facilitate the limited monitoring and executive influence we do have. If we had real knowledge of the mind, we would recognize that its mechanisms are to a degree autonomous. We would see that what control we have is exerted through effective alignment with those mechanisms rather than any simple volitional direction of them. Conscious learning is an ideal, not a reality.

Discovery of a Hurt Friend

When Jared sat on the curb outside the diner and thought about his friend, he had no clue that Andrew had tried to make him a gift of the stolen transmission. His experience was at first dominated by his own egocentric issues; his effort to be principled and not steal conflicted with his desire to have a solid transmission for the car he is so fond of. As he moved on to class and was fully captured by infatuation, any thoughts of Andrew were far from his conscious mind.

Later that day, however, as Jared sat in his room and was supposed to be working on a history term paper, Andrew suddenly reappeared in his awareness. Very quickly Jared's thoughts progressed—developed—from sensing something was curious about Andrew's offer of the transmission to realizing that Andrew had intended it as a gift. The price was so low that it could be only that. Andrew knew Jared needed a transmission that had a working fifth gear. As this realization dawned on Jared, he became embarrassed that he had been so insensitive to Andrew's intention. He replayed the sound of Andrew's voice as Andrew grumbled and left the diner; Andrew had complained because he was hurt.

This realization of Andrew's good intention is one more facet of Jared's reality that he must try to integrate within his concept of self-in-world. For our purposes, the example illustrates how the multiple threads of experience are woven within the unconscious as well as conscious domains of an adolescent's mind. Jared took in the information about Andrew, but because it was merely a base curiosity about what was happening, it was information only superficially. At first it did not do much informing within Jared's conscious apprehension of the situation. As this potential information incubated, it became integrated within Jared's semantic reference frame, fully unconsciously, as the day progressed. By the time he sat down to his homework, the information had formed a strong enough pattern of meaning to break into his awareness the first time his thoughts were idle.

Jared's learning during this time was unconscious. Over the several hours of his afternoon experiences it was being consolidated in the background of his awareness. It then manifested itself as a powerful new pattern in consciousness, a concept, an effective instance of being informed. And Jared was changed.

The next chapter considers the process through which this happens. The brain's consolidation of memory proceeds. As it does, experience develops, both above and below the surface of awareness.

4
Motivated Experience

In this chapter we will look for the processes of motivation that give life to the mind. We have seen that learning—in both animals and machines—must be controlled. In the mammalian brain, the controls involve expectancies, and these are rooted in bodily needs and desires. The control of certainty and uncertainty thus revolves around motivation. The result of this control decides whether raw data becomes meaningful information.

We have seen that the actual mechanisms of the mind are not easy to bring to conscious awareness. This limitation—the opacity of introspection—is a major barrier to psychological analysis of the mind. Introspection is particularly ineffectual in understanding the mechanisms of motivational control.

Motivation is an obvious question when we consider the behavior of other people. When your friend announces she is going by herself on a vacation to Hawaii, you are immediately curious about her motives. Why Hawaii? Why now? And, of course, is she really going alone?

Interestingly, however, when you consider your own experience, the question of motivation often evaporates. The perspective of the naïve view of the mind seems to cause motivation to be less apparent, even invisible, in the process of subjective experience.

Observing Experience

One assumption we often make without noticing it is that experience is passive. Subjectively, it seems as if things just happen to us. Our own psychological processes seem to run on their own, without the need for an active, internal motivation. For example, when you are on a diet, food looks and smells really good, and even the simple idea of eating it is compelling enough that you may feel you need to have it immediately.

Of course, when it's someone else who is on a diet, it's easy for us to recognize that the person is being driven by hunger. Furthermore, it becomes obvious that the person's current orientation to food is biased by this hunger. But subjectively, motivation changes the appearance of reality so integrally that what the dieter passively recognizes is the unitary phenomenal effect of the great and overwhelming attractiveness of food.

This invisibility of the active influence of motivation is not, of course, absolute. Because people eventually learn what hunger does to their intentions of self-restraint, they may even make it a personal rule not to shop for food when they are hungry. However, to follow this rule, one needs a logical inference and a good deal of objectivity. Motives do their work so close to the basis of conscious experience that we are subjectively captured by their influence. We then need explicit, objective reasoning, like personal rules about shopping, in order to remind ourselves to deliberately infer their workings. In the naïve view of the mind, we often see reality changed by our motives, but we think it is just reality.

Naïve Agency

Another default perspective of the naïve mind may be related to the limited insight into motivational mechanisms. This is the assumption that we voluntarily control our minds. The identification of the subjective mind with voluntary control is taken as a patent and incontrovertible fact in our intellectual culture. Whereas much of experience just seems to happen passively, without need for internal motive, naïve subjectivity assumes the mind has an active, integral agency, in other words, a voluntary capacity to determine what happens. From this naïve view, we do not need to search for the motives of our actions because there is an inherent explanation for their causes. We create them, de novo, by naked will.

Most people would agree that free will is something we should have, but it sounds better in the abstract than when you try it at home. It does not take much study of the psychological function of real people to conclude that free will is an ideal that fades quickly in the face of powerful implicit motives such as hunger, lust, or even sleepiness, which regularly overwhelm the will. The assumption of agency turns out to be another default assumption (and often a mistaken one) that accompanies the naïve view of the mind. This *is* a default with a purpose. Even if the assumption of voluntary control of the mind is a delusion, it may be a necessary one.

When a person with schizophrenia loses the assumption of agency, the result is a predictable search for external causes for the activities of the mind. If I do not control my thoughts, then somebody else must. From clinical work with people with schizophrenia, we know the result will be a bizarre but predictable control delusion. In this delusion, the breakdown of the personal sense of agency causes them to believe that their mind is being controlled by the devil, radio waves, scientology, or whatever occult power is now salient in the popular culture. A kind of agency seems to be implied by the experience of the subjective mind. If the sense of an internal agency fails for any reason, then a causal vacuum arises, a void whose presence is subjectively sensed. To fill this void, a replacement agency spontaneously comes to mind.

And when the mind's activity is caused by an active agent—whether it is the self or aliens—we feel little need for a motivational explanation. Because they are willed, the mind's actions need no ascription of an underlying cause. Even in the face of overwhelming evidence of its inadequacy, the implicit will manages to survive critical examination. Even when we are abandoned by the will, the naïve mind remains motive blind, spontaneously fantasizing an external agency to fill the void.

The Magic of Creativity

There are times when, even for a sane person, it is obvious that the will is weak. For example, any time new learning is required. Old learning, such as work activity that is familiar, engages a practiced expertise and runs so well on its own that we just need to show up. However, learning new skills is difficult, so difficult, in fact, that the inadequacy of volition overwhelms the naïve assumption of agency. The weakness of the will becomes a palpable fact of experience.

For example, you may have been skilled at reading, writing, and various forms of problem solving while in school, but you never really learned algebra. When you had to use it, it was so hard to struggle with this new problem domain without clear instruction and with no familiar cognitive skills that you wanted to give up. A struggle like this is repeated in any new problem domain. While the struggle is active, it gives serious pause to the assumption of simple volitional control of the mind. When the mind is challenged, what is required is not just a pattern of intelligence but also the motivational energy to organize and develop this pattern. The creative person finds the motive effort. Most people find ways to avoid it.

Avoiding the struggle is a great release of the anxiety of being lost without a solution in a strange and novel domain. To be creative, we must not only tolerate but also engage the anxiety of uncertainty. That way lies information. Even for the creative person, the requisite motive is no simple act of will but requires hours, days, or even a lifetime of effort. Motivation is the engine of the creative struggle, and it often operates in ways we do not understand. Because the mind's mechanisms are largely unconscious, motive blindness may extend even to creative demands for which we exert our best conscious effort. For example, in a critical stage of problem solving, you may become overwhelmed with the facts of the problem. Looking to where the solution should appear, you find only painful confusion. You experience the anxiety and struggle—the emotional wake of the motivation—but you cannot see results.

Objectively, psychologists who study creative problem solving know that the mind is indeed productive during this stage of subjective emptiness, but its products are not subjectively accessible because they are still forming at a preconscious developmental level. This period of problem solving is called, appropriately enough, *incubation.* When the creative results do reach a level of development sufficiently robust to appear in awareness—in a moment of reflection or in a dream—we experienced them as a gift, an instantaneous insight. The creative results then seem far removed from the conscious, effortful motive struggle that organized their latent intellectual roots. The results appear foreign to the self, willed by some magical agent, perhaps a creative muse.

Scary Magic

Because they are integral to the self and yet uncontrolled, the mind's hidden motive mechanisms may seem scary. I have seen this in my work as a psychologist with people who are emotionally confused. In their confusion, many people become afraid of their own mental processes. They lose the illusion of being in control and then may question their own sanity.

I have also found that the fear of the mind is often readily apparent in people who do not know I am a psychologist. There is usually a normal conversation, such as at a party or reception, and then someone mentions I am a psychologist. For some people who did not know this, there may be an immediate facial expression of panic and embarrassment, as if they had just discovered someone observing them in the nude and they do not look so good. Acutely self-conscious, they divert their eyes to the side as they try to remember what they might have revealed in the recent conversation, or they admit the fear of transparency outright in a comment such as "Oh, no! You must be analyzing what I'm thinking!" In reality, of course, I cannot tell what they are thinking. And even if I could, I normally would not be that interested, although, because they became so self-conscious, I cannot help but wonder what they are hiding.

What do these reactions mean? It could be that even normal people have a constant stream of such lewd and unusual thoughts that they are always on guard against being found out. These reactions happen so often that I am starting to wonder. However, here we may gain a deeper insight into the normal state of the naïve mind. Even our own minds seem magical. Because we are untrained in the mechanisms of the subjective mind, mental processes happen in ways we do not understand. Where we are untrained as subjective observers and unprepared to apply rational analysis, awareness may be quickly overwhelmed by magical thinking. Consciousness is but thin ice on deep waters.

If you were a scientist at the end of the 19th century, such as William James or Sigmund Freud, it would be logical for you to think that science was poised to provide insight into the human mind. Science had just shown us our evolutionary origins, so we could now expect equally profound revelations on human nature—not just insight into the mind as an academic or even a medical topic but insight into the mind as it appears to everyday people. In fact, in their different ways, both James and Freud reasoned about psychology in practical terms that people could apply to their own lives. Freud, especially, believed that scientific analysis could give us insight into the workings of the unconscious mental mechanisms that shape our lives.

However, somehow science failed to inform the subjective knowledge of the mind. Academic psychologists lost track of the real-world questions addressed by James's psychological studies, and Freud's psychoanalysis eventually failed both in the clinic and in philosophical inquiry. Sometimes I think the failure of psychoanalysis may have been one of the significant events of intellectual life of the 20th century, leaving us split across the chasm of the two cultures. Many people, especially artists and literary intellectuals, were fascinated with the psychoanalytic promise of insight into the unconscious motives that shape personal choices. But other people, especially the hard-nosed scientists, denied that this was a viable topic for science. Some even seemed threatened by discussion of the unconscious mind.

The result was that, in the domain of subjective experience, science has left us intellectually naked. We have only the innocent beliefs of children, with no training in scientific analysis of subjective experience. We are left trudging through life, as it were, in our phenomenological birthday suits.

The Neuropsychology of Motivation and Mind

Even with Freud's considerable genius, psychoanalysis could not illuminate the mind's motivational basis through introspection alone. What is required is a scientific analysis, built methodically on evidence that can clarify the vagueness of personal introspection.

Motivation did become a central question in academic psychology in the mid-20th century. It was seen most prominently as a component of stimulus-response behaviorism. The goal was to use learning as a way of creating more complex behavior, starting with the motives of simple drives such as hunger or thirst.

Then, however, with the cognitive revolution, motivation seemed to be misplaced. After cognitive psychologists were ridiculed by behaviorists as unscientific, they soon returned the favor by ignoring fundamental behaviorist issues, such as drives and motivations. Within an information-processing analysis, computers do not have motivations. If the mind is like a computer, cognitive psychologists assumed, we do not need to worry about motivations, either.

Later in the 20th century, when social and clinical psychologists eventually addressed questions of motivation, it was with the prevailing cognitivist explanations. The premise was that our objective cognition (how we think about things) causes our subjective emotion (how we feel about them). Motives could be engaged, actualized, or inhibited, all depending on how they are cognitively represented. This set of strict cognitivist assumptions remains the basis of the only psychotherapy to receive research attention today: cognitive behavior therapy.

As we have seen in examining animal behavior (Chapter 2), many of the findings of learning effects cannot be explained by a simple behavior reinforcement theory. Rather, they require explanation in terms of an animal's cognitive representation of expected rewards or threats. Yet the cognitivist interpretation—that things can be fully explained by the form of the representation—does not work in its cold, dry form either. The expectancies that guide animal behavior are valued and motivated. The theory we need to understand such expectancies is one in which the expectancies themselves are charged with motivational significance. How can we understand the integration of motivational control with cognitive representation in this way? Could we align the principles of cognitive representation with those of behavioral control?

The brain is instructive in this regard. Unlike a mentalistic theory, the brain cannot be understood in the abstract alone. Rather, it has a concrete plan that manifests its biological, evolved origins. Looking into the plan more deeply and searching specifically for the motivational basis of memory and cognition, we will find that this motivational basis emerges from the bodily basis of homeostatic self-regulation.

Biologically, this should be no surprise. Psychologically, this is not only a surprise; it is also disturbing. What we find is not a vague reference to the physiology that must be there (because after all we know we are biological) and that we can then politely ignore. Rather, what we find is that motivation is integral to the most fundamental of the brain mechanisms of experience—those responsible for creating memories.

Without memory, the structures of the mind are gone as soon as they arise. Within memory, the experience of a lifetime is organized, sorted for its personal significance, and understood, more or less. Acting to direct the very mechanisms of working memory that support consciousness, our motives shape experience so integrally that their effects occur before we know it. Research is now revealing the motivational processes that are essential to the process of memory consolidation itself, which selects the events of perception and cognition that will become integral to one's retained personal history and thus to the self. By studying these mechanisms, we can see how information emerges in experience from the primordial dialectic of data and value.

We start our review by considering the basic mechanisms of bodily control of behavior. These are the visceral and arousal processes that represent physiological drives and needs in order to regulate internal states and motivate our actions. At first this may seem more a biology lesson than a philosophical inquiry into the substrates of experience. However, I maintain that this biology is the essential foundation for understanding the bodily basis of the mind's activity.

The key insight into the neural mechanisms of motivation is *vertical integration:* All levels of the neuraxis, from the primitive circuits of the brain stem to the cortex of the telencephalon, must be coordinated in the control of neural activation. We need to pay careful attention to our biology lesson just to visualize this hierarchy of neural systems, forming as it does the evolutionary architecture of the mind.

Studying vertical integration leads us to a deeper, even more surprising realization. The brain's activation control systems are essential for the embryonic development of the brain itself. Although the general form of the brain's anatomy is determined by genes, the fine structure of its networks is determined by the patterns of activity in the networks themselves. The vertically integrated motive control systems are necessary to stimulate and regulate this activity. Motive controls stimulate the first awakening of our movements in the womb, and in this way they provide the essential engines for neuroanatomical differentiation.

In higher animals, motive regulation of neural plasticity does not stop at birth but continues throughout life. To understand how neural plasticity allows memory, we will consider the process of memory consolidation, organizing experience within the brain's synaptic web. We will apply the principles of representation and control found in Chapters 2 and 3 to the consolidation process within the specific architecture of the human cortex; in doing so we will find that consolidation appears to integrate the representations of the visceral core of the hemisphere with those of the sensorimotor shell. This integration is always motivated—continually excited and modulated by the vertically integrated activation control systems.

The study of the brain's motivated anatomy will thus yield clues to the ongoing formation of memories, organized as they are over hours, days, and years. Through this study we will be better able to understand expectancies—the concepts of the future that are perhaps the most important products of the memory system. Coalescing both past and future biases, we find memories to be actively exercised—instantiated within the stuff of experience—through dreams, fantasy, and other forms of reflection. By understanding the neural activation of these processes, we may find ways to look beyond motive blindness to find clues to the motivational controls that act—consciously or not—to direct the organization of experience.

Mechanisms of Motivation

Let us again take up the principles of information control, this time in the context of the body. Essential bodily states, such as temperature, are regulated by immediate physiological reactions, such as sweating or shivering. The analysis of control in these physiological systems proved fundamental for several fields of science, including electronics and computers. What physiologists discovered was how the body self-regulates and makes adjustments necessary for the various metabolic and physiological states necessary for life.

An integral function in this process is sensing the internal state, such as body temperature. Another function is effecting the adjustment, such as sweating. Together, these processes—monitoring and adjusting—allow the body to achieve homeostasis, the maintenance of critical internal states within physiologically acceptable ranges.

In the mid-20th century, electrical engineers faced increasingly complex system design problems. For example, they needed to aim antiaircraft weapons on the basis of radar information, or they needed to adjust a new invention, the transistor, so it would not oscillate out of control. These problems were eventually understood to be problems of control theory, which purports to explain how interactions among the elements of a system could be designed to allow the system to self-regulate. In fact, engineers had designed closed-control loops for many years, but it was from the physiologists that they learned the principle that explained the process. Armed with this information, the engineers became skilled technologists of control. The key principle was homeostasis.

An example of a well-known control loop is a thermostatic home heating system. The thermometer senses when the air temperature of the home drops below a set level. At that point, a switch is automatically tripped to turn on the furnace. As the temperature rises above the trip level, the thermostat (the thermometer and switch combination) shuts off the heat. With this simple closed loop, a feedback relation is established between the room temperature and the control of the heat. Through similar principles, feedback relations in the body allow the autonomic nervous system to maintain multiple bodily parameters—oxygen, temperature, blood pressure—in ranges necessary for physical health. This is causation, but not a simple, linear, one-way cause. Rather, it is a continuous, circular causation in which sensing and effecting are linked in a tight functional loop.

Recognizing the close parallels in the control problems faced by electronic and physiological systems, Norbert Wiener in the 1940s formulated his principles of feedback and control, an approach he called cybernetics. Over the next several decades, cybernetic reasoning was an essential foundation for the development of electronic data processing (EDP).

In the early years, a small group of intellectuals believed that cybernetics would herald a fundamental advance in many areas of science, including psychology and brain research. Certain important theories, such as Gregory Bateson's ideas on psychopathology and family interactions, were explicitly formulated in cybernetic terms. Psychotherapists readily adopted Bateson's ideas from communication and control engineering in describing human interactions. By the 1960s, the term "feedback" had migrated from cybernetic theorists through psychotherapy circles to become an integral concept in popular culture.

Emotion as a Cognitive Construct

However, the gap between body and mind proved a formidable barrier for cybernetic theory. It may seem a natural extension of the physiological principle of homeostasis to consider human behavior as guided by the monitoring of need states and the corresponding engagement of motivational control processes that then shape experience and behavior. But these principles seemed too difficult for psychologists to apply to the mind's internal mechanisms. Certainly the behaviorists wanted nothing to do with internal mechanisms of any sort. And for the cognitive psychologists, the computer metaphor did not seem to require either emotion or motivation. Even when psychologists developed specific theories of emotion, their hypotheses seemed paralyzed by motive blindness.

For example, the most popular academic psychology theory of emotion in the latter 20th century was a variant of an idea William James had suggested more than half a century earlier. In its modern form, the theory proposed that the physiological responses of emotion are not only visceral (autonomic) but also diffuse and nonspecific. What makes an emotion, according to this cognitivist theory, is the cognitive interpretation that we apply to the physiological feelings that arise in the particular situation.

It is true that many of the autonomic responses to emotional situations are diffuse. The sympathetic branch of the autonomic nervous system engages a common set of visceral actions, such as adrenalin release and heart rate increase, in response to varied psychological contexts from fighting to fleeing. The psychological theory proposed that emotions are created when a person interprets the physiological response (termed "arousal") in light of the psychological context.

For example, you are unexpectedly asked to say a few words at a large banquet you are attending. When you hear this, you immediately feel your heart race, your knees go weak, and your breath catch. The interpretation you place on these physiological responses then shapes the emotions that ensue. It will be one thing if you interpret these bodily feelings as evidence that you are challenged, excited, and ready to give it a try. It will be another thing if your interpretation is that you are having a panic attack and that you have not only lost the ability to speak but will also soon lose both consciousness and bowel control. Bodily responses are thus open to interpretation.

Psychotherapists know this and have become skilled in training their clients to frame their emotional reactions in ways that are adaptive rather than catastrophic. However, academic psychologists went further and concluded that emotions *are* cognitive constructions. This was a strange turn of mentalism. What happened to motivational influences that we know must shape the behavior of biological organisms? Did academic psychologists *themselves* succumb to motive blindness?

Apparently so. The naïve view of the mind may take hold even for those whose job it is to explain things like emotion and motivation. Cognition—the mind—is assumed to operate objectively, free of emotional influence. Continuing the tradition of philosophers since the Middle Ages, this magical, noncorporeal mind then applies an interpretation to the physiological response such that the body then figures in experience only through its peripheral physical sensations. By failing to recognize that emotions and motives act on cognition directly, academic psychologists indeed succumbed to motive blindness. As a result, once again the mind, as we know it, escaped the body.

Experience Embodied and Constrained

This abstraction of cognition to an objective realm may be forgivable for philosophers and experimental psychologists (and perhaps linguists), but clinical psychologists should know better. It is their job to cope with all of the impolite stuff that real people bring up. They should know that the mind is shaped by motivational influences.

People who are affected by strong emotions such as anxiety and depression provide unavoidable evidence that emotions constrain both experience and cognition in predictable ways. The emotion (the input or affective side of feelings) inherently engenders the motivation (the output or effective side). The anxious person seems to anticipate threat, as if the state of anxiety biases attention toward danger. Now, we know this is a two-way street: The cognition of anticipating danger is certainly enough to make anyone anxious, but in many cases the cause is clearly the emotion, and the effect is seen upon the cognition.

Because anxiety is often a chronic trait, it can be difficult to sort out which comes first, the anxious cognition or the anxious emotion. However, depression is in some cases quite acute, and in some people it may alternate rapidly with mania. In these cases, we have seen that the person's cognitive bias predictably follows the emotional state.

In these things, people are not that different from rats. When a person is depressed, the world looks terrible. If depression becomes severe, the self is experienced as worthless. Yet if the depression is lifted—either spontaneously or through a treatment such as drugs or electroshock therapy—the person's cognitive outlook improves in proportion to the mood change. For those who cycle to mania, the cognitive bias is distorted positively, in exact proportion to the mood swing, shifting not only to optimism about the world but also to a grandiose conception of the significance and power of the self.

Importantly, the emotional constraint of severe moods is embodied. Although it is commonly assumed that the depressive bias is a psychological trait, clinical observation shows that it comes and goes with the depressed mood state, even when this state is manipulated physiologically (as with drugs or electroshock therapy). Furthermore, the variance of optimism-pessimism with mood state is true for normal as well as pathological depression.

When experiencing failure, for example, it is normal to become slightly depressed. The negative bias that ensues may be adaptive, causing us to reappraise the context of failure in a negative light. The transitory pessimism is then a cybernetic, self-regulatory mechanism that is automatically recalled when we encounter this context again.

As with many emotional influences, these mood state influences operate under the cloak of motive blindness. People who experience a strong mood change do not experience themselves as being influenced by a motive process. Instead, they think the world changes.

Even those who frequently cycle between depression and mania fail to recognize the transient nature of the motive influence in the depressed state. Instead of experiencing a mood of depression as a temporary, constraining influence on perception, their experience is one of a world gone bad and a self without value. Instead of recognizing that emotional influences come and go, they cannot remember being happy—even if they were pathologically manic just a few days ago. In the depressed cycle, the subjective perspective becomes rigidly hopeless in the face of a felt eternity of having no value as a human being. When subjective reality is masked by motive blindness in this state, even suicide can appear to be a logical choice.

The implication of mood biases for a scientific analysis of subjective experience is clear. Not unlike the rats that are prepared to expect big rewards—or that collapse into learned helplessness—mood states modify our views of the world. They do this not as an external influence on the mind that is apprehended as such but as a force that biases our received reality without a trace, operating within the pervasive shadow of motive blindness.

Cybernetics of Moods and Drives

How do these influences of moods and drives operate? When your hunger causes the smell of food to be so pleasurable that you become alert and excited by the prospect of an approaching meal, what are the mechanisms that make this happen? How are the visceral and hormonal signals relayed to the brain in ways that can shape the networks carrying out perception and subjective evaluation? When you become elated after achieving success with a difficult problem, whether it is a major challenge at work or a simpler one like finally organizing the closet, how does this happen? How does the cognitive evaluation engage the arousal mechanisms that achieve the bright affect and increased alertness of your elated mood? These are questions of self-regulation, of extended forms of homeostasis. In neural terms they require answers in terms of multiple motivational mechanisms, from brain arousal, to elementary attention, to specific constraints on working memory.

From studies of neurophysiology in animals and from clinical observations in humans, we know the general outlines of these motivational mechanisms. As with most questions of brain function, the results are surprising primarily because any given function is found at multiple evolutionary levels of the neuraxis.

Let us review more carefully the levels of brain organization that we first examined in Chapter 2. This time the focus is on the representation of bodily and motivational control systems, the multiple evolved systems of self-regulation. This is not an easy review because to develop a functional model— to get the big picture—it is insufficient to simply point out a brain part or two. Rather, the theoretical perspective we need requires us to consider how motivational control is organized across the entire neuraxis, the levels of nervous system from spine to cortex.

Certain primitive reflexes, such as the withdrawal of a limb from an abrupt, painful stimulus, are found in the spinal cord. Although one might not consider these reflexes as motivational, they are fully cybernetic in that they close the loop between the stimulus and the response. Furthermore, the architecture of the reflex is informative. Each segment of the spinal cord shows an architectural pattern that is maintained at higher brain levels.

The anatomy of the spinal cord gives a simple but fundamental plan for understanding the organization of the neuraxis (Figure 4.1). Dorsal (toward the back) regions of the cord receive nerves with sensory input. Ventral (toward the front) regions send out nerves with motor output. Both sensory (dorsal) and motor (ventral) areas have two major functions. One is somatic, where the connections are to and from the skin and muscles. The other is visceral, where the connections are to and from the internal organs. For both types of functions, reflexes are achieved through interneurons that connect sensation with action.

For the visceral (autonomic) functions, the cybernetics are what may be described in engineering terms as closed loop: there are automatic (autonomic) controls that link a viscerosensory input (such as sensing a lowering of core body temperature) with a visceromotor output (such as constricting the blood vessels of the skin to conserve heat). Both the energetic (emergency or sympathetic) and the conservation (rest-and-digest or parasympathetic) divisions

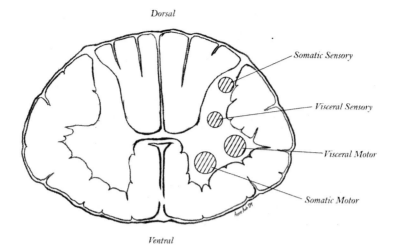

Figure 4.1. Schematic of the spinal cord. Somatic motor neurons are found in the ventral (body front) horn of the gray matter. Somatic sensory neurons are found in the dorsal (body back) horn. Visceral (autonomic) gray matter for viscerosensory and visceromotor functions is found in the intermediate zone between the dorsal and ventral horns.

of the autonomic nervous system operate in this closed-loop fashion, without the need for attention or voluntary control. The cybernetics of each division and level of the neuraxis provide control properties that become integral to more complex mechanisms of motivation and self-regulation (Figure 4.2).

Although there have been many opportunistic twists of fate in vertebrate brain evolution, a consistent principle is that each higher brain level maintains the segregation of sensory and motor functions. It also maintains the segregation of somatic and visceral functions. The segregation and the interaction of these elementary functions remain organizing principles for the brain—the physical embodiment of mind—at each of the multiple levels of vertical integration.

Neural Control

When we examine the functional systems of the nervous system in this way, we see levels and bodily functions. We also see the residuals of vertebrate evolution in the somatic and visceral functions at each level of the nervous system. The theoretical challenge is to understand how the integrated psychological functions of a person arise through the coordination of the neurophysiological functions across these multiple levels.

In terms of control theory, when we look to the brain we find multiple control influences with interlocking patterns of balance and influence. Thus, physiological homeostasis is achieved not through one control process (as with a thermostat causing a furnace to start) but through multiple influences,

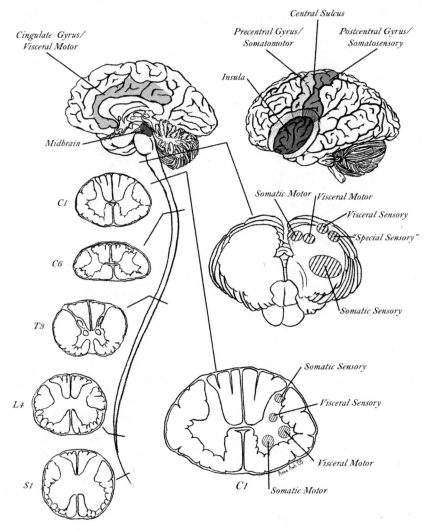

Figure 4.2. View of levels from the spinal cord through the medulla and brain stem to the cerebral hemispheres. At each level the somatic sensory and motor pathways are segregated from the visceral sensory and motor pathways. At the level of the cerebral hemispheres, the visceral sensory and motor networks are represented at the limbic core of the hemisphere, whereas the somatic sensory and motor networks are represented in the neocortex on the hemisphere's lateral surface.

including controls on visceral activity at the hindbrain level, controls on the brain's alertness from the midbrain level, and controls on patterns of motivational drive at the hypothalamic level. At the limbic and cortical level, the brain's cognitive processes are engaged, with influences both to and from the more elementary subcortical levels.

In sexual arousal, for example, there are specific and predictable autonomic controls on bodily function, including peripheral vasodilation (flushed skin) and genital vasocongestion (getting wet or hard). These are accompanied by motive influences on both perception and behavior, including heightening of the hedonic value of tactile stimulation and energizing of actions to increase that stimulation. Although the interlocking network of influences is most stereotyped and thus most easily traced at the elementary physiological levels, it extends to the mechanisms of attention and memory as well. Through them, focused adaptive influences are exerted on experience.

Although most of the neurophysiological studies of sexual motivation have been done with animals, one well-known experiment was conducted with electrical stimulation of limbic circuits in a human patient. R. G. Heath implanted electrodes in the limbic system of a woman with a mental disorder in order to study the effects of electrical stimulation on her disturbed experience and behavior. When current was turned on to the electrodes implanted in the septal area (part of a limbic circuit related to sexual behavior in many animal studies), the woman showed a specific and consistent tendency to describe her interest in sexual activity during the psychiatric interview. When the current was turned off, the references to sexual interest stopped.

Although the medical utility or even ethics of this experiment may be questionable, it provides a concise and memorable demonstration of a key theoretical question: the direct influence of limbic activity in motivating cognition. Human cognition continually integrates motivational influences that apply dynamic biases on perceived value. These motivational influences not only affect hypothalamic and simple visceral functions but also apply both motive and evaluative controls on the content of experience.

Motivated Anatomy

The question for understanding everyday human experience is how psychological function emerges from multiple levels of the neuraxis. Even to explain the coherent function of these multiple levels in scientific terms is a considerable theoretical problem. It can be described as the problem of *vertical integration*—how higher levels of brain systems have evolved from lower ones. At the same time, we must understand how the higher levels are able to subordinate and integrate the function of those lower levels. This was the problem that John Hughlings Jackson addressed more than a century ago with his evolutionary account of the human brain. Jackson speculated that this is perhaps achieved through hierarchic integration of multiple—what I am now calling vertical—neural levels.

The evolutionary order that Jackson recognized in neuroanatomy is difficult to reconcile with everyday experience. Perhaps in some abstract way we can accept that mood states are important to experience. Maybe we could even admit that our biological urges arise from primitive brain parts, but we do not experience or feel levels, certainly not in relation to personal experience with problems of mental control. The brain's multiple levels of organization are hidden from the naïve view of the mind, latent within the implicit unconscious.

Nonetheless, we can follow Jackson's example and reason from evolutionary principles, much as we did in considering the animal learning mechanisms discussed in Chapter 3. Higher levels of the brain may support memory processes that impose an increasing delay between stimulus and response. Within the simple networks of the spinal cord, reflexes are carried out with direct neural connections mediating between sensory input and motor output. At each higher level of the brain, there is not only greater complexity in both stimulus evaluation and response preparation but also increased time that can be spanned to organize this complexity in mediating between input and output.

Psychologically, the processing over time may be reflective, as when ruminating on the significance of recent events provides mental structures that can prove useful in the future. Or, as rats taught us so well, the processing may be explicitly anticipatory, preparing responses in advance of the relevant environmental events. In each case, the multiple levels of the neuraxis must be recruited to play their roles in extending the neural processing, spanning broader intervals of behavioral and experiential time.

Understanding vertical integration is challenging for the theorist, and it poses a major barrier to the layperson who wants to understand the brain.

For example, Jared's friend Kim shows a high degree of intelligence as she tries to understand the lessons in her history class. While she struggles to apply her intelligence effectively, it may help her to know more about the mental resources she brings to the task. How could we teach her about the vertical integration of neural mechanisms underlying her intellectual skills?

Vertical Integration in Personal History

In my work as a university professor, I have found a way to give students the required perspective on the development of the neuraxis: Have them study the embryonic development of their own nervous systems. Incredibly, the residuals of vertebrate evolution can be seen, recapitulated in general outline, in the small, real process of our own embryonic development. As we have learned from psychologists such as Jean Piaget and Heinz Werner, embryology is also an excellent model for understanding psychological development. Furthermore, I find a certain philosophical perspective comes from contemplating the state of your nervous system at the stage of your life when your tail was longer than your torso.

To convey this perspective, we could teach a student like Kim about the history of her brain in the first 2 months of her (postconceptional) life (Figure 4.3).

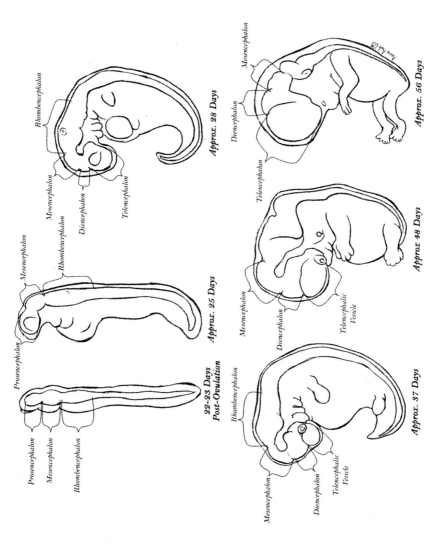

Figure 4.3. Embryonic development of the nervous system, showing the progressive elaborations of the levels of the neuraxis.

Kim's primitive neural tube first differentiated three bumps, vesicles around the cerebral spinal fluid in the interior of the neural tube (22–23 days). These bumps developed into the rhombencephalon (hindbrain), the mesencephalon (midbrain), and the prosencephalon (forebrain). At about 28 days after conception, the prosencephalon differentiated into the diencephalon (interbrain) and telencephalon (endbrain). By the time of Kim's birth, her telencephalon had expanded laterally and front-to-back, wrapping around and covering much of the brain stem (hindbrain, midbrain, and interbrain) (Figure 2.1).

The Principle of Functional Differentiation

This developmental progression organizes the human neuraxis under genetic control. This is the vertebrate evolutionary plan, tweaked by the myriad selection accidents of our ancestry to create an individual human brain. The evolved architecture shows the need for vertical integration. In human evolution (phylogenesis), each level gained its functional capacity through subordinating and elaborating the functions of the previous level. In individual development (ontogenesis), some kind of recapitulation of this genetic unfolding occurs in order to recreate the fully organized human form of each child's brain. In order for Kim's brain to function in the course of her psychological development, it must achieve an effective functional integration across all of the levels of her vertically organized neuraxis.

The insights from embryology do not stop at birth. Research of the last several decades has shown that the mechanisms of embryonic differentiation of the brain's networks continue to operate even after the transition from fetus to neonate. In fact, in an amazing accidental strategy of mammalian evolution, the embryonic neurodevelopmental mechanisms have been retained as key mechanisms of psychological development after birth.

First, let us examine the way the fetal brain differentiates its connectivity. It forms connections through a process called *subtractive elimination,* which refers to the strategy of making numerous connections early in development, then subtracting the ones that are not used. Although the gross connectivity of the brain's networks is under genetic control, the fine connectivity is achieved through this process of functional sculpting.

For example, nerve fibers must connect the frog's retina to its midbrain. Certain chemical gradients induce these fibers to grow toward the retina, but the fine mapping (allowing the brain to reflect the spatial structure of the visual world) would be very difficult to organize with something as crude as chemical gradients. Instead, evolutionary selection happened on a somewhat gross but effective strategy: Make a *lot* of retinal connections with the midbrain, then keep *only* those that make a good map (i.e., the ones that reflect correlated activity in the visual field). Only the effectively patterned connections survive.

In everyday language, this can be described as a "use it or lose it" rule. In technical terms it is a form of subtractive elimination through *activity-dependent specification*. Functional activity (in this case correlated visual input) is what specifies the fine structure (visual map) of the brain's representational networks. As the brain grows, it differentiates its architecture on the basis of the correlated functional traffic in the network. Neurons that fire together wire together.

Motive Control of Neural Plasticity

Of course, this process of functional differentiation requires *function*. This is where motive control comes in. There must be some activity in the network so that the correlated activity can be selected for retention, and yet the fetus is still in the womb, protected from interface with the world. Early studies of embryonic neural differentiation showed that the baseline of activity necessary to achieve functional sculpting of the networks is naturally exhibited by the vertebrate embryo. At a certain stage of development, the embryo begins a spasmodic pattern of motor activity. The ancients described this as the *quickening*. This self-generated action causes the patterns of sensation (feelings) within the embryo's brain that then create the correlated activity necessary to differentiate the connections of the brain's networks.

The brain's own activity is thus a primary agent of the anatomical differentiation of network architecture. Neurodevelopment is, in large part, *self-organizing*. The controls on brain activity that achieve this are distributed across the multiple levels of the neuraxis. Of course, these controls are responsive to the information traffic with the world, so that functional differentiation can be described as *experience dependent*. However, at each level of the neuraxis, unique controls motivate the process of functional differentiation. As we study the function of each level of activity control in the next section, we can then question its contribution to the psychological mechanisms of experience.

We know abstractly that the evolution of more complex forms of motivation and self-control has produced increasing capacities for spanning between stimulus and response. Rather than reflexes producing actions from fixed stimulus patterns, higher mammals are able to mediate their actions with increasing evaluation of environmental data. What we are now figuring out is that this mediation between stimulus and response—this capacity for representation and organization of information—is achieved not just in mental space but also in the physical space of the brain's networks. The more complex activity-control systems that allow memory to span between stimulus and response also motivate the functional sculpting of corticolimbic anatomy.

Thus the evolution of intelligence drew upon an accident of development that caused the fetal differentiation (morphogenesis) to continue after birth. In this way, the brain could incorporate the information of the world within its neural structure. Kim's memory *is* her fetal growth continued. The activity-dependent sculpting of neural connectivity in embryonic differentiation is now only the beginning of a momentum of adaptive information structure that continues throughout life. As it continues, activity-dependent sculpting forms the fine architecture of cortical anatomy to encode the distributed representation of memory.

The principles of distributed representation and control that we examined in the previous chapter are thus woven in an intimate pattern within the developing child's neural architecture. Motives are the organizing forces, operating across the vertical levels of the neuraxis and providing the neural excitement that allows the sculpting of information patterns. Experience is thereby consolidated into personal patterns of neural connectivity. Through their work in directing anatomical differentiation, the motives of the self operate as the engines of experiential structure.

Thus, through radical neoteny (maintenance of fetal form), the human brain remains plastic (adaptable). We mediate increasing spans between stimulus and response. As a result, the uniquely human capacities of conscious appraisal and voluntary control have become necessary to direct the increasingly deep reflection on the meaning of past events and the increasingly involved planning of future actions. If we could understand how this occurs, we might see how human consciousness itself has evolved as the emergence of a powerful memory capacity allows us to apprehend an increasing span of experiential time.

A Challenge for Self-Control

We can consider the roles of both unconscious structure and conscious direction of the intellect in the context of a typical problem of everyday life, one in which motivations are mixed and contradictory, and yet the person must apply intelligence quickly and effectively to make an important decision.

In her 17th year, Kim has been interested in Jared, a quiet, good-looking guy in her history class. After class on Friday she is pleased and excited when he asks her out to a party at a friend's house that night. After school she calls her best friend, Tara, to tell her. They laugh and squeal with delight as they reflect on this new development. They are both aware of the new social status Kim will have with their friends if she dates this cute boy. As their boisterous conversation unfolds, activating a series of associations in memory, their adolescent hypothalamic mechanisms readily modulate working memory with motive influences. Tara embarrasses Kim by joking that now she can get into Jared's pants.

That night before the party, Jared picks Kim up in his cool little car, and they have fun talking and laughing as they drive through town. When they get to the party, Kim is fully engaged in her new role of being Jared's girlfriend. The self-image she is realizing tonight, of being a senior and being in this relationship, is one she has anticipated for a long time. Her self-image becomes the defining context for the mix of motives and values that cross her mind through the evening's flow of conversations and personalities. The mental structures in her memory—her images of herself and her life—span not just the immediate moment but this entire segment of her life as well.

Kim has time to reflect on things now because Jared has left her alone in the kitchen as he talks with his friend Andrew. He was attentive to Kim earlier in the evening, and he seems to like her a lot. Still, she is perplexed and a little irritated to see him drinking beer with his buddy while she stands by herself.

As it gets late, Kim notices the effects of Jared's beer drinking. Even through her infatuation, she recognizes that he is getting drunk and uncharacteristically loud. The glow of her perfect evening is now shattered by anxiety. She experiences a sinking feeling in her stomach as she realizes that this drunk boy is her ride home.

Kim then knows that she must make a decision: whether to confront Jared and not let him drive or to go along with him and risk trouble with her parents, trouble with the police, and even a serious accident. If we analyze the forces operative within Kim's mind in the next few minutes, we can infer the contributions from multiple levels of her nervous system.

In neural terms, Kim's experience is at first dominated by her own visceromotor (gut-level) responses. Her sinking feeling is an emergency signal of something disappointing and maybe dangerous. Recognizing this threat diffusely,

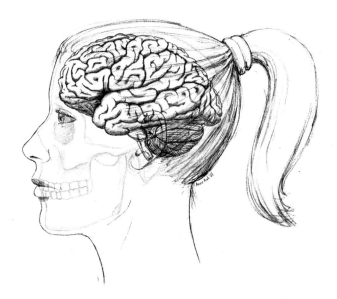

Figure 4.4. Lateral view of Kim's left hemisphere.

Kim must call upon the brain stem arousal mechanisms integral to her anxiety, energizing not only the primitive fear of her limbic circuits but also the vigilance and focused attention recruited through multiple corticolimbic pathways. If she is successful, she will achieve vertical integration of the midbrain, interbrain, and endbrain systems that have guided her neural development. In the process, she will focus and balance the interlocking feelings, motives, and reasons that will converge on her decision.

Level One: Hindbrain

At the hindbrain (rhombencephalon) level of the lower brain stem are not only sensory and motor pathways but also visceral control centers for internal bodily functions such as respiration and blood pressure. Because these hindbrain circuits are critical to survival, they have been strongly conserved in evolution. This means that genetic mutations that affected brain stem circuits were likely to be fatal. The result was that little room remained for evolutionary variation and selection at this level of the neuraxis. After hundreds of millions of years, the architecture of the human hindbrain still follows a very rigid, primitive plan.

As her anxiety collides with the disappointment in the loss of her valued evening, Kim has multiple autonomic (visceral) responses that are mediated by the lower brain stem. As she worries about what might happen, she blushes with the embarrassment and shame of the expectancy of the social conflict that may soon occur with Jared, her dad, the kids at the party, and who knows who else. The collision of her anxiety with her former good mood causes an intense, abrupt disappointment that spreads down through the visceral controls in her lower brain stem. The result is a sinking feeling in her stomach that is so strong she feels sick.

Level Two: Midbrain

At the midbrain (mesencephalon) level, both somatic and visceral sensory data engage a network—reticulum—of nerve connections. These integrate multiple inputs on the status of the body and on external stimulation within the *reticular activating system*. When these multiple inputs are processed in the mesencephalic reticular architecture, they can influence the arousal of the entire cerebral hemispheres of the brain through neurochemical projection systems. Both street drugs and therapeutic psychiatric drugs influence arousal level and mood states through their actions on these mesencephalic reticular projection systems. The high of a stimulant drug or the alertness of a panic attack depend heavily on these neurochemical roots. At this level, we can see the integrating functions of the midbrain reticular networks that monitor bodily and sensory data to regulate both psychological and physiological arousal.

For Kim, the elation associated with the social arousal of the party was actualized within her mesencephalic reticular activating system, making her experience and actions bright and upbeat. However, this component of midbrain control is now overwhelmed by anxiety and vigilance as she considers the threats, both social and physical, that are posed by Jared's intoxication. Her sense of alertness has now galvanized into a need for safety.

Level Three: Interbrain

At the top of the brain stem are the diencephalic (interbrain) structures. For mammals, the division between somatic and visceral control systems is clear at this level. The somatic sensory and motor traffic to each cerebral hemisphere is funneled through the thalamus, a more or less round structure on the left and right sides at the top of the brain stem (Figure 4.5; see also Figure 2.4). In the thalamus are specific nuclei that act as centers for each sensory modality and

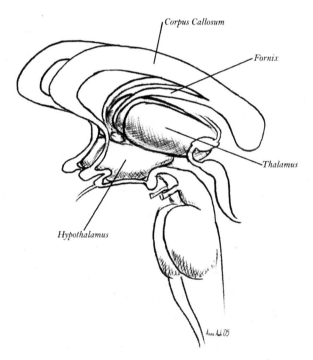

Figure 4.5. View of the thalamus and hypothalamus at the top of the brain stem. The thalamus is the gateway for sensory input to—and motor output from—the cerebral hemispheres. The hypothalamus monitors and regulates the internal bodily states.

for regulating aspects of motor control. These specific thalamic nuclei connect the relevant somatic traffic with the neocortex of the cerebral hemisphere. In addition, the thalamus has nonspecific nuclei that are important to regulating the activity level of the cortex that receives and sends the specific somatic projection signals. Together, the specific and nonspecific thalamic nuclei are important in the regulation of the sensory and motor traffic between the cortex and the world. In the cognitive domain, these same mechanisms may be important to the control of attention for both perception and action. Kim's attention is riveted to Jared's every move as she fine-tunes her own behavior in preparation for a showdown.

The second major division of the diencephalon is the hypothalamus (Figure 4.5), which is so called because it is located below the thalamus (the prefix "hypo-" means "under" or "beneath"). This integrated group of nuclei is the central executive control of the autonomic nervous system. It regulates the visceral milieu (the internal landscape of bodily organs) through homeostatic controls. Aligned with its role in the visceral functions, the hypothalamus has direct influences on somatic behavior through primitive motivational drives such as hunger, lust, panic, and rage. For Kim, sympathetic arousal prepares her body for confrontation, whereas panic and rage tug her in opposite directions. Now higher levels of the neuraxis are needed to orchestrate an effective response to this confusing situation.

We thus find a clear division of function at the diencephalic level. The somatic function (i.e., the traffic with the environment), whether through external sensation or motor actions, is regulated through the thalamus. The visceral function (i.e., the traffic with internal bodily organs), including both viscerosensory and visceromotor controls, is regulated through the hypothalamus.

In today's cognitive neuroscience, we typically think of the diencephalic (interbrain) circuits as providing primitive support functions for the more fully elaborated telencephalon (endbrain). However, as we shall see, the separation of *somatic* sensory and motor organization (in the thalamus) from *visceral* controls (in the hypothalamus) may be a key to understanding the active, motivated organization of experience.

As Kim tries to collect her thoughts, the cognitive representation of her situation leads to visceral responses, and these augment particular cognitive representations. She reasons through the implications of Jared's intoxication with capacities of attention and working memory that reflect the control influences projecting from the thalamus to guide the cognitive representations that are forming in her extensive cortex.

The result of her thought process is a neural representation that is not just a dry, objective idea of the danger of her situation. Rather, it is a vertically integrated construct that is propagating across all of the levels of her neuraxis. It engages a strong visceral response with direct influences on both hypothalamic drives and lower brain stem autonomic responses. Through its multiple levels of representation, this visceral response is experienced. Kim feels sick, and simultaneously she knows something is seriously wrong.

As the control influences course up and down the neuraxis, there is order to the neurocybernetic mechanisms. Because the functions of the telencephalon must elaborate their basis in the diencephalon, the diencephalic segregation of somatic sensorimotor (thalamic) and visceral (hypothalamic) functions provides order to the control of Kim's telencephalon and to her experience.

Level Four: Endbrain

For more complex motivations, the mammalian hypothalamus is itself heavily regulated by the structures of the telencephalon (endbrain). These include the basal ganglia (nerve groups at the base of the brain) that are important for motor organization, including initiation and sequencing of actions and action patterns. The telencephalic structures also include the limbic circuits, engaging the hippocampus, amygdala, and septal nuclei together with the limbic cortices forming the border or limbus of the core of the cerebral hemisphere (Figure 2.4).

The massive networks of the telencephalon, which are distributed through the cortical mantles of the left and right cerebral hemispheres, provide capacities of memory representation for mammals that are not possible for simpler creatures. As we will see shortly, the most important physiological processes of the cortex are those that support this representational capacity, consolidating memories within the sensory and motor regions of the cortex.

Consistent with their roles in weaving functions across the multiple levels of the neuraxis, the limbic circuits not only send and receive projections to the cortex of the cerebral hemispheres but also project back down through the diencephalic structures. The way this happens, creating the pattern of connections from cortex to the interbrain, gives us a major clue to the organization of the cortex.

The best way to understand this clue is to apply the classical evolutionary-developmental reasoning worked out by neuroanatomists such as C. Judson Herrick: Look to the developmental roots. In this case, these are the evolutionary developmental roots of the telencephalon found in the diencephalon, showing the clear division between somatic and visceral functions that is found at every level of the neuraxis. Whereas the somatic/visceral division can be difficult to sort out within the large, wrinkled, and highly evolved cortex, it is a clear and simple distinction at the interbrain level. Somatic control is thalamic. Visceral control is hypothalamic.

The thalamic projections are preferential to the sensory cortices (including visual, auditory, and somatosensory or touch) and to the motor cortex. Together, these areas are responsible for the interface between the organism and the outside world. Here we find the *somatic function* of sensory and motor interface with the world represented at the highest level of the neuraxis.

In contrast, the hypothalamic projections to the cerebral hemispheres are focused on the limbic networks at the core of the hemisphere. These extend the *visceral function*, including the monitoring and control processes of the hypothalamus, to the telencephalic level. Within these projections is a further separation of viscerosensory monitoring (in the ventral limbic and insular cortex networks) from visceromotor control (in the dorsal limbic or cingulate gyrus networks). See Figures 2.1 and 2.9.

To explain Kim's experience in the physical terms of her neural organization, the complex representations of the telencephalon are the end points of vertical integration. Although she does not know it, of course, her experience has an implicit structure formed by the visceral and somatic boundaries of her neurocognitive apparatus.

Kim's conscious appraisal includes her awareness of the sinking feeling as she faces her decision. Unless she gets really sick, however, this overt visceral sensation is only peripherally important to the motives that shape her experience and actions. She also experiences the conflicts among several feelings tonight: her fear, her anger, her desire to be in control, and her need for acceptance. These motives have an important basis in the visceral representations of the limbic networks. Yet, like the sinking feeling, these motives are only elements of her emotional landscape, and none of them are definitive.

Not unlike the rodent subjects, Kim's reward anticipations were very high tonight. Her learning capacity is now strongly affected by the discrepancy between her valued expectancy and the situation she now perceives. Although Kim is clearly influenced by primitive mood states, she is also influenced by the more complex experiential integration of herself in the context of her situation. We could call this her *contextual self-concept.* It is the most fundamental framework for her understanding of the evening's events. This is the memory structure that spans time, draws from her personal history, and is able to project forward, creating a "memory" of the future—an expectation of what is to come. The capacity for self-representation, certainly one of the more complex products of the telencephalon, forms a unique reference frame for human cognition. Shaping this reference frame and embodied at the multiple levels of the neuraxis, Kim's self-concept is now shifting between multiple unstable and incompatible states.

On the one hand is the new but already prized and familiar image of herself as the popular girl who is Jared's girlfriend. On the other hand, she recalls—over a few minutes of quiet reflection now as she stands in the kitchen—self-concepts that have been important to her for many years. Her present experience resonates with her images of herself as a trusted daughter, an assertive young woman, and a realist. She clearly remembers the pact she made with Tara two years ago that they would never risk a stupid death on the highway.

Vertical integration is essential to adjust the neuraxis to achieve harmonic resonance with each of these concepts of self-in-time. Motives and emotions that are rooted in the lower levels of brain organization are important, but they do not operate out of context. The defining context for Kim's feelings and actions is framed by the complex psychological organization of her experience of herself in the present situation. The multiple components of the neural hierarchy play their roles as they are recruited and shaped by this complex psychological process.

Vertical integration is essential to this story if Kim is to think effectively now. She must recruit multiple levels of the neuraxis to contribute in a coordinated fashion if she is to make the right decision. In psychological terms, this requires several things. She must remember the most important facts and concepts that are relevant to her choices. She must anticipate and then evaluate the most probable outcomes of each choice. Finally, she must organize the context of the various elements of her self-image to articulate—consciously now—the values that will shape her decision. As we will see next, her representations of self-in-context have a clear neural structure. Her sense of self is rooted in the motivational base at the visceral core of each hemisphere, and her image of the world is articulated in the neocortical networks forming the sensorimotor interface with the world. Her concepts of self-in-context span and integrate these levels.

Kim, of course, is unaware of neural structure. She is acutely self-aware but not of thalamocortical projections or limbic elaboration of visceral responses. Rather, she is conscious of the events around her and of her images of herself. Her perception of events and her articulation of action plans are formed in the sensorimotor networks of the cortex. Her internal feelings, as well as the values and self-references derived from them, are formed in the corebrain limbic networks. Between these representational bounds (the anchor points of the visceral core and the specification of experience at the somatic shell), Kim's defining concepts of self-in-context must emerge. This is the confluence of sensation and need. The result is information.

Thus, as we analyze it in psychological terms, Kim's experience illustrates what is perhaps the most remarkable and defining feature of vertical integration of the human brain: the multiple constraints on the representation of memory. In humans, as fetal mechanisms of neural self-organization have been maintained to direct an extended psychological development, the mechanisms of the neuraxis have extended the mediation between stimulus and response to achieve ongoing processes of conceptualization and expectancy. These processes form the structures of the self and are continuations of the anatomical self-organization of embryogenesis. Through these processes, Kim can not only recall the defining concepts and experiences of her past but also project her experience into the future. By integrating the motives and urges of her visceral limbic core with the articulated perceptions and actions of her somatic neocortical shell, Kim can remember what is important, reason and imagine how things may go, and evaluate the probable consequences of what happens next.

Architecture of Memory

Imagine a society in which adolescents were not just left to cope naïvely but were trained to understand the mechanisms of motivation that shape the new freedoms and decisions of adult life. If we are to design a curriculum for intelligent and disciplined users of the mind, one of the required courses must be neural architecture. As the preceding review of computational neuroscience shows, the principles of distributed computation are beginning to provide insight into the way information is represented across massively parallel networks. With these principles, we can begin to interpret neural networks such as those that Kim must synchronize as she tries to understand the immediate future unfolding before her. Now we come to questions about how complex psychological structures—of the self and personal history—are organized within the telencephalon. To explain how the mind acquires information adaptively, we need to understand the network architecture responsible for forming memories.

What we find is complex and at times confusing, crafted not by principles of organization but by defaults. These are the defaults that conferred function upon the residuals of myriad accidents of mammalian evolution. To explain the control of the mind's dispersed representations, this architecture is highly informative. As we learn to infer function from distributed connections, we are learning to read the patterns of cortical architecture. In these patterns we find the story of the adaptive control of experience. Each of the connectional patterns that we studied in Chapter 2 provides structure to intelligence. The most important pattern for motivational control may be that formed by the core-shell differentiation, between the limbic core at the medial center of the hemisphere and the neocortical shell on the lateral surface. This differentiation provides theoretical insight into the close (and perhaps inseparable) links between motives and memories.

We should begin with the basics, the evidence that the connections between core and shell are necessary to form memory.

Boundaries of Memory Formation

Research with both animal experiments and human clinical cases has shown that laying down memory traces requires some kind of interaction between the limbic regions at the core of the brain and the neocortical regions on the lateral (outside) surface of the cerebral hemispheres. To understand the nature of this interaction, we examine two fundamental clues.

The first is that if the connections are severed between limbic and neocortical regions, new memories cannot be formed. Importantly, these connections are not limited to the cerebral hemispheres of the telencephalon but extend vertically to integrate the diencephalon and even the midbrain and lower brain stem as well. This finding implies that some corticolimbic process, often described as *consolidation*, must occur in order to lay down memory traces across the distributed networks of the neocortex.

The second clue is that, even if the critical diencephalic-limbic-cortical connections are damaged, old memories may be accessed, depending on the connections that are retained. This suggests that a memory is not localized in one site. Rather, once it is laid down in tissue, a memory is widely distributed across the networks of the neocortex of the cerebral hemispheres.

Corticolimbic Architecture

Our understanding of the architecture of the human cortex has undergone major advances in the last several decades with the development of quantitative methods for describing the connectivity between regions of the cortex. Without quantifying the connections, we can find connections among many regions of the cortex in such a confusing pattern that it seems as if there is no order. However, once the connections are counted, as in the careful studies by Pandya, Barbas, and their associates, the pattern emerges. The density of connections reveals a distinct architecture.

Let us review the basic core-shell architecture of corticolimbic pathways studied in the last section of Chapter 2. The posterior cortex includes the sensory pathways for vision, audition, and touch. Although there are many specialized regions that can be separated, the general pattern of connectivity breaks down into four networks for each sensory modality: the primary sensory cortex, the unimodal (one sense) association cortex, the heteromodal (multiple senses converging) association cortex, and the paralimbic (visceral and motivational) cortex.

Refer again to Figure 2.10. This figure (at right) shows the major pathways and the four major network divisions for the visual system. Visual input from the eyes through the optic nerve is first processed through the thalamus and the midbrain (showing the residual parallelism of the evolved, hierarchic neuraxis). The most specific visual pathway projects to the primary visual cortex (denoted by the shaded section of the arrow at the back of the brain, on the right of Figure 2.10). The processing sequence is then an orderly one. The primary visual cortex projects to the visual association cortex, which projects to the heteromodal association cortex, which finally projects to the paralimbic cortex.

This fairly simple overall pattern is complicated by two subdivisions of visual cortex, the dorsal processing stream targeting the dorsal limbic areas (cingulate gyrus and hippocampus) and the ventral processing stream targeting the ventral limbic areas (insula, anterior temporal, orbitofrontal cortex, and amygdala). The four corticolimbic levels—limbic, heteromodal association, unimodal association, and primary sensory neocortex—are seen within *both* the dorsal and ventral visual processing streams, shown by the top (dorsal) and bottom (ventral) arrows in Figure 2.10.

The same corticolimbic network levels are seen for both the somatosensory (touch) and auditory (hearing) senses (Figure 2.8). The primary sensory cortices for these senses also receive specific information from the thalamus, and this information is processed and projected to the sensory association cortex, then to the heteromodal association cortex (where it is integrated with other senses), and finally to the limbic cortex (where it is integrated with motivations and emotions).

Roots at the Core, Mirrors at the Shell

As we have seen in our study of Kim's brain, the major principle for understanding the integration of sensory data in the cortex of the telencephalon is the organization of the data paths through the diencephalon (i.e., the thalamus and hypothalamus). As Herrick taught us, we find our roots in history. Through evolutionary developmental reasoning, we seek out the organization of the cortex by studying the base of its roots in the diencephalon. The sensory input (from the eyes, ears, and skin) is relayed from the thalamus to the primary sensory networks of the neocortex, where it is processed and "associated" with other information in a linked set of four networks. Because this network set is linked sequentially, the dominant order of processing must be primary sensory, then unimodal association, then heteromodal association, and then limbic.

Just as in Herrick's reasoning the thalamic projections help to define the sensory and motor functions of the neocortex, so the hypothalamic projections help to define the sensory and motor functions of the limbic cortex. The difference is that the thalamic connections compose the environmental interface (the shell of the organism), whereas the hypothalamic connections constitute the interface with the internal organs, the organism's core. The thalamic projections are *somatic*, dealing with the skeleton and musculature of the body, as well as the receptive surfaces of the skin, eyes, and ears. In contrast, the hypothalamic projections are *visceral*, forming the roots of corticolimbic operations—the viscerosensory and visceromotor functions of homeostasis with the internal organs, autonomic nervous system, immune system, and hormones.

At the somatic shell we thus find mirrors of the outside world. For the thalamic projections to the shell, the external surfaces of each sense organ represent or mirror the world in a certain way. These representations are then modeled in detailed neural networks by the primary sensory neocortices. The visual cortex, for example, is mapped retinotopically, meaning that its network topography corresponds to that of the retina, the back surface of the eye. Thus, the upper visual field is separated in visual cortex from the lower visual field, and the left side of space is separated from the right side in an orderly, retinotopic map of visual space. The somatosensory cortex is somatotopic, with the lower body mapped on the medial wall of the hemisphere and the upper body on the lateral wall. This is not an exact mirror but a kind of carnival mirror that distorts space as a function of sensory importance. Large areas are devoted to the sensitive surfaces of the lips, tongue, and fingers (Figure 2.7).

Remarkably, the auditory cortex appears to be *tonotopic*, organized with similar frequencies (tones) mapped next to each other. Apparently the "surface" of the ear that is mirrored by the auditory networks is not an interface with the space of the surround but with its frequency spectrum.

For many years, the maps of the sense organ surfaces have provided tempting clues to the functional mapping of cortical networks. However, it has proven difficult to build meaningful psychological theories of the general network structure. We had localizationist ideas on the one hand and vague holistic ideas of mass action of the nervous system on the other. Both were demonstrably wrong, but we did not have good alternatives. Our ignorance was partly as result of not knowing the actual connection pattern (the quantitative connectional anatomy) and partly a result of having no metaphors—no theoretical models—for how connections imply function.

Now, thanks to modern neuroanatomists such as Deepak Pandya and Helen Barbas, we have quantitative maps of the connections. Moreover, thanks to many creative connectionist theorists, we have computational principles that are derived from the theory of distributed artificial neural networks. We can therefore reason explicitly for the first time about the physical structure of mind. Functions are not localized in a simple one-to-one fashion with neurons but rather are broadly distributed across the connectional patterns of entire networks.

Figure 4.6 is a schematic of this structure in one sensory pathway, such as vision. In the previous figures, the cortex has been shown as a wrinkled, two-dimensional sheet, but this sheet has thickness—a third dimension. Stratified across that thickness are the cortical layers, which are organized rows of neurons that seem to have different functions.

In the separation of the regions of a limbic-to-neocortical pathway into sections, the layers of each region of cortex are separated by the granular layer of cortex (shown by dots in Figure 4.6). Layers above this are called *superficial* or *supragranular* layers, and layers below it are called *deep* or *infragranular* layers. The function of the granular layer of primary sensory cortex can be

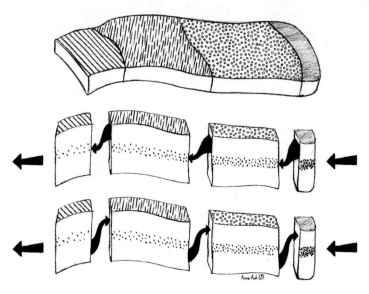

Figure 4.6. Patterns of connections across regions (or levels) of a single corticolimbic pathway, such as the visual pathway, as unfolded in Figure 2.12. The cortex represented by the pathway is separated into primary sensory (shaded, at right), unimodal (stippled), heteromodal (dashed), and limbic levels (striped, at left) (after Pandya).

inferred by its connections, which are primarily the input connections into sensory cortex from the thalamus. Thus, when the visual thalamus sends connections to the primary visual cortex, these go to the granular layer (see Figures 2.8, 2.11, and 2.12).

The dominant flow of connections (shown by the arrows in Figure 4.6) follows a regular pattern with the "incoming" sensory data (e.g., from the primary visual cortex toward association areas) progressing from the supragranular layers to the granular layer of the target region (as in the top of Figure 4.6). The unimodal visual association region, for example (the second region from the right), receives its granular layer input from the supragranular layers of the primary visual cortex (the first region on the right).

This is the primary direction of neural information traffic that scientists normally consider. The data comes in from the senses to primary sensory cortex and then gets processed in some unknown way by the "higher" brain regions, the association cortices. By this logic and from what we know now about cortical connectivity, the "highest" regions in the hierarchy are the limbic cortices.

This is a somewhat disconcerting implication of the findings of modern anatomy. The limbic areas have always been recognized as the *primitive* areas of the brain, regions of the cortex we share with even simple mammals. In fact, we call the nonlimbic cortical areas "neocortex" because they are thought to be newer than limbic cortex. Finding that the limbic cortex is positioned at the center of hemispheric traffic has been something of a shock.

Not only does the anatomy give us clues to the global corticolimbic framework of memory consolidation, but the nature of the connections between adjacent corticolimbic rings implies something about the consolidation process as well. The sensory data appear to be processed in each of the adjacent networks, within the supragranular layers, before they are passed on to the next network (i.e., to that network's granular layer). However, as Figure 4.6 shows, there is also substantial traffic in a back-projection direction, going from limbic out to neocortex. Furthermore, this feedback traffic has its own layered pattern from deep layers back up to the superficial layers of the previous network.

Great Limbifugal Expectations

A natural first assumption has been that the order of processing across the networks of the cortex starts in sensory cortex and then goes in some way toward cognition. However, we have just seen that the neural connections also exist to support back projections, going from limbic toward neocortical networks (Figure 4.6). Just as spinning a weight on a string results in an outward, centrifugal force, so this direction of processing could be called "limbifugal" in that it proceeds from the limbic core outward. Because the networks are linked adjacently, the outward limbifugal order would still be sequential. The order would be limbic, then heteromodal, then unimodal, and then primary sensory (Figure 4.6).

Can we interpret this bidirectional neural architecture as demonstrating a bidirectional operation in the mechanisms of the mind? The first naïve idea we have of cognitive processing is that cognition starts with raw sensation; then it is elaborated in perception; and then it is integrated with attention, memory, concept formation, and other more complex mental modules. This was the key assumption of British empiricist philosophy, that the mind could be known from analyzing its basis in the elementary evidence of sensory perception. How could cognition go backward from the visceral core out to the sense organs?

Although this backward direction of information processing is unintuitive to the naïve view, it was anticipated by psychological theory. Careful experiments with human perception in recent decades, such as those by Roger Shepard, have suggested that memory often serves as the defining context for the development of a percept. Shepard's experimental studies led him to this conclusion. Even in the simplest acts of perception, experience appears to be interpreted. We see what we expect to see. Shepard underscored the radical influence of memory and expectancy when he described human perception as "hallucination constrained by the sensory data."

Shepard's views have been popular in neither cognitive psychology nor the derivative discipline of cognitive neuroscience. Experimental psychology in the United States took its assumptions from British empiricism. One of these assumptions is that perception is an objective process (the dogma of immaculate perception), as if we must assume objectivity in perception if we are to have an objective (and, by implication, clean) state of mind.

Yet the bidirectional traffic in the corticolimbic pathway seems to require a psychology as disturbing as Shepard's. The density of neural pathways is as great going from the limbic core toward the neocortical shell as it is from the shell to the core. Memory reaches out to sensation as much as sensation gives the evidence of perception.

Studies of neural activity in the cortical regions of monkeys have confirmed not only that processing is bidirectional but also that it reverses predictably. When memory dominates the corticolimbic pathway, then processing is *limbifugal* (outward). When sensation dominates the pathway, then processing is *limbipetal* (inward). In these studies, a monkey is trained to make a perceptual discrimination of an external signal for a reward. Under this control of an external signal, the first neuroelectric response is seen in sensory cortex. Then there is a response in limbic cortex. However, when the monkey is trained to remember the signal in order to get the reward, the first response is in limbic cortex. Then there is a response in sensory cortex. This back projection—in a limbifugal direction of processing—suggests that the monkey is imagining (hallucinating) an internal recollection of the sensory signal. The sensory response is then driven by the remembered image.

It is as though the mind breathes: in, then out, weaving a confluence of distributed representations, weaving visceral meaning to arbitrate with external reality.

Supramodal Coherence

Although the dominant connections are within-pathway (within the adjacent, linked corticolimbic network levels from limbic to primary sensory cortex), there are also important connections that are cross-pathway, as in the concentric rings of cross-pathway connection lines in Figure 2.12: Even with a loose, metaphoric interpretation of the connection patterns, there are fairly clear implications of how function must arise from the architecture created by the pathways and their cross-pathway links.

For each of the sensory pathways, the pattern of connectivity suggests a high degree of *specificity* in matching the thalamic model (i.e., the input to the cortex from the thalamus). This contrasts with a greater *generality* at the corebrain root of the pathway (i.e., the visceral-limbic base). For the primary sensory neocortex, the interconnections are limited to the adjacent unimodal association cortex. There are no connections to other primary sensory regions, to motor areas, or to anywhere else. With the exception of their adjacent neighbors, the primary sensory cortices are virtual islands within the connectional seas of the massive networks of the cerebral hemispheres (Figure 2.12).

Proceeding from the sensory module toward the limbic base, the degree of interconnection among cortical regions increases with each adjacent linked network; as a result, the greatest interconnection is seen for the limbic networks. They send projections not only to themselves but also to widespread regions of the motor pathways, as well as to all of the sensory modalities of the hemispheres.

This connection pattern implies that the integration of cortical operations—the functional coherence that extends beyond a specific sensory modality—must be greatest for the limbic base of the network hierarchy. This is a profound implication.

Medieval philosophy held an idea of the "common sense," the region of the mind that would receive the convergence of the specific senses. Drawing on the somewhat more specific metaphors from computational modeling, the anatomy of the primate cortex reveals an architecture in which the densest interconnection (and thus the integration of the common sense) must be achieved at the visceral, limbic base of the hemisphere. In biological terms, this makes sense. The limbic core is under hypothalamic control, providing motive direction of memory integration.

But is this structure of neural connections relevant for experience? Subjectively, can I sense that the differentiated features of sensory experience are articulated in relative neocortical isolation? In contrast, is there some phenomenological trace of the integration of my mental process, assembled in holistic fashion in the densely interconnected, motivated, visceral networks at the hemispheric core? Could we imagine a philosophy in which common sense is found at the gut level?

Cross-Pathway Spheres of Influence

Another connection pattern appears within the interregional or interpathway connections. As we have seen, the major connections are between the linked adjacent networks of each sensory modality (primary sensory → unimodal association → heteromodal association → limbic—and back again). However, a coherent pattern is also evident in the connections that cross from one pathway (e.g., vision) to another (e.g., somatosensory or motor). This pattern is described by another key anatomical principle: the tendency for the connections from a given level (for example, the unimodal association level) to stick to that level. Thus, quantitative analysis shows that the projections from visual unimodal association cortex tend to align with the auditory unimomodal cortex. Similarly, the projections from heteromodal cortex from one region tend to go to the heteromodal cortex from another region. These cross-level connections are illustrated in cartoon form by the concentric "rings" of connections across similar network levels in Figure 2.12.

This pattern implies that each of the four "levels," from limbic to primary sensory neocortex, operates as a coherent computational unit. We have seen that, computationally, connection implies function. If so, then each level of connection must exhibit a certain coherence of function as a result of its neural architecture.

What does *this* structural constraint imply for the structure of perception? Are there four levels of structural influence latent within my conscious experience? Discrete waves of network coherence through which embedded domains of emotional and sensory qualities differentiate out of the common, gut-level sense?

But Where Are the Cognitive Modules?

The modern quantitative studies of the primate cortex have thus revealed a highly structured architecture of memory representation that is remarkable in its specificity. Yet this order is not what we would expect on the basis of cognitive theories. Where are the modules for the mental faculties we have always assumed would be carried out on sensory data? Where are the faculties portrayed by the box-and-arrow theories of cognitive psychology, such as spatial memory or verbal reasoning?

What we find are amazingly intricate networks, each embedded within the other and each linked to its neighbor by highly stereotyped connections. These must imply—in some way we do not understand—the nature of their functional relations. Do the forward projections—toward the input layer (layer 4) of each more internal network—imply that each more exterior network "stands in" for the input from the sense organ? If so, does this imply the evolutionary history of the linked networks? Each more differentiated neocortical network seems to process the sensory data in a more complex way. Furthermore, each one seems to have evolved to interpose itself between the thalamic sense data input and the outer layer of an existing network. As it evolved within the "primitive general cortex" of early protomammals, each new network seems to have created another "shell" of mirrored processing that was interposed between the sense organ (or the thalamic relay of it) and the corebrain limbic processing of visceral significance.

Do the back projections of each network toward its more superficial neighbor imply that the neighbor is embedded in the deeper, more internal network? These projections seem to take the processing results of the more internal network and feed them back to modify the intrinsic processing of the more external network. Is this the way that memory shapes perception? Does the core network project its output, its digested information, back to the intrinsic processing of the shell network in order to set a template for what is expected as new input data? With four of these major networks or levels linked together, should we conclude that memory is constructed through a functional architecture of nested processing levels, each operating in a kind of patterned rhythm and each more specific than its embedding level?

Structure of Mind

Until we have more sophisticated tools for physiological network analysis, we can only theorize, and rather vaguely at that, on the functional significance of this linked structure of corticolimbic networks. However, even a preliminary theory takes on interesting form when required to be anatomically correct.

Start with the most parsimonious assumption: What you see is what you get. If we reason from the representational cybernetics of Chapter 3, we conclude that the mind must be achieved by the connectional architecture of the brain. There could be hidden cognitive modules in there somewhere, but so far we cannot find any. Rather, what we find is the human mind instantiated in a highly patterned and differentiated architecture. As exquisite as it is, this architecture is hardly unique but rather a predictable variant on the venerable mammalian plan. In fact, we now know how to read the evolutionary plan in the embryogenesis of each child and to see this continued in the activity-dependent plasticity that shapes that child's development of a self.

At the apex of the vertical hierarchy of neural structure, rooted in subcortical circuits and with strong control from the hypothalamic visceral mechanisms, is a densely interconnected limbic core of the telencephalic hemisphere. Branching out from this core are three additional network levels, each progressively more isolated and specialized in capturing its unique sensory channel, mirroring in this way the critical information from the external world.

What if experience were isomorphic (matched in form) with this architecture? Within the anatomy, we begin our search for function by reasoning with computational principles implying that memories—the base substance of mind—are stored not in any privileged location but distributed throughout the networks. Conceptual elements or representations must then be assumed to follow the stacked structure of the corticolimbic connectional pattern.

At the core must be the most integrative representations, formed through the fusion of many elements through the dense web of interconnection. This fusion of highly processed sensory and motor information (abstracted through three previous levels of network processing), together with direct motivational influences from the hypothalamus, would create a *syncretic* form of experience. Meaning is rich and deep, with elements fused in a holistic matrix of information charged with visceral significance. Emanating outward from this core neuropsychological lattice are the progressive articulations of neocortical networks. Finally, at the shell are the most differentiated of networks, differentiated internally with the finest of the network architectures in sensory cortices and differentiated globally through their isolation from the general hemispheric architecture. The most differentiated networks of the hierarchy are the most constrained by the sensory data, forming close matches with the environmental information that is in turn mirrored by the sense receptors.

What if these are indeed the bounds on the architecture of experience? If so, they would be defined by bodily constraints. At the core are the demands for internal, visceral self-regulation, not unlike the motive expectancies that shape the stimulus evaluation of our rodent cousins. At the shell are the constraints of the somatic, sensorimotor interface with the world. At each boundary, the structure is constrained in a way that must make its representations concrete. Sensory data are inherently concrete, bound as they are to specific present information. Visceral influences are concrete in their own way, pressing the internal needs and physiological states onto the networks to bias feelings and actions.

It must be between these bounds—within the intermediate networks between shell and core—that abstraction, on a good day, takes form.

Thus, when forming a perception, we may not only process the sensory data but also start with expectancies formed at deeper network levels. Perhaps, as in the hopeful rats, these motivated expectancies are what give value and significance to the somatic data. At the limbic core are visceral representations that capture the significance of past experiences. These implicitly form expectancies that anticipate new good things, in the case of hope, or bad things, in the case of anxiety. As bidirectional processing percolates up and down the corticolimbic hierarchy, the reentry process begins to articulate patterns in the intermediate level that negotiate effectively between expectancy and the sensory data. The visceral core may begin with certain biased states, but—in its resonance to the significance of the sensory data—the visceral core may become strongly altered in state itself, further shaping and constraining the limbifugal dialectics.

The result of this process would be a psychological construction—a concept. Formed in this anatomy as a physical process, the concept must be fully bounded by bodily constraints on both its internal and external borders.

The Visceral Basis of Action

In outlining the corticolimbic architecture, I have concentrated on the sensory structure of the posterior brain, but a similar architecture holds for the motor pathways of the anterior brain. Although the latter is somewhat more complex to analyze because we cannot separate the sensory modalities, the architecture of the frontal lobe is in many ways more significant for psychological function than that of the posterior brain. This is because of the massive elaboration of the heteromodal regions of frontal cortex that seems to have been a defining feature of human evolution.

The four corticolimbic levels are readily identified in both pathways of the frontal lobes, the dorsal processing stream from the medial limbic (anterior cingulate) cortex to the medial and dorsolateral frontal lobes, and finally to motor cortex, and the ventral processing stream from the orbital limbic cortex to the orbital and ventrolateral frontal regions, and finally to motor cortex. Both pathways include prefrontal (analogous to heteromodal) cortex and finally primary motor cortex.

What is remarkable in the human brain is the extensive elaboration of these motor pathways, especially the heteromodal regions that make up the prefrontal (in front of motor cortex) lobes. From the mid-20th century, electrical stimulation of the frontal cortex has shown that, at the base of the frontal lobes, the stimulation produces visceral responses. Stimulation of the dorsal pathway produces visceromotor responses, and stimulation of the ventral pathway (including insular cortex hidden under the temporal lobe) appears to produce viscerosensory responses.

The implication of these findings is that the roots of the frontal lobe must be found in the motivational mechanisms of the hypothalamus and associated subcortical circuits. This anatomical fact has been the cornerstone of perhaps the most informative theories of frontal lobe function, those based on the motive control of action.

If we put together the network architectures of perception with those of action, an interesting story of adaptive behavior emerges. As a result of experience, expectancies are held in the most embedding limbic networks. These are valued expectancies, predictive models for how the world should be—how it should feel. Through resonance with the environmental data patterns in sensory networks, perceptions are continually arbitrated between the internal (visceral) and external (somatic) interfaces.

Actions are engaged by these motivational and visceral expectancies. On the basis of the anatomy, we must infer that the organization of action is achieved through an inversion of the network pattern, progressing from the limbic core out to the motor cortex. In this limbifugal process, the output of more internal (limbic frontal and heteromodal frontal) networks cascades up to modulate the processing of the more external action networks in primary motor cortex.

In the structural network constraints on this process, motive templates and adaptive goals form an embedding context for the progressive articulation of actions within the external motor cortices. There is evidence that the microstructure of actions recapitulates this architecture literally, as global postural movements of the trunk musculature, charged with visceral constraints, are first formed as the basis of actions. Then limb movements are recruited within the pattern established by the axial, postural muscles. Finally, more discrete actions are articulated with fingers, lips, and tongue as the fine structure of behavior.

This process of action formation emerges in child development just as it does in the moment. At both time scales it begins with gross, visceral generality and progresses toward fine, somatic specificity. However, as with perception, backward projections from primary motor cortices to heteromodal and limbic cortex complete a feedback process, reshaping core motives and intentions according to what is available and possible. And as before, concepts—in this case concepts for acting on the world—emerge through waves of arbitration between these two poles, constrained by the concrete requirements of the body and the world.

Perhaps this progression achieves not only motor products but ideational products as well. Because we can find no mental modules in the brain, ideation can perhaps be achieved only through the extant network architecture. If this is so, then the active basis for thought must be formed on the motivational infrastructure of the visceral core of the brain. It is then progressively articulated into a more differentiated, actualized, somatic embodiment through four linked networks and fed back to the motivational structures at the limbic core before ideational products—concepts—are fully formed.

Apparently, as greater complexity and temporal span accrued to behavior, evolution selected for neural algorithms that were effective not only in simple motivation of action but also in the support for delayed gratification, for extended planning, and optimally for the vision to anticipate the future. We must be careful, therefore, in assuming that the motive controls for the frontal networks are the simple needs and urges of the limbic circuits and hypothalamus. Rather, because of the extensive encephalization of the human brain, the motive controls for thought must provide forms of energy, direction, and force that are no longer concrete motives. The motive systems now provide new forms of cybernetic support for complex processes of ideation and planning. In this process, higher brain systems subordinate lower ones, but at the same time they remain dependent upon them for elementary functions of arousal, memory, and attention. The core-shell arbitration that allows memory consolidation is energized and regulated by the same activation control systems that allowed activity-dependent specification of cortical anatomy in embryogenesis. Experience thus requires continual recapitulation of the embryonic journey. In each instance, the coordinated function of adaptive behavior requires that vertical integration—coordinated across all levels of the neuraxis—must be recruited to support the ongoing motivational context for unfolding action plans.

As Kim realizes her predicament, the functional patterns of her brain are structured both by her brain's mammalian anatomy and her specific personal history. The sinking feeling in her stomach is a visceral sign of the syncretic representation formed in her corebrain networks. This primitive concept includes a context of mood that then shapes her specific thoughts. Her previous mood (supported by widespread projections from brain stem activating systems) was expansive with confidence and optimism. With the recognition of her danger and the sinking feeling that accompanied this recognition, her mood shifts to an unusual and unpleasant mix of anxiety and disappointment. This new mood then spawns congruent, emotionally distressing ideas and images—rippling out from the limbic base to take their more articulated forms in the nested neocortical networks of each of Kim's cerebral hemispheres. Images take form, achieving conscious animation though the sensory and motor networks that are engaged, each in turn, by the deeper, latent internal patterns of Kim's memory.

We do not know exactly, of course, how functional patterns are distributed across the corticolimbic networks. Judging from the way memories are activated, these network patterns often seem to resonate and recruit new patterns, as if forming complex waves of meaning. For Kim, her abrupt anxiety unleashes a rush of images and ideas from the sense of unfairness that her fine evening is ruined, to the fear and embarrassment over what might happen, and even to visual images of disapproving looks from her friends. However, these feeling states also unleash a stream of anticipated arguments—with Jared and perhaps with her parents—arguments that are still sublingual but beginning to spill over into words at the tip of her tongue.

These emotional controls involve subcortical mechanisms of motivation. For example, the processes of anxiety and vigilance engage both the amygdala of the limbic system and the associated activating system in the brainstem. Yet these primitive mechanisms were initiated by Kim's evaluation of the situation, which required complex and abstract concepts supported by the full network of hierarchies of both cerebral hemispheres. The two-way interaction between the abstract cognitive representational abilities of the cortex and the more elementary motivational controls of subcortical circuits is illustrated by the mix of feelings (e.g., anxiety) and ideas (e.g., anticipation of disapproving looks).

The sinking feeling itself resonates with the patterns of her corticolimbic hierarchy, activating fragments of her historical self. One fragment is a long-forgotten memory of a previous sinking feeling. That memory quickly rushes back, escalating into a cascade of reconstructed images. When Kim was 9, sitting in the back seat of the car, her dad ran over a dog. She saw the little poodle run out into the street, and she heard the screech of the tires and then the thump, thump as both front and back wheels hit it. When that happened, the sinking feeling came as she imagined the violence done to the dog by the force of the car. She sensed immediately and viscerally the transformation from live, cuddly animal to disfigured carcass. She felt sick.

Kim's cognitive representation is organized within her linked set of core-shell memory networks, and these reflect not just her current thoughts but also her personal history of feelings and understandings. At the party, Kim is occupied with this flashback for only a few seconds as she stands in the kitchen before quickly putting it out of her mind, but her resolve is galvanized nonetheless by the visceral response and its semantic resonance. She will not risk such violence on the road with a drunk boy.

Kim's mood and conceptualization of self-in-context has vacillated wildly in the last few minutes, and it will shift twice more within the next few seconds. At first, she resolves to take control and is assertive and even angry about having to deal with this situation. Then, however, she shifts her mental frame again to try to hold on to the fun and promise of the evening while still asserting control. The neural structures of mind are not simply passive constraints on her experience. They are the vehicles through which Kim is able to take control of herself. With the magic of fiction, we can endow Kim with a highly effective intelligence that enables her to regulate her own motivational processes for optimal coping under these difficult circumstances.

Kim's neurocognitive frames, her bodily structures of mind, are implicit—barely perceptible in her preconscious experience. Even if Kim cannot verbalize her understanding, her awareness of these meaning-frames is formative for the unfolding of her experience and actions over the next interval of time. When she feels assertive, she also feels capable of taking control. When she feels optimistic about holding on to the good feelings of the evening, she also feels capable of creative risks. Kim's concepts emerge from a developmental order, formed first from a base of feeling and postural preparation in the visceral limbic networks. They are then articulated through the specific patterns of connectivity within her corticolimbic hierarchy into percepts and concrete actions that take shape within the somatic, sensorimotor networks of her conscious thoughts and actions.

Kim looks around the kitchen and takes a deep breath. The vertical integration of her neuraxis now shifts into a new mode. Whereas the strong responses in her limbic and hypothalamic regions have recently been reactive, responding to the recognition of her situation and the meaning it has in relation to her personal experience, she now takes an active role in her circumstance. This shift is marked by multiple transitions. Her sense of agency and her confidence in being in control take form at the limbic base of her frontal lobe networks. Within a few seconds, multiple patterns of Kim's nervous system become aligned with this awareness of agency and control. There is a new level of parasympathetic tone in her autonomic nervous system, balancing and attenuating her earlier sympathetic stress response. Her posture shifts slightly, reflecting her motive bias toward action preparation.

Kim briefly but consciously consolidates her resolve as she remembers her pact with Tara—that they would never take stupid risks in cars. At this point, the formative influence in Kim's nervous system becomes an embedding context, a memory structure that serves as a new, global context for the many specific choices and decisions she will make in the next few minutes. With its primary representation within the limbic networks, this emerging concept organizes the multiple subcortical and brain stem systems, including the autonomic and postural attitudes, as well as a shift in the brain stem arousal projections that shift her mood away from anxiety and toward elation. With the corresponding confidence of mood, Kim forms a contextual frame within which she will both reconstruct her good mood about the evening and ensure that she will be the one who drives home. The expansiveness of her self-generated optimism allows her to be creative, even playful, framing a coherent attitude of mind that integrates the contributions from the multiple levels of the neuraxis.

Kim walks up behind Jared and presses her body against him. This is the first time she has touched him like this, and she knows they both will remember it. As she does, she reaches her left hand into his left front pocket and her right hand into his right front pocket and deftly extracts his car keys (which were in his right-hand pocket). As she turns away, she reminds herself to tell Tara: She got into his pants after all. As Kim walks toward the door she looks back and says, "I'm taking you home with me tonight, Buster."

The kids at the party take in this interchange with wide eyes and raised eyebrows. Jared, his testosterone-addled adolescent brain easily overwhelmed with sexual imagery, follows Kim out to the car. Even in his demented state, Jared wonders what just happened. Within seconds of encountering the cool night air, his sexual imagery fades. He slowly realizes that next week in school his new name will be "Buster."

Kim still has a lot to do tonight, figuring out how to drive Jared's car and later calling her mom on the cell phone to get a ride from Jared's house, all the while keeping Jared both happy and under control until she gets him and his car home. Nevertheless, taking charge of her world seems doable now that she has taken charge of herself.

Kim's capacity for effective intelligence in this situation began with her skill in self-regulation. In psychological terms, this meant dealing with her emotions enough to bring optimal motivational patterns to bear on her experience. She faced with honesty the fear of a real physical threat, and it shocked her out of one context into a radically new one. The fear augmented her alertness and sharpened her thinking. She coped with a rush of negative feelings—embarrassment, anxiety, and past trauma—and used them as signals that coping was needed, rather than signs that she was being overwhelmed. She even used her ability to get angry at the unfairness of her situation to forge the resolution to assume control. Her anger enhanced the limbic motivation of her thoughts, allowing the persistence and confidence to take action. Finally, in an unusually skilled maneuver of self-regulation, Kim rekindled her optimism for this promising evening in parallel with her resolve, enabling her to draw upon not only her playfulness but also her sexuality as she engaged in a creative path to her goal, finding a way to turn a dangerous scene into one that was both safe and fun.

In neural terms, Kim's brief tour through motivational space engaged a series of modes of vertical integration of her nervous system. Her mind was embodied in the multiple frames of bodily structures. In each transition of her experience, there were shifts in arousal control from the brain stem, shifts in the configuration of motive control from the hypothalamus, and, perhaps most important of all, multiple limbic activations of concepts in memory that were able to arbitrate between visceral, motivational influences and the constraints of sensory and motor reality.

Thus, from a theoretical analysis of the nested anatomy of the mammalian corticolimbic networks, we can speculate that Kim's series of shifts in psychological frame through these few minutes were associated with specific patterns of memory activation. These created expectancies that were, in turn, constrained by the connections possible within her corticolimbic architecture. At the limbic core, highly charged and syncretic patterns served as the defining framework. These embedded the more specific and articulated—more or less conscious—images, ideas, and actions that then took shape within her sensory and motor networks. The result was an interface, linked across the four networks of corticolimbic architecture and achieving a cybernetic negotiation between Kim's internal self and the reality of her world.

5
Visceral and Somatic Frames of Mind

The framework for the mind must extend from the most elementary levels of visceral meaning to complex, hierarchical patterns of abstract understanding. As we gain increasing understanding of the brain's physical representations of information, we find opportunities to grasp the patterns and structures of the mind more directly. When we apply this theoretical analysis to the problems of mind that affect real people, we see the need for a theory of experience as a whole. If it is to be fully understood, the mind cannot be parsed into discrete faculties. In biological terms, the mind is bounded by the body, with motivational constraints at the core and sensorimotor constraints at the shell. Between these constraints and continually shaped by them are the structures of experience.

The mind's embeddedness within the whole organism is shown by both its spatial and its temporal extents. Spatially, the mind is distributed across the visceral and somatic frames of bodily structure, arbitrating between the body's motivational base and its contact with the world. Temporally, the mind's ongoing operations extend well beyond the appearance of present awareness, as these operations continually integrate the span of life experience. In psychological terms, the mind is a function of the whole of personal history. The mechanisms of memory consolidate the ongoing representation of events, not in isolation of a static present but in the dynamic context of an unfolding life.

Assembling Experience

As mentioned earlier, the English word *experience* has two meanings. It refers both to the present workings of consciousness and to the accumulation of skills and knowledge from history. Because these things seem different to us, we need two definitions for this word in the dictionary. In typical usage, we infer one or the other meaning from how the word appears in context.

This distinction may arise because we think of consciousness as operating in the immediate present, whereas memory operates on things in the past. However, as the story of Kim's situation at the party illustrates, conscious experience is fully dependent on memory processes. If your memory could extend no more than a second, your consciousness would become not only ephemeral but also highly superficial. We might say that the depth of consciousness is determined by the depth of its mnemonic roots, which give experiential access to personal history. Once we understand this, we can see how the consolidation of personal history generates expectancies that frame the momentum of consciousness.

How, then, does a memory become activated? How does it stay alive? How does it become associated with other relevant memories to form an integrated context for a decision? From a scientific analysis of brain mechanisms we can learn a number of specific facts about memory. In the preceding chapters we studied the architecture of memory—its spatial organization. Now let us examine the processes that integrate it in time. Although concepts of brain function do not provide simple answers to psychological questions, they can be organized to create a set of theoretical tools. With them we can then reason in interesting ways about the physiological mechanisms that provide the latent infrastructure for both conscious and unconscious mental processes.

The Excitement of Kindling

Before long, the technical advances in imaging brain physiology will enable us to visualize the dynamic processes of memory formation and conscious experience in each person's brain. Though at the present time we have only crude clues, they are intriguing.

One clue is given by the way the cortex responds to electrical stimulation. If an electrical stimulus is applied to the cortex, it causes neurons to discharge. In a gross way, this stimulus excites the natural electrophysiological activity that is integral to the function of cortical networks. For example, if someone's visual cortex is electrically stimulated, that person will see flashes or patterns. If the motor cortex where the control of the arm is localized is stimulated, the person's arm will twitch.

Years ago, neurophysiologists found that when they electrically stimulated the cortex, it became more sensitive to further stimulation. Because of this *sensitization* effect, additional responses could be elicited more easily, even using lower levels of electrical current. This excitable reaction was called *kindling* (like starting a fire).

Remarkably, once a kindling reaction was sufficiently developed, the cortex would often continue to discharge, sometimes going into extensive seizures.

The afterdischarges continued even when the stimulus was no longer applied. This implied that an intrinsic property of the mammalian cortex is the ability to sustain activity once it is elicited. Clinically, neurologists and neuropsychologists are familiar with similar spontaneous forms of electrical activity in the disorder known as epilepsy.

Theoretically, the phenomenon of kindling suggests that the cortex may be wired in some way to adapt itself to (i.e., to "remember") the effects of stimulation. Remarkably, this principle operates within the highly specific anatomy of memory circuits. When researchers studied the anatomy of kindling, they found that the limbic networks of the cortex—at the core of each hemisphere—are the most likely to kindle. Regardless of where the electrical stimulus is applied, the kindling response is most likely to be seen in the limbic cortex at the core of the hemisphere.

Two facts of the anatomy of the limbic cortex may help explain its reactivity. First, as we have seen, the limbic cortex is at the core of the hemisphere networks, weaving the hemisphere's neocortex with the major limbic-diencephalic circuits.

In relation to epilepsy, this observation fits with clinical experience; epilepsy tends to focalize (or become concentrated) in the limbic networks of the temporal lobe. In relation to psychological function, the anatomy is similarly revealing. Kindling occurs across the same pathway of connections—between the limbic and neocortical regions—that is essential for memory formation.

Could these neurophysiological dynamics provide clues to the psychological dynamics of memory activation? When Kim considered her situation at the party, her cognitive evaluation of it was constrained by her patterns of memory activation. She might have remained transfixed by the images of her new, valued role as Jared's girlfriend. Instead, she resonated more directly to the dangers of the situation. The result was a strong visceral reaction. Did the activation of her sinking feeling and her subsequent memory of the dog occur because her ongoing corticolimbic processing kindled the relevant patterns in limbic networks? What determined which memories captured Kim's attention? Are certain memories more likely to be activated under emotional stress?

When animals are kindled with an electrical stimulus, researchers have found that changing the excitability of the brain, such as with stimulant or depressant drugs, also changes the efficiency of kindling. Importantly, emotional stress also sensitizes kindling. This means that the brain's natural response to stress can increase the kindling response, just as if additional electrical stimulation had been added.

The implication of this evidence is that stress causes a change in some form of neural arousal, which then causes kindling to be more effective. These are very different influences on neural tissue—emotional stress, drugs that increase arousal, and electrical kindling—yet they appear to act on common neural mechanisms that regulate the brain's memory system.

These effects are clues to vertical integration. If kindling is sensitized by stress, could it also be that Kim's memory access was sensitized by the stress inherent in her predicament? Did this lead her to focus her thinking more clearly just because her corticolimbic processing was sensitized by the brain stem arousal of her acute anxiety? How close was she to excessive brain stem modulation, leading her to become overfocused, overwhelmed, and unable to grasp the complexity of her situation?

Even more important for theoretical purposes is the observation that, according to Pavlovian principles, the kindling response can be classically conditioned. In the jargon of the learning theory considered in Chapter 3, this means that an animal can be trained to associate a sensory stimulus, such as a tone, with the kindling response (i.e., with the electrical discharge of the cortex). When the tone is presented, the animal's limbic system discharges just as if it had been electrically stimulated. Experimentally, this is an interesting trick. We can then train the kindling response simply by applying Pavlovian learning principles.

Theoretically, this may be a profound observation because it implies that the corticolimbic kindling response is under the same form of neurophysiological control as the normal processes of learning and memory. Kindling suggests a conceptual model for the physical processes that allow memories to be activated. Given the ongoing nature of memory consolidation, these are the processes of neurophysiological excitement that allow memories to be assembled, organized, and distributed across the extensive networks of the cerebral hemispheres.

The Hidden Weaver

Thus experience—if it is significant enough to be captured in memory—engages the processes of neurophysiological excitement that integrate the events of that experience within the brain's network patterns. These processes are not conscious. Rather, they are triggered more or less automatically by experiential events and sensitized by internal needs that are typically unconscious, cloaked in motive blindness. In addition, they may evolve in time, in the background of the unconscious mind, for hours, days, and even years.

Neurophysiologist Sir Charles Sherrington described the brain, with its intricate pattern of interconnections, as "the enchanted loom." If we take up this metaphor, then many findings suggest that consolidation, the weaver, is always at work within the unconscious mind, out of sight yet tirelessly integrating each life episode within the tapestry of personal experience. Remarkably, there are interesting clues that the process continues during sleep, taking unique forms during each stage of dream and dreamless sleep.

The mechanisms of consolidation remain enigmatic. The evidence we have shows that research on these mechanisms will offer key insights into the life of the mind. From computational models, we can see how recursive, reentrant processing could result in complex representational patterns that are distributed across widespread networks, such as occur in the mammalian cortex. From the modern theoretical analysis of animal learning, we have seen that animals form valued expectancies that guide their evaluation of events. These are motivated expectancy structures, Freud's "motive memories." The evidence on consolidation implies that our motive memories operate out of awareness (but persistently and effectively) to organize the implicit structure of experience.

Studies of consolidation in animal learning have provided what is essentially negative evidence of consolidation—by applying a stimulus, such as electroshock, that disrupts it. When a seizure is induced, the ongoing mechanisms of consolidation are disorganized, and associations that have recently been acquired are then lost. The animal or person has amnesia for events for hours or days *before* the shock was given. The implications are twofold.

First, the mechanism of consolidation cannot be maintained when the electrical storms of seizure discharges are set loose across corticolimbic networks. This is not surprising since we know that the normal interactions among neurons are electrochemical in nature and are thus similar in kind to kindling and seizures. Seizures interrupt the ongoing electrophysiological processes that are required for consolidating memory. These are physical processes of the mind.

Second, the consolidation of a memory trace is not an instantaneous effect but requires some interval of time. In fact, even after the animal has completed all of the perceptions and actions that will be associated in the learning process, consolidation still takes time. If it is disrupted, even after many minutes (or even many hours), the memory will be lost.

Remembering and Revising Personal History

Organizing experience is thus an ongoing process of the unconscious mind. This fact requires some rethinking of our naïve assumptions about learning and memory. It has always seemed obvious that some activities of neurons support sensory processes, while others support motor processes, and it would be logical that some *association* or learning processes assemble these into coherent patterns of behavior. But did we ever guess that the mind's assembly would be so extended and dynamic? The mind's weaver appears to be tireless, continually working—and apparently reworking—patterns of experience in the background, as other sensory events and behavioral efforts occupy the foreground of awareness.

Neuropsychologists who deal with patients with memory deficits often observe the importance of consolidation, primarily through seeing it disrupted. One important piece of evidence is the retrograde amnesia that occurs after a head injury. If a person is knocked unconscious, even if the brain damage is minor, some degree of amnesia—loss of memory—almost always occurs for events immediately preceding the insult. The more severe the injury, the more this retrograde amnesia extends backward into the person's immediate past. The effect seems exactly parallel to the blocking of consolidation that has been systematically studied in animal experiments. If consolidation is disrupted, then the memory traces that were active, still being integrated into storage, may be lost.

An interesting and unexpected effect of the consolidation process is that a memory that is re-activated, even after being integrated into storage, may become more susceptible to interference than one that is not. This effect has been demonstrated in animal experiments in which a drug is given that impairs the chemical basis of the connection between neurons. The result is that memories that are being consolidated at the time of the drug's action are susceptible to interference.

The unintuitive result is that things that normally strengthen the memory (because they make it active at that time) also make it more vulnerable to disruption by anything that interferes with the consolidation process.

Some evidence suggests that a similar process occurs in human consolidation. Practice in learning—exercise that should strengthen a memory—may also make that memory more susceptible to interference. This interference may not be limited to chemical blocking agents but may include the inherent interference that is encountered in parallel-distributed memory systems (i.e., through the stability-plasticity dilemma). If so, then when a memory is activated, such as through being cued or recalled, it may become more susceptible to interference by new, similar material than if it had not been recalled at all.

These findings have practical implications for understanding memory recall. When a person is questioned about an event, such as a crime, we assume, within the naïve view, that the person's memory is a veridical record of the event, like a video that can be played back. Instead, the memory is a reconstruction, and the distributed cybernetics of the memory representation cause it to be susceptible to interference by correlated data, such as alternative interpretations that an investigator may suggest. Each recall, each access of the memory, is thus an intervention. Memory seems to be altered by each effort to address it, not unlike the elementary particles of quantum physics.

These findings may also have philosophical implications. Our histories are contexts for current perceptions. They form embedding semantic frameworks (implicit expectancies) for creating meaning (information) from the sense data. In the process, the act of perception itself opens our histories for reevaluation and—because of the distributed nature of the representation—for revision. As a result, the record of history itself is altered. To the extent that we learn from experience, the process of meaningful perception is one of self-transformation.

Unraveling the Tapestry of Experience

Most of what we know about consolidation suggests it is critically important over the span of a few minutes and then results in memory storage that is enduring. However, certain findings imply that experiences are consolidated within long-term memory over not only hours and days but also years.

These findings have come from studies of the consolidation disruption that occurs in people as a result of electroshock therapy. Electroshock, or electroconvulsive therapy (ECT), is an effective short-term treatment for severe depression. However, an important side effect is the disruption of the ongoing consolidation process, which thereby produces memory deficits for recent events. These deficits are similar to the amnesia that results from brain damage. In many cases, the primary memory deficits are short lasting and are an acceptable side effect in order to achieve improvement in a severe depression.

On one occasion a number of years ago, I worked with a depressed patient who had a series of five ECT treatments, one every other morning. I was a young clinical psychology intern covering for a psychiatrist who had prescribed this treatment but was called away. My job was to interview this woman to determine the effectiveness of the treatment. Because I had a full caseload of my own, I had to interview her in the mornings before her treatments. What I soon realized was that this timing meant that I talked with her during the time before each treatment—the time for which she would have no memory. The treatments were effective in improving her depressed mood; as a result, in each interview she was progressively more able to engage her life issues with appropriate energy and emotional responses. My interviews documented her progress in her medical chart. However, because the induced seizures disrupted her consolidation, each morning she greeted me as a stranger and had no recollection of our previous conversations.

This was for me an impressive demonstration of the volatile nature of memory. A similar and more permanent demonstration is witnessed by anyone who spends time with a patient or relative with Alzheimer's disease. Other observations with the memory deficits of ECT have been less direct, but they have highly significant theoretical implications. This research suggests that on any given day, within domains of the mind that are fully unconscious, we are consolidating events from the past, potentially over many years of personal history.

The question that guided this research was whether remote as well as recent memories may be disrupted by ECT. Searching for events within our popular culture that would be arrayed systematically in time, researchers made the obvious choice of old television shows. Because the public usually tires of TV material within a few years, a question about a particular show or character can pinpoint the memories that were formed within a specific window of time.

By testing each patient's knowledge of a couple of decades of television material, the researchers were able to calendar the memory deficits systematically. They found that ECT is most disruptive of recent events, as expected. However, it also impaired memories for remote events. Although this effect might be explained by a diffuse impairment of mental capacity, the analysis of the data showed that the deficit was organized in a regular progression of the severity of impairment retroactively in time. Instead of a recency effect (more impairment of more recent memories) measured in minutes and hours, this study found a graded retrograde deficit extending over many years.

Recursive Assembly of the Self

The consolidation of experience is thus ongoing in the background of the mind, assembling personal history in space and time. In the mind's spatial extent, our analysis of the architecture of memory has shown that the consolidation process reverberates throughout it. The process engages corebrain networks of motivational regulation, as well as the more exterior cortical networks of somatic articulation. Because of the distributed nature of network representation and the dense interconnections of corebrain networks, the integrative core exhibits very little modularity. This implies that, particularly at the evaluative base of experience, the whole person may be affected by each meaningful experience.

In the mind's temporal extent, the events of a given hour or a particular day are not isolated but are woven into memory structures that reach over broad spans of personal history. This implies that personal experience—the implicit historical self—becomes an organizing influence in selecting what is retained from events. Conversely, the experience of significant events becomes incorporated with the memory that makes up personal history. The self is thus *active* in the sense that it operates as the implicit agent of experience and *recursive* in the sense that it transforms itself through each active encounter.

The Self Reflected

The core and the shell networks seem to play different roles in representing and organizing memory and thus the self. Yet modern psychology has taught us that, perhaps somewhat paradoxically, the self is not separable from the social context in which it develops. Through extensive analysis of clients over years of therapy sessions, the psychoanalysts who have studied how personality develops have concluded that the self is formed of "object relations," in other words, relationship templates that become vehicles for perceiving social reality.

Thus, some people have a need to be included, and they therefore approach everyday interactions as opportunities for inclusion. Others, because of their unique developmental history and the genetics of their temperament, need autonomy. These people will vigorously exercise autonomy, even in a social context in which this behavior is surprising to everyone else and may not be in their own interests. The embedding context of social relations is a critical context for neuropsychological theory. As we conduct our analysis of the bodily mechanisms of experience, we must recognize that the roles of visceral and somatic structures are important not just in the internal operations of the mind but also in the self that is discovered through interactions with others.

As their brains progress into adult maturity, adolescents gain a capacity for abstract thought that allows them, more than ever before, to observe the self as the agent of life experience. Children have an implicit self, but adolescents become self-aware. In examining the core-shell dimension of the structure of intelligence in Chapter 2, we reviewed the example of Jared's understanding of his conversation with Andrew. Jared interpreted the surface features of Andrew's speech and actions by mirroring them within his own sensorimotor networks. An important theoretical question was how Jared's corebrain structures of motivational and emotional control would contribute to his understanding of Andrew's emotional state of mind. In Chapter 4 we saw how the corebrain motivational networks exert pervasive influences in consolidating memory and cognition. It may be useful to consider the example of Jared's understanding again, this time recognizing the integrated function of the nested set of core-shell networks. In doing so we will discuss how the integrated function of the corticolimbic networks causes each act of memory to be an operation of the self.

As Jared sits on the curb outside the diner after lunch, several things occur to him. His understanding of the interaction with Andrew, including both his present interpretation and the memory that he will retain from this episode, is framed by what might be called operations of the self. He is perhaps most aware now of having made an easy decision—not to take a stolen transmission—that at first seemed like it would be hard. He reflects on the meaning of his actions and thereby gains new insight into himself.

In the way of adolescents, this is the first of several major insights Jared will integrate this week, leaving him a different person by the time the week is out. Contrary to the naïve idea of self-knowledge, in which we assume we know ourselves through direct introspection, Jared's self-understanding is not apprehended directly from subjective knowledge of his own mind. Indeed, operating from the naïve view as we all do, his subjective access to the processes of his mind, including his core values, is limited. He must learn of himself by observing his own actions and seeing himself in the eyes of other people. In this process, the external sensorimotor networks mirror the world, allowing Jared to interface with it. At the same time, his corebrain motivational networks allow him not only to apply the evaluative processes inherent in his own values but also to achieve empathic resonance with those who enter his experience. In each domain we find an interpersonal basis for the bodily structures of mind.

The decision on Andrew's offer seemed hard at first because Jared really wanted the transmission. This desire was itself framed by Jared's personality. Jared identifies with his car and defines himself in relation to his skill in fixing and driving it. Yet, surprisingly, it took little thought for him to tell Andrew he didn't want the transmission. This choice was not a result of logical reasoning, although Jared could have given reasons if asked. It was rather a felt judgment, an intuitive certainty, and an expectancy implicit in the values Jared has developed. He has faced the ethical issues of car theft on several occasions with his friends. Previously it has been hypothetical, but he has exercised his judgment enough that the conflict between his principles and his desire is now readily dominated by his principles. Both the principles and the desire are elements of the self, conceptualized in the organization of his identity.

Although he has no conscious reflection on childhood experiences during his thoughts just now, Jared's attempt to understand himself operates upon long-standing concepts—network patterns of the brain—that have been organized and reorganized throughout many years of his life. His self has taken new form in the last few years, highlighted by his acute adolescent self-consciousness. Still, the integrated organization of Jared's personality reflects his entire childhood history, incorporating the residuals of the many interpersonal interactions that he has internalized in various forms.

Jared is aware of the currently salient elements of his identity, both his sense of principles and his desire for the new transmission. Sitting here on the curb, idly watching the traffic go by, he is impressed by how these things operated just now. He always thought that acting out of principles would be difficult, yet just now it was not. He learns about himself by observing his own behavior. The processes of the mind are largely unconscious, but he can see their products in his experience.

Jared wonders about Andrew's principles. Does Andrew rationalize stealing the car? Is he really involved in stealing? Does he know the risks? As Jared gets up to walk to history class, he thinks about how upset Andrew was when he refused the transmission. What's up with that? It doesn't make sense for Andrew to hassle *him* when it is *Andrew* who's taking the stupid risks.

Jared tries briefly to understand Andrew's perspective, but at this point he sees the situation through his own egocentric viewpoint, the default state of the naïve mind, and thus fails to understand how Andrew felt.

As Jared was thinking about this interaction, some information patterns took form. Others remained incomplete because they are discrepant with his contextual understanding. The observation of himself acting easily on principle was a novel one and informative, and it is now becoming integrated within a revised self-image. The observation of Andrew's distress, however, did not make sense. In other words, it did not fit with Jared's expectancies. The resultant feeling of discrepancy, this significant uncertainty, will spawn an extended process of consolidation. This will be a dynamic information process, continuing over the next several hours to structure the relation of need and data. Importantly, this memory operation goes on in relation to the interpersonal basis of Jared's construction of self.

Research with problem solving in laboratory experiments has shown that people remember the problems they did not complete more than the ones they did. So it is with Jared. In the next couple of hours of our story, he will rekindle his infatuation with Kim with such enthusiasm that he seems to completely forget about Andrew. Yet this forgetting is true only for his conscious experience. The discrepancy of Andrew's unexpected distress will trouble him and come back to haunt him later that night.

When he talks with Kim after history class, we also see operations of Jared's self. He is certainly attracted to her in the way young men are to young women. But his attraction also has features that are unique to Jared's personality. He admires the way she explains things in class, and he is self-conscious about his own limited ability to articulate his ideas. When she seems impressed by what he says after class, his confidence grows. In the coming hours and days, he will see himself differently, now through the eyes of someone he admires and who seems to like him. It will be a gradual yet transformational experience for Jared, as memory structures of value and image resonate until they achieve a permanent reorganization of his self-image. This self-image will be guided by the interpersonal exchange of information, in this case by seeing his reflection in Kim's eyes.

However, his self-image will also be challenged by Andrew's reaction. That night, as he thinks back over the events of the day, the interaction with Andrew revisits him. Through no obvious association with what he was thinking, the question of Andrew's distress reappears. This occurs because of its autonomous activation, its ongoing consolidation and semantic organization, in his unconscious mind. Jared suddenly realizes that Andrew must have intended the transmission as a gift. This must be because Andrew values Jared's friendship. That is why Andrew was so upset.

We can interpret the neural structures of Jared's sudden insight into Andrew's intentions. The mechanisms of consolidation, weaving patterns across the core and shell networks, must have kept this issue alive in Jared's unconscious mind. The consolidation continued because the experience was both incongruent (uncertain) and emotionally significant for Jared. Information comes from the confluence of uncertainty and need.

This was not just a static memory in the sense of a snapshot of experience. Rather, the representation was a multileveled idea that included interpretations as well as perceptual data. The consolidation process was not just holding a fixed idea but was organizing it. As a result, a new version or interpretation of the concept took form in Jared's unconscious mind. How did it then break through the barrier to consciousness?

One way to explain this is vertical integration. An idea that gains enough personal significance also gains limbic resonance—motive kindling. This supports background, unconscious consolidation in the corticolimbic networks. In the process, the limbic resonance of the idea may become sufficient to reach down to brain stem arousal systems and call up enough arousal to wake up the brain, activate the idea, and launch it into consciousness.

Once the idea is in consciousness, it receives explicit attentional control and reasoned elaboration. As he recognizes Andrew's intention, Jared then reasons through the implications and quickly realizes that he was blinded by his own egocentrism. As he asserted his principles and resisted temptation, his perception of events was sufficiently constrained by operations of the self that he failed to perceive Andrew's intention.

Jared now has an important new perspective on himself to try to integrate within his identity. This is the knowledge of the insensitivity he can display when he is preoccupied with himself. Jared's self is consolidating not just personal but also interpersonal significance. To the extent that he fully integrates this realization, Jared will become a more complex person who is able not only to respond with confidence to others' appreciation of him but also to observe his own natural insensitivity and perhaps—in a future situation—to buffer his natural egocentrism with an improved capacity for self-monitoring.

Bodily Self in Social Context

The human mind is so strongly embedded in social interactions that its composition can be understood only in terms of those interactions. Indeed, the social context is important for all levels of the mind's structure. In addition, neuropsychological organization has interesting implications for the structure of empathic response to others.

A few years ago researchers in Italy were recording the activity of neurons in a monkey's frontal lobe as the monkey performed certain movements. The goal was to study the neural patterns related to the control of movement. Unexpectedly, they noticed activity in these neurons when the monkey *observed* movement, such as by a person in the laboratory. The same neurons of the frontal lobe's motor system that were engaged when the monkey picked up an object were also active when the monkey watched a person pick up an object. It was as if the monkey were organizing its perception by imagining what it would be like to perform the action.

A number of studies using functional magnetic resonance imaging (fMRI) technology have suggested that a similar process—covertly mirroring actions during observation—occurs with human subjects. When we first considered Jared's interpretation of Andrew's angry outburst in Chapter 2, we reasoned that the sensorimotor networks of Jared's cortex would be important in allowing him to interpret what Andrew said, mirroring the sound of Andrew's voice. The findings on "mirror neurons" add a new element to this reasoning, suggesting that Jared's observation of Andrew may have been paralleled by activity in Jared's own motor system. This activity unconsciously and implicitly mirrored what it would be like to raise his voice and storm out of the diner as Andrew did.

The important question becomes whether the mirroring occurs throughout the corticolimbic hierarchy. Does mirroring occur only in the sensory and motor cortices (the articulated shell of the exterior interface), or does it take place in limbic cortex (the visceral core) as well?

Although the evidence is still preliminary, fMRI studies suggest that when people observe someone else experience pain, they do so with limbic responses that parallel their own emotional response to pain. If these findings turn out to be a general indication of how the brain works, then we may expect that the evaluative mechanisms of corebrain networks are integral not only to the motivational constraints of the self but also to the empathic motive influences that allow us to resonate emotionally with the experiences of others.

Thus Jared's brain activity negotiates experience between a motive core and a sensorimotor shell not only in the internal operations of the self but also in the interpersonal interactions that frame the self within the social context. In this analysis we can see how the consolidation of experience achieves an adaptive structure that is alternatively *egocentric* (with the self as center) and *allocentric* (with the other as the center of attention). Jared's challenge in this story has been to exercise and actualize the self while not losing his sensitivity to his friend.

In both cases we must understand the bodily frame of experience. The sensorimotor networks are the gateway between brain and world, whether constrained by self-manifestation or synchronized with mirroring another. The corebrain motivational networks provide integrative and evaluative controls, whether establishing the needs of the self or resonating with those of another. In both cases, significant experience is organismic, engaging the entire vertical hierarchy of the organism's neuraxis rather than modular or isolated into discrete mental faculties. Furthermore, the neural structures of experience—both the visceral and the somatic functions—frame the bounds of the consolidation process. They are boundary constraints on the process of the mind, whether the result is an assembly of a more coherent identity or the elaboration of a more empathic resonance with the experience of someone else.

Learning Across the Layers of Identity

Jared's thought processes are operations of the self. His experience involves not only projecting the self to shape interactions but also incorporating the reflections from interactions to modify the self. Adolescents undergo remarkable growth in both their bodies and their minds. They achieve new capacities for abstract thought, for questioning accepted values, and—through their new conceptual skills—for asserting a new identity. The process of maturing into adulthood is one of learning across all levels of neural organization and engaging both the somatic and visceral boundaries of cerebral network function.

However, a remarkable imbalance in the capacity for learning occurs across the core and shell networks. For many years we have known that the sensory and motor cortices of the somatic networks—at the lateral or outside surfaces of the cerebral hemispheres—become mature earlier than other cortical networks. Many brain researchers have recognized that this maturation is associated with decreased plasticity—reduced capability to adapt to learning experiences.

In recent years, however, other researchers have discovered that human corebrain limbic networks have an opposite bias, retaining a fetal-like plasticity that allows them to become reorganized with experience to a remarkable degree well into adulthood. The result of this imbalance in network plasticity is that learning—and we can infer the organization of an identity—continues with childlike spontaneity within the integrative, visceral-evaluative base of the mind.

This contrasts with the somatic interface layer, which, stabilized relatively early in childhood, remains relatively untouched in supporting an individual's characteristic style of perceiving and acting on the world. How do we recognize this early maturation of the shell? An important indication of neural maturation is the formation of the *myelin sheath* of fatty cells around the axons (fiber tracts) of neurons. When this forms, it facilitates the speed of nerve conduction, allowing mature neural networks to operate more efficiently than juvenile ones. Classical observations of the postmortem anatomy of child brains by P. I. Yakovlev showed that the first areas to myelinate early in childhood are the sensory and motor networks of the neocortex. This myelination implies that the sensorimotor areas become mature while the association (and limbic) areas are still developing.

Although these observations remain fundamental to our understanding of human brain development, new anatomical findings in recent years have suggested a new interpretation. The traditional interpretation was that the primary sensory and motor networks were the most primitive brain areas, whereas the association areas were responsible for more abstract function. According to this view, the early maturation of the more primitive sensorimotor networks would be consistent with the principle of recapitulation of evolutionary progress (phylogeny) by individual developmental progress (ontogeny). A child would first grow the primitive networks and then the more recently evolved (association) networks of the cortex.

Although this principle of ontogenetic recapitulation of phylogeny is highly informative for understanding the process of development, in brain development there are many exceptions that are themselves as remarkable as the rule. This aspect of cortical maturation is one of them, and it may be useful to consider the evolutionary roots to help explain the biological context of plasticity and learning.

First, because the cortex of the sensory and motor networks is highly articulated in discrete layers (whereas that of the limbic corebrain areas is more diffuse and primitive in its architecture), some anatomists have concluded that the evolution of the cortex may have proceeded from the limbic core (as the primitive general cortex of reptiles) toward progressive differentiation of the neocortical layers. According to this view, differentiation would have begun with the limbic, progressed through heteromodal to unimodal association, and then continued to the primary sensory or motor cortex.

Whether you believe this evolutionary story or not (and many neuroscientists do not), the recent evidence shows that, apparently complementing the early maturation of the sensorimotor shell networks, there is a *late* maturation of the corebrain limbic networks. In fact, in humans the corebrain networks may remain in a juvenile state well into adulthood. Helen Barbas and her associates have found that the limbic cortex of mammals manifests neurochemical properties that are retained from fetal development, as if maintaining the developmental plasticity of embryogenesis.

We know that humans are highly *neotenic* mammals. This means our development is retarded. Whereas normal mammals become sexually mature and behaviorally competent in a few years, we remain hairless and playful, with a childlike incompetence and dependence on parents that can persist for two decades or even longer, depending on the length of postgraduate education.

In relation to development and evolution, the evidence on cortical maturation is paradoxical. Mammalian development seems to have subverted the recapitulation principle through some fortunate accident of genetic selection that allowed differential *heterochrony* (mixed timing of the rate of development) for the corebrain and interface shell networks. However this process relates to the evolutionary structure inherent in the genome, the result is fixity at the shell and plasticity at the core. The genes that control the timing of development seem to have been selected to allow the sensorimotor shell networks to move quickly toward maturation, even as the corebrain networks remain in a juvenile state.

We can speculate on the adaptive significance of this developmental heterochrony. Perhaps the sensorimotor patterns must be established early to allow mammals to sense and act in consistent ways. Conversely, with the extended juvenile period allowed by effective mammalian parenting, the regulatory and integrative networks at the hemispheric core are left immature, so they remain capable of learning throughout the extended juvenile period.

Whatever the evolutionary explanation turns out to be, the effect is to shape the developmental course of the architecture of memory. To be sure, the consolidation process traverses the corticolimbic networks. Moreover, we know that even adult mammals are capable of modifying the sensory and motor networks to some degree. Nevertheless, the substantially greater plasticity at the core implies that each learning experience operates strongly on the integrating motivational networks while leaving those of the sensorimotor interface in their preformed states.

Growth and Rigidity of Language

We can see the concrete effects of this differential core-shell maturation in a cognitive domain where plasticity is expected and where its absence is noticeable: language. Children acquire language with remarkable efficiency, yet we fully expect that adults can learn a new language if they apply themselves sufficiently. However, soon after puberty, the sensorimotor skills of language—both the expression of a language without an accent and the accurate comprehension of the unique sounds of that language—cannot be learned. Most adults can learn the meaning of a new language—or even several languages—quite effectively but not the surface articulation and perception. An adult will always have the accent of the native (childhood) tongue in learning a new language and will be unable to perceive the full range of the speech sounds of the new language.

In contrast, a child learns a second language with all of the proficiency of a native speaker. This developmental plasticity appears to be explained by the differential maturation of the core and shell networks. As the neural networks of the sensorimotor shell mature, they become highly resistant to modification by experience. By puberty, they have locked in; as a result, even the most intelligent adult is unable to develop fluency in the perceptual or motor articulation of a new language.

At the same time, a childlike plasticity remains in the corebrain networks. This allows values and integrative concepts of the self to remain fully open to the effects of whatever experience one encounters. Helen Barbas, who discovered this anatomical imbalance, has wondered whether the phenomenon might explain the human vulnerability to mental disorder.

Human adolescents may be particularly good examples of cerebral network heterochrony. Their bodies become sexually mature, and their sensorimotor networks gain a level of maturation that will be modified only slightly as they continue into adulthood. But their corebrain networks of emotional self-regulation remain relatively childlike (labile and reactive) as they attempt to integrate a stable adult identity. Consistent with Barbas's speculations, many mental disorders have their onset in adolescence and early adulthood. Yet, even as the plasticity at the core leaves us vulnerable, it may be an essential mechanism to allow the continuing transformations of the corebrain self-regulatory and self-representational structures. These extended transformations, no longer complete after even a decade of adolescence, may be necessary to achieve an identity that is complex enough to match the reality of life in the Information Age.

Sitting in his room after school, Jared tries to understand the meaning of his insensitivity to Andrew. In this reflection, his sensorimotor networks help him represent the situation and the interaction, mirroring events in his imagination. However, they are relatively unaffected by the information processing that reverberates across the reentrant corticolimbic networks. In contrast, his corebrain networks are fully modifiable, with the result that a rich mixture of feelings and concepts compete for control of the enduring network architecture. He is both surprised and embarrassed by his insensitivity. These spontaneous emotional responses do their work to change his corebrain representations. The result is an immediate, automatic, and unconscious shift in the pattern of his identity.

However, there is a less spontaneous reaction operating at this evaluative level as well: curiosity about himself. Jared recognizes that he was so intent on acting with principle, maybe even a little self-righteous about it, that he failed to notice Andrew's feelings and did not even realize what Andrew intended. It was a spontaneous realization when Jared first recognized that Andrew intended the transmission as a gift. But now this conceptual structure progresses into a volitional process, a willed intention to recognize his failure in this situation and to try to make amends to Andrew.

The result of these few hours of experience is a complex set of changes in Jared's identity. These changes operate primarily at the corebrain level, where Jared's neural plasticity remains most fluid and where he organizes his understanding of the meaning of events through his feelings about them. Most of these changes were unconscious, resulting from the background consolidation of the day's events in relation to his previous history. Still, Jared exercised an important instance of conscious self-regulation as he reflected on his insensitivity. He formed the intention to be more conscious of this fault of his nature and to try to be more aware of how his friend must feel.

In this process, Jared's intention is another form of expectancy, another active process emergent from the architecture of his memory. Operating within the corticolimbic architecture, this intention to integrate the experience in a certain way must draw on the adaptive, corebrain base for its motive energy, and it must have its greatest influence on the corebrain levels, where plasticity is greatest. Because it emerges from the corebrain level, the intention has little articulation and therefore takes only vague form in Jared's consciousness. Yet the intention operates dynamically, kindling the consolidation process across the linked corticolimbic levels. The effect is to organize a new concept of himself that incorporates a more appropriate humility, offering Jared the chance to develop more sensitive and perceptive responses to the people in his life.

Abstraction in the Body Dialectic

If the visceral and somatic neural structures are the foundation of such examples of mind, we must consider how they are elaborated into complex, abstract cognition. Inherently, the representations at both core and shell are concrete. Sensations, actions, and motives are organized into simple, elementary forms, and it is obvious that abstract cognition must be elaborated at a level above this.

In fact, many philosophers and psychologists assume a fundamental separation between the mind (the true domain of human intellect) and the mere sensory and motor operations that we share with animals. For a neuropsychologist, of course, this assumption shows the naïve view at work. The mind is no ghost in the machine. Rather, it *is* the machine. If we study its mechanics, we can understand its function. The mind must have some structures that are close to perception, others that are emergent from the networks responsible for concrete action, and still others whose activities correspond with the body's needs and desires.

Even though the mind emerges from its bodily roots, it cannot be reduced to those roots. The visceral and somatic frames are concrete bounds defining the conditions that structure the mind's operations within the intervening levels. To understand the more complex products of human intelligence, we must comprehend how abstraction emerges from the concrete boundaries at both the shell and the core.

At the shell, it is obvious how the operations are concrete. Perception must interact with specific sensory data. Actions must be effected through the muscles. Even as I try to explain my intention in this paragraph, I must navigate from my intentions to the movements of my fingers on the keyboard. At the core, concreteness is similarly enforced by bodily reality. Needs and urges arise from neural control of the organs. Through the fog of motive blindness, needs and urges achieve their effects on the mind, and in our experience, reality is changed. If complex concepts are formed between neural networks with these concretizing bodily functions—and we know they must be—then we need to understand how.

Abstraction allows concepts that extend over broad semantic domains. An abstract idea stands above any specific fact or event and thereby captures the meaning of many specifics. For example, you may have a concrete idea of the car you rode in when you were a child. This might include the memory of the way it sounded, the way the windows framed the world going by, and maybe even the texture of the wear on the seat. You can recall these concrete features of your experience with this specific instance of "car."

However, you are also able to form the abstract concept of "car" that applies to many cars that you have known. You can even understand the notion of "vehicle" that includes other means of transportation that might not even have wheels. By capturing many exemplars within a category and not just one concrete form, abstract concepts create a hierarchic structure of meaning that is essential to sophisticated intelligence.

The importance of abstract thought may best be seen in its absence. The loss of the abstract attitude is one of the clinical signs of brain damage. For example, you probably have never seen black snow, but if you were asked to imagine it and then to describe what it would be like, you could. You might even be able to deduce that it would melt more quickly than white snow because it would absorb more sunlight. But if you had suffered brain damage, this "as if" frame of mind would be very difficult to achieve. Patients with frontal lobe damage will say, "Snow can't be black. It's snow!" Their concepts are concrete, linked closely to perceptual experience. Their experience cannot become hypothetical, disembedded from the sensorimotor qualities.

So, too, does concreteness arise from visceral constraints. When you are very hungry, food looks and smells quite good. Your experience is concretized by your visceral state. In that state, food is not a hypothetical idea. It makes you drool. Motives are integral to the visceral basis of mind. They typically operate out of awareness and cause the core representation of the self to be unavoidably egocentric.

Encephalization of the Somatic Function

Consider how sensory and motor patterns could become elaborated into more sophisticated concepts. At one level is a motor sequence, a program specification to innervate muscles to achieve an action. An example could be Jared's adjusting the steering of his car. This is invariably integrated with a sensory monitoring operation. For example, Jared feels the inertia shift and watches the progress of the car changing into the next lane. At a somewhat higher level is a more abstract representation of the motor plan—say, Jared's intention to leave the freeway. At a still more abstract level is Jared's idea of safe driving. Although this idea may not be conscious very often, it is of course an essential guide for his actions, and we hope it forms an implicit, embedding context for his experience as a driver.

However, how are such levels of abstraction represented in the brain? An important theoretical idea is *encephalization*. We saw earlier that this concept refers to a newly evolved brain structure taking over a function that was previously served by a lower (previously evolved) structure. For example, most birds have excellent vision, and so do most primates. But the visual capacity of birds is achieved by neural algorithms at lower levels, processing visual data in the midbrain and even in the retina. In contrast, in primates not only diencephalic but cortical visual networks as well have evolved to provide more elaborate processing of the information from the retina. As a result, primates have a more flexible visual system. Instead of specialized pattern detectors, the primate visual system engages more generalized operations in which percepts have properties of concepts.

To avoid the naïve interpretation of evolution in terms of human values, modern biologists point out that the visual system of birds is only "lower" in some arbitrary scale of evolutionary complexity. To serve their adaptive niches, the visual system of a hawk or an owl is highly refined and fully adequate, yet the vertical integration can be traced clearly across the phyletic order. Vision in the primates includes the same general structures of retina and midbrain as in birds, but these structures are superseded in important ways by the "higher" (terminal addition) structure of the cortex. Vision in the primates is encephalized to the extent that people with damage to visual cortex may have blind spots (large areas of the visual field that conscious visual attention cannot access because the visual function is no longer complete in the lower structures of the human brain), even though the prior forms of these structures were complete visual systems for our evolutionary ancestors.

Interestingly, people with damage to the visual cortex have residual visual function provided by the remaining subcortical visual networks. These patients may be able to reach accurately for objects in their blind spots, in the phenomenon known as "blindsight." They can do this even as they report that they see nothing in those areas. For the conscious interpretation of vision, the encephalized visual system of humans has now distributed essential functions to the cortical visual networks. This restriction of consciousness to the cortical level of representation is interesting, given that important elementary functions of orienting vision in space are still carried out by subcortical levels of the vertically integrated neural hierarchy.

Another example of encephalization is the cortical control of the motor system. In humans, we identify the motor cortex with the control of movements. If I suffer a stroke in my left hemisphere, I may lose control over the right side of my body. If the motor area of the cortex is severely damaged on this left side of the brain, I will never regain function of the right side of my body. Yet if the same cortical damage were to happen to my cat, Toby, his motor impairment would be transitory because the encephalization of motor function within cortical networks is substantially less in felines than in the big, neotenic primates. Within a few months of complete cortical ablation of the motor strip, Toby would exhibit more or less intact motor functions.

In the vertical hierarchy created through the terminal additions of vertebrate evolution, the higher network thus takes over some aspects of control from the lower ones. Yet the lower networks are in no way vestigial, and the integrated operation of the entire vertical hierarchy remains essential to the behavioral function. If a lower level is damaged, the effect can be catastrophic. This implies a corresponding shift of function in the lower levels, with the result that they transition from being complete control centers to becoming subordinate auxiliaries that support the more differentiated function of the recently evolved neural level. In theoretical terms, we can say there must be a kind of *coevolution* of the multiple levels of the vertically integrated neuraxis, all doing their part to support neuronal teamwork.

Thus, in the cerebral representation of mind, the first direction of encephalization that we need to understand is that of the somatic function. This is the elaboration of complex, abstract, and unbounded conceptual abilities from the simpler, concrete, and stimulus- or response-bounded representational abilities of the sensorimotor networks. Somehow the consolidation process, when negotiated across the four linked networks from core to shell, allows the creation of hierarchic, abstract mental representations. These representations—such as words, images, or symbols—must be linked to sensorimotor patterns in some way because they can be interfaced to the world only through such patterns. Yet abstract concepts, as they are woven by consolidation into representations that span the multiple corticolimbic networks, are no longer dependent on any particularly sensorimotor structure.

The extended juvenile period of humans engages a developmental process that allows a level of unbounded representation in which information becomes highly elaborated in novel forms. As a result, we can imagine sensory qualities we've never seen, like slimy trees or librarians in tuxedos. More importantly, we can think about actions without acting. We can imagine classes of action, such as going fishing or reading fiction, without much concrete detail of the action plans. The categories can be abstracted from the action. These flexible, diffuse representational skills allow us to achieve abstract concepts that are removed from sensory qualities, such as good investments or unrequited love.

Encephalization of the Visceral Function

Although it is indeed an abstract concept, the example of unrequited love takes us closer to the visceral than the somatic function. In working to understand the theoretical implications of neural architecture, I have concluded that abstract intelligence may owe as much to the visceral core as to the sensorimotor shell. This conclusion is suggested directly by the physical structures of the mind, the mammalian corticolimbic networks organized in the very specific core-to-shell pattern. But the value of this idea—for a theoretician, of course—is not just that it is consistent with the evidence. It is also fun. It has implications that are surprising and interesting.

It is a fairly obvious and well-known idea that cognition must involve a more abstract elaboration of sensorimotor function. Encephalization of the visceral function would mean, at the basic level, that the mammalian cortex has also taken over some aspects of control of the internal homeostatic functions, including both sympathetic (energizing) and parasympathetic (conserving) divisions. The control of homeostasis must include both viscerosensory and visceromotor functions. However, encephalization of the visceral function must also mean that control systems that initially evolved for elementary mediation between visceral states and somatic control have now become elaborated to achieve general-purpose control capacities.

Thus, in an elementary visceral control, feeding-related reflexes may become primed by a state of hunger. Or, your orienting response—as you quiet your visceral processes while you look or listen intently—may be primed by an unexpected stimulus. But in a complex animal, in which hierarchic representational memory systems have evolved to interpose delay between stimulus and response, the visceral controls may allow more deeply motivated control of mental representations.

Through experience with food, for example, an animal learns that certain contexts make for good foraging. The sensory and motor experiences with these contexts then become represented hierarchically, above the sensorimotor level, to represent abstract features of these good feeding opportunities. At the same time, however, these abstractions take on visceral-evaluative functions. The hope for food becomes represented more abstractly as a regulatory influence, serving as an integral motive for that context-concept. This is not a direct visceral feeding reflex but a more complex, hierarchic representation of motive for significant (usually good) foraging contexts.

In the mammalian cortex, the greatest connection density occurs within the limbic—visceromotor and viscerosensory—networks at the hemispheric core. These networks receive direct hypothalamic control, meaning that they are closely regulated by the elementary visceral functions. In humans, by encephalizing the visceral functions, the corebrain networks come to provide a kind of evaluative conceptualization of qualities of significance that may be as important to abstract thought as are the sensorimotor functions.

It is not just that the visceral function is a more abstract version of biological motives. Rather, it becomes an evaluative and controlling influence for intellectual activity itself. For the artist, the beauty of the application of a certain method is not only emergent from the sensorimotor experience with the method but is also energized by the motive excitement that becomes integral to the abstract sense of beauty. For the theoretical scientist, the excitement of a new concept is a quality of experience associated with realizing that the concept fits certain validating evidence and suggests unexpected new ways of thinking. For the heavy equipment operator, the confidence in moving over a section of ground becomes an abstract motivating influence that, coupled with the current project goal of this section of the work, shapes the specific sensorimotor plans for applying power to the tracks, sensing the center of gravity, and setting the bite of the blade.

Dialectics of Abstraction

Thus some ideas, because of their emotional significance, gain a purchase in the life in the mind. They achieve an extended consolidation, whether conscious or unconscious. An attractive object becomes emotionally charged, and the memory of that object becomes similarly charged. When Kim became interested in Jared, the sight of him at school was exciting. His appearance and mannerisms were attractive to reflect upon, and she soon found herself daydreaming about him.

It is an important clue when a desired object appears in awareness, unannounced. It may appear in a fantasy or an idle reflection with no apparent association to what's happening in one's thoughts or surroundings. This object may be a sexually attractive person, a thing one wants to buy, or a juicy bacon cheeseburger. Subjectively, the appearance of this thing in awareness—for no apparent reason—signals that it must be important. Objectively, we can infer that within the background of the mind, within the unconscious, memory operations have been engaging the representation of the object, most likely because of the emotional significance in limbic networks that responded to the representation of this object. Once multiple network levels are linked with sufficient activation, the integrated representation appears to resonate as if in a multilevel harmonic. The representation—the idea—may then achieve sufficient activation that it visits conscious experience.

A *hedonic cathexis* (Freud's term for a positively charged attraction) can be a powerful force in the mind, but it pales in comparison to a hostile cathexis. As the old adage goes, friends come and friends go, but enemies accumulate. This is because hostility binds memory with unusual strength and tenacity. A hostile encounter spawns motivational reverberations that are powerful enough to distract you from focusing on anything else. You may wake up in the middle of the night, suddenly conscious of the threatening object.

The mind is thus shaped—guided in its ruminative consolidation—by the emotional significance of experiences. This shaping is not always conscious. The research on consolidation shows that much of the integration of personal experience goes on in the background. The emotional significance of a memory acts as a motive engine, keeping the idea excited.

Because the dynamics of memory mechanisms are played out within a highly constrained architecture, that of nested corticolimbic networks, we can infer the structure that is integral to these consolidation dynamics. The tissues that are most excitable and that seem responsible for linking concepts to their emotional and motivational significance are those at the limbic core. These are the cortical networks that form the integrative base of the sensory pathways. They also form the generative base of the motor pathways. In relation to the overall architecture, the limbic networks manifest the visceral functions, the internal needs and drives centered on the hypothalamus. It makes sense, then, that the self-regulation of memory consolidation—the background organization of the mind—is guided by motivational controls.

At the limbic core, concepts not only are diffuse and syncretic but also incorporate holistic representations of visceral significance. Certainly this significance must have a strong instinctual basis that is linked to the body's internal needs and states. Yet human development is so extended and socially complex that we must also assume that personal significance is not just biologically wired but also highly shaped by childhood experience. Intelligent social animals have a rich set of needs for attachment, dominance, curiosity, and mastery. These complex social and intellectual needs must be represented at the motivational base of the corticolimbic hierarchy—in some cases with their own hormonal and neuromodulator controls—as prominently as the feeding and reproductive urges. Moreover, in those rare creatures who have the capacity to think about themselves, such as human adolescents, the motive controls must include needs for self-definition and self-actualization.

At the corebrain level, the excitable limbic networks may represent one-fourth of the full architecture. They make up the corebrain base of the four levels. The concepts at this level must be holistic, infused with affect, and regulatory in the sense that their excitability is able to control the consolidation and processing at the other levels. They could be called feelings, though not in the sense of emotions that have only subjective import. Rather, judging from the architecture of the memory mechanisms, these feelings are the engines of conceptualization.

At the heteromodal and unimodal association cortex levels and finally at the sensory or motor cortex level, the representation patterns become progressively more differentiated and more articulated in relation to the concrete sensory and motor operations at the somatic interface with the world. Weaving patterns of thought across these levels, consolidation achieves a recursive self-modification of representations across the corticolimbic hierarchy. Through this process young people may learn to construct representations that are not reduced to either somatic or visceral bounds. In this way, they are able to form abstract structures of mind.

It is merely a description of fact to say that abstract concepts are not concrete on either visceral or somatic boundaries. A more interesting theoretical idea is that abstraction—as a hierarchical, generalized representation spanning multiple network levels—may actually be created through the tension between the differing constraints arising in the core and shell networks.

The reality interface of the sensorimotor shell must impose concrete constraints. However, perhaps these somatic networks are able to offer structure that serves the representational needs of the abstract idea, such as words with adequate connotation or images with sufficient precision. Similarly, although visceral needs may typically collapse conceptual structures to the lowest (gut-level) common denominator, the corebrain networks may develop the capacity for more intellectual motives. Examples may be the curiosity of an open uncertainty or the disciplined anxiety of logic. Intellectual motives like these could be productive of hypothetical, hierarchic structures of mind.

Bodily Frames of Abstraction

Putting together the dynamics of corebrain recruitment and kindling with the structure of the cortex's nested, multileveled architecture, we can formulate a physical theory of intelligence and experience. The mind's contents are continually assembled and reworked, mostly unconsciously. The structures of thought are increasingly differentiated as they progress through limbifugal, core-to-shell transformations. In the opposite direction of recursive processing, the same conceptual structures are now combined, integrated, and evaluated for significance as they are processed through limbipetal, shell-to-core transformations. The visceral core provides one bodily boundary, discerning the self-regulatory meaning of the linked set of patterns. The somatic shell provides the other bodily boundary, framing the constraints of the world as they are instantiated at the body surface. The world is met at the body surface, at the somatic shell, either through the sense data or the information patterns required for taking concrete action.

In this dialectical arbitration of opposing constraints, abstract concepts appear to emerge through recursive, bidirectional exchange across the linked, hierarchic, and embedded networks. At the core, concepts may take the form of feelings and intuitive, preconscious constructs. We can surmise that, through coupled representations across the linked corticolimbic hierarchy, our concepts differentiate toward specific and concrete forms—images and movements—as they are actualized at the sensory and motor interfaces with the world. The progression from diffuse core to differentiated shell is a process in which each concept becomes represented in reentrant fashion. This means that the network at one level has point-to-point contacts with the network at the adjacent level. Furthermore, it means that the processing at the adjacent level is able to feed back or forward recursively to the previous level. The overall corticolimbic organization is hierarchic in that the articulated sensory and motor patterns are nested within progressively more integrated patterns at the intermediate levels and finally embedded fully within the holistic, affectively charged limbic core.

This nested hierarchy of cortical networks may be implicit within the hierarchical structure of abstract concepts themselves. Each layer of conceptual abstraction—from "cow" to "animal" to "organism"—is a concept in its own right, with both a sensorimotor boundary and a visceral boundary, providing both the detail and the gist that all concepts require. The nesting of a visual image of four-leggedness and the soft texture of cow fur within a global sense of a warm and friendly creature allows the concept of "cow" to be both elaborate and meaningful. As a result, this concept can join the ranks of semantic forms that are built up in layers of increasing abstraction through the child's cognitive development. Concept formation takes a lot of work, each concept requiring its own process of core-shell arbitration, but concept retrieval takes less than a moment, allowing intelligence to flow with mercurial speed and grace.

Because it must find some resonance at the limbic core in order to receive the consolidation that gives it life, each concept or idea—each element of experience—must be motivated. Because it directs the structures and dynamics of memory, human motivational control must be integral not only to the practical operations of the mind in everyday life but also to the most abstract and sophisticated achievements of intelligence that have been laid down over years of development.

For an illustration of how abstract thought might arise from the bodily basis of mind, we can again consider a story of Kim's experience, this time as she applies her intelligence to the problem of understanding her assignment in history class.

When she gets home from school, Kim is preoccupied with thoughts of going to the party that night with Jared. She immediately calls Tara to tell her about it. However, Kim also has homework that is becoming a major problem. She has a history term paper, due next week, on the causes of the Civil War. Even though she has checked out and read several library books and found a very good website on the Civil War, she has not been able to find ideas for a term paper. She knows that slavery was fundamental to the South's economy but not to the North's, but this does not explain what happened. She cannot understand how people could kill each other in their own country. She cannot comprehend how the North and the South could have grown so different in the short history of the nation.

As Kim sits down to her homework now, she is frustrated with the blank space in her mind—the void where concepts should appear. She is in the advanced placement history class. She has impressed the teacher with her careful preparation and critical questions, and she expects to get advanced placement for college. She has learned to trust her intelligence. However, now she seems unable to say anything more about this war than what is obvious from her textbook. If this is all she can do, both she and her teacher will be disappointed.

Earlier in the week, Kim had completed her reading and internet search and tried several times to outline her ideas, but now that Jared has asked her to the party, her thoughts are captured by this new relationship. She fantasizes that she and Jared are living in the South and that they are going to get married. She rapidly elaborates on this fantasy. There is a big plantation wedding and sex in an interesting location. Then they take a trip into town in an open buggy. There they meet their family at the big restaurant on the square and have animated discussions about their state seceding from the union.

As Kim reflects on her brief fantasy, she recognizes that meeting their family at a restaurant is something people do now but probably did not in 1860. In fact, Kim found it difficult to imagine what they would do in their lives on the plantation (other than sex, of course). When she now thinks about it again, there was not much else to do but farm work. Kim actually feels isolated as she imagines living out on a plantation with no cell phone, no instant messaging, no internet, and no cable TV.

She realizes, somewhat incredulously, that even as people talked about the war, they would not know what was happening even in the next state. The only news they would get would arrive days or weeks after events happened, and it would be brought in on the stagecoach or riverboat and dependent on highly unreliable stories in newspapers or word of mouth. Even radio did not exist.

At this point Kim is idly reflecting on her fantasy, not actively fantasizing. She had given up trying to find an angle for her paper. Then suddenly it occurred to her. In 1860 people *were* isolated. Without our modern means of communications to keep in contact with the rest of the country, their only direct knowledge was of what was going on in their town. *This* could explain how different parts of the country could grow so far apart over time.

Kim immediately saw the approach of her paper taking shape. She started a new outline from scratch, seeing how the textbook causes of the war could be made more interesting by contrasting the access to information for people then and now.

Motive Engines of Information

In this story, as often happens in real life, intelligence takes form through what at first seems like idle reflection. In this case, Kim entertained a fantasy to elaborate her imagination of life in the context of a historical period. The concrete fruits of her imagination then formed the basis for an abstract concept, an insight into the way that communication among people forms the basis for culture. The concept was general, not limited to any specific examples that it explains. Yet the concept does link to the evidence, which is formed by concrete facts about the reality of the times then and now.

The concept was complex, largely unconscious, and motivated. There were multiple personal, visceral motives that directed Kim's fantasy, much in the way that psychoanalysts have described. But the most formative motive for her abstract concept was curiosity: her interest in understanding people in those times. This motive took on a complex function that she was conscious of only implicitly. It integrated representational as well as energetic qualities. Kim's curiosity was structured, beginning with the specific question of how people could become confined to local perspectives. Formed as a directed, significant question (a vectoral, valued uncertainty), Kim's motive-laden representation became not only a structure of memory but a directional, expectant influence that consolidated specific facts as well. Organized by the vehicle of her fantasy, Kim's unconscious mind consolidated these facts into the cognitive structure of a hierarchic, general insight.

Primary Process Cognition in Fantasy

It might seem that this is just a story about a schoolgirl's fantasy, and of course it is. How can a schoolgirl's fantasy and her capacity for imagination be relevant to a theory of neural mechanisms of intelligence? Actually, the scientific literature on creativity provides many examples that suggest that imagination and even playful fantasy may be integral components of creative intelligence.

Albert Einstein, for example, developed his concept of relativity theory from "thought experiments" in which he imagined himself in a certain relationship to physical qualities. He might imagine the feeling of being in an elevator dropping through space or what the world would look like riding on a light wave. These fantasies allowed him to visualize and otherwise mentally represent the physical properties that he tried to understand. His capacity for representation led him to imagine what turned out to be highly novel concepts. It may be significant that Einstein was aware of the bodily nature of his imagination, for he reported that he often formulated theoretical ideas first in visual and "motorific" images, well before he could find a mathematical structure to capture their essential elements.

Even before we consider more complex forms of intelligence, it is important to understand the experiential process of Kim's fantasy. As Freud realized, fantasy is a fundamental form of thought, a process of the mind in which needs and desires operate as unconstrained motive engines. Events and scenarios are then created in the imagination with the sole purpose of satisfying those need and desires. This is an operation of memory—a process of the cognitive apparatus—in which the constraints of actual sensorimotor implementation are subjugated by the power of motive impulses.

Fantasy thus suspends the practical limitations of reality. The motive roots of the visceral core are then given unusually wide rein. Freud saw fantasy as the primordial form of thought. He believed that a similar mechanism of wish fulfillment—imagination driven by desire—could explain the mental life of dreams. Freud described the motivational control of fantasy and dreams as *primary process* cognition.

We can consider Kim's fantasy in relation to the corticolimbic architecture that supplies the physical structure of her memory and cognition. In fantasy (and maybe in dreams), the hierarchy of nested brain networks becomes responsive to the limbic core. The excitable limbic networks resonate to the imagination of events and actions that are relevant to the needs and urges that they represent. As we have seen, the visceromotor and viscerosensory limbic networks are in turn highly responsive to hypothalamic drive—the control influences from elementary needs in primitive subcortical centers. In terms of information, fantasy is charged at the visceral level, and yet it is unconstrained at the somatic (reality interface) level. The result is an unbalanced kind of information. Concepts and images can be formed with only one purpose: to make you feel good. The cybernetics are then loose and free, so that concept formation is rapid, productive, and even somewhat unpredictable.

The result may be a pure motive kindling: maintenance of things in memory because they resonate with subjective values. There is an approach to abstraction here, but it is lopsided, weighted strongly toward visceral, hedonic constraints. As Kim imagined living in the 1860s, her imagination operated under limbic drive. Images arose that fit her values, her vague intuitions of attractive things. Her imagination resonated with not only general human motives, such as sex and power, but the motives of her personal history, such as her enjoyment of debates with her family, as well. Her imagination was limbic driven, but it still took form across all of the linked networks of the cortex, giving rise to visual images of people and scenes. She imagined feelings of touching and being touched and smells of food cooking in the summer kitchen.

It must be that in fantasy, just as in more realistic forms of thought, recursive—repeated and self-organizing—interactions are organized across the four network levels, thereby linking memory contents with the motives that sustain them. We can infer from the anatomical pattern of connections that the recruitment of motive drive from limbic areas involves a reentrant, cyclic process through which ideas are generated, possibly in diffuse, vectorial, gistlike form. Then, kindled in limbic roots, the ideas have the persistence in memory to become elaborated more fully in sensory and motor pathways. Through such dynamic mechanisms, Kim spins a rich and complex fantasy. She entertains patterns of experience that borrow from realistic images of the outside world but subjugates them to fit her immediate subjective values.

Secondary Process Cognition in Action

Freud believed that the challenge of the mind is to mediate between primary process representations and the realities of a demanding environment. This mediation—between urge and world—he described as *secondary process* cognition. For Freud, secondary process cognition forms the basis of the ego, the agentic self that takes charge of mediating between the urge and the world. Primary process cognition is shaped strongly by unconscious motives (shrouded in motive blindness), whereas by mediating with the immediate sensorimotor contact with the world, secondary process cognition operates more squarely in the domain of conscious experience.

Freud's account of secondary process cognition was intended to describe the tension between bodily needs and the social context. A child must learn to cope with internally based urges on the one hand and the demands of the social environment on the other.

This is a more elaborate psychological model than the one we are drawing from corticolimbic anatomy. Perhaps, however, Freud was wrong in thinking that the constraints of social appropriateness are the fundamental basis of secondary process cognition. Perhaps the secondary process function is a more elemental constraint, emerging from the requirement for actualizing a desire in the somatic contact with the world. The structure of the constraints at the somatic surface, of physical as well as social reality, may be the primordial embodiment of secondary process cognition.

This reinterpretation would not contradict Freud's observations but rather suggest a more elementary neural basis for them. Social demands would certainly present the reality context for a child's behavior, yet the more fundamental constraint may be the requirement to cope with reality of any sort and to work out sensorimotor structures that allow desires to be met.

Let us consider again the structure that unfolded in Kim's skilled maneuver at the party with the car keys. This was a clear challenge of interfacing personal motives with environmental demands. Preconscious psychological concepts, even if experienced only as vague moods and intuitions, served as the foundation for her action plan. She also had conscious goals in mind: to keep Jared happy and to take control. She had assembled enough confidence and optimism to adopt a playful, risk-taking attitude in forming her plan. Even to explain Kim's simple action of reaching for the keys, we must start with the psychological context for her motor plan, the feelings and attitudes that formed its base. Then we can consider the developing representation of the motor program across the linked levels of her corticolimbic networks.

As Kim pressed against Jared, she had multiple goals for this action in mind: positioning herself, distracting Jared's attention through bodily contact, and then raiding his pockets. We can infer that the general frame for this action and the motive goals that sustained it were organized in limbic regions. Actions must start with postural settings, effected with the large axial (torso) muscles, which are closely aligned with motive influences. Kim placed her feet squarely to position her body and support her reach, and she leaned into Jared while keeping her balance. Strong visceral responses from the feeling of being against him were generated in multiple limbic and autonomic centers. Her action then took more specific form, implementing the more discrete movements of her arms around Jared, then differentiating the angles of her wrists as she plunged her hands into his pockets, and finally articulating the fine actions of her fingers as she felt in his pockets and closed on the keys found by her right hand.

In understanding this action sequence, it may be useful to consider an interplay between primary process and secondary process cognition, reinterpreted in the light of functional neuroanatomy. The constraints are framed simply between Kim's visceral motives and the physical demands of her body's somatic interface with the immediate environment. Her goals were represented by general concepts (playfully engaging Jared's attention and getting the keys) and by associated regulatory feelings (the confidence of being in control, the sexual quality of bodily contact, and the sense of balance, coordination, and competence) as she executes her maneuver. These goals and feelings are not adequate unless they are actualized through the secondary process function—movements that interface precisely with the world—in this case, Jared and his pockets.

At the core of Kim's brain are syncretic, visceral feelings. Linked across multiple corticolimbic networks are more abstract goals that imply the directions that actions should take. The task of organizing the action is one of establishing motive and postural frames at the limbic level, then progressive recruitment of more specific proximal (close to the body) limb movements in the intermediate premotor and motor cortical levels, and finally the actualization of the most articulated, distal (far from the body) wrist and finger movements in the primary motor cortex. Importantly, this progression is also refined by back projections to the postural and affective limbic settings from the detailed representations of the sensations throughout Kim's motor systems, from her torso to her fingertips. The constraints of her internal needs are represented at the core, and the constraints of the world are mirrored by the sensory and motor networks of the shell; Kim's task is to organize an action sequence that arbitrates recursively across these constraints, back and forth from need to reality.

Although the constraints at multiple levels must be coordinated through reentrant, recursive links across levels, the fundamental progression is a wave of action from core to shell, from the holistic postural-affective base to an increasingly differentiated specification of discrete movements. This may be the most fundamental framework for behavior, between mental activity that is driven by need and desire at the visceral core and mental activity that is constrained by sensorimotor contact with the world at the somatic shell. It is at the world boundary that the secondary process must be achieved.

Embodied Abstraction and the Concrete Self

Is this developmental framework just for physical actions? Could we generalize the linked corticolimbic framework engaged by Kim's action plan to understand the physical, bodily mechanisms of the more abstract intelligence she is able to apply to her schoolwork?

When the imagination in fantasy is closely directed by needs and motives, it remains concrete. Bodily urges (hunger, lust) and ego needs (belonging, dominance) act directly to shape the images and scenes. Usually that is the point of fantasy and also the beauty of its simple purpose: to exercise the imagination for satisfying urges and needs.

In Kim's story, however, as sometimes happens in reality, the fantasy proved productive intellectually. It formed the basis for a more refined motive: to understand the fragmentation of the country in those times. This motive was able to arbitrate with the available evidence and thereby generate an abstract concept: the role of communications in bonding a culture. This concept was then highly productive for Kim's external, real, adaptive demand—to understand this history problem. The primary process cognition became a generative, creative influence for meeting secondary process constraints.

Kim's memory may have worked at consolidating patterns on multiple levels simultaneously. Her concern with not having an idea for the paper would have been a latent pattern in her semantic (meaning) network space. Her fantasy may even have been motivated intellectually in some latent or unconscious fashion and not just by her primitive desires for social and sexual gratification. Because she has strong intellectual motives, she wants to understand life in the 1860s and specifically the regional factionalism. The effect of these implicit constraints, acting unconsciously through her primary process cognition, was to create an abstraction integrating key data on specific events in relation to key desires. The result was information, forged in fantasy out of the elements of need and fact.

Thus we can theorize that the creation of this abstraction had to be accomplished within the exact structure of Kim's mammalian corticolimbic architecture. Drawing from the computational principles of connectionism, we can infer that, as Kim reflected on her fantasy and began to understand the isolation of people in those times, the abstract representation of this insight must have required linked patterns across multiple networks, from the global representations of value and need at the limbic core (her need to understand this problem) to the specific articulation of sensations and actions at the neocortical shell (the facts of life in those times). This insight was a product of the mechanisms of the self. It became abstract only because it transcended the concretizing influences of the self, including both the egocentric values of the core and the literal mechanisms of the shell. In this example, information emerged from uncertainty and was arbitrated at the intersection of need and reality.

Combining neuroanatomy with computational models, we now have a more discrete architecture for the mind than was available to Freud. By examining the physical structure of the brain, we can entertain a more literal interpretation. Primary process cognition is driven by the visceral controls at the core of the brain, and secondary process cognition must fit the somatic constraints of the sensory and motor interface with the world. This is an interpretation that fits all mammalian brains, and yet it must produce the most complex abstractions of the human mind. Complex psychological functions must arise from bodily structures. There is no other source for them.

There are no brain parts for abstract faculties of the mind—faculties such as volition, insight, or even conceptualization—that are separate from the brain parts that evolved to mediate between visceral and somatic processes. When Einstein developed his intuitive understanding of the relative frames of inertial systems, he did so by means of bodily mechanisms that had become capable of abstract representation. This was a progression of abstractions the likes of which had never been seen. Einstein understood the relativity of acceleration and gravity by imagining that he was in an elevator. He formulated the continuity of mass and energy by visualizing what it would be like to ride on a beam of light. He imagined manipulating the forces of nature with his motoric images. His syncretic sensory constructions and implicit action programs transcended the concreteness of everyday perceptions and actions. The resulting abstractions in fact transcended the known laws of nature. Even though he knew the current laws of physics, Einstein's secondary process function failed to constrain his fantasies the way that Freud said it should. As a result, his primary process function was allowed free rein, and it imagined a very different world. Of course, now we all have to live in it.

If we assume that a nested structure of concepts must take form across the nested structure of the neural networks of the corticolimbic hierarchy, we can then specify the structure of abstract conceptualization. This is a structure of mind based on bodily forms. Abstraction has one level of representation emergent from syncretic, evaluative, and regulatory representations at the limbic core. It has two intermediate levels in the association areas of the cortex that very likely not only provide greater structural differentiation but also act as a buffer from the concretizing effect of visceral demands. Abstraction then has a necessary articulation in specific sensory and motor forms, often in verbal constructions, in the primary and sensory motor networks that allow communication with the outside world.

With concretizing influences at the level of somatic articulation and concretizing influences at the level of visceral affection, it is no wonder that abstraction turns out to be a rare and precious construction of human intelligence, a fragile creature of the body dialectic.

Perhaps the most important lesson from the brain's corticolimbic architecture is the implication that concepts (especially abstract ones) cannot be captured at one level of network representation. Rather, they must be organized through linked structures spanning the full corticolimbic hierarchy. Kim's ideas of communications and cultural identity will be articulated in the specific sentences of her paper, but no one sentence or even paragraph will fully capture her insight. She must build a context of meaning for her ideas. Moreover, if the reader is to find it interesting and effective, this context must be motivated. Kim's writing must give voice to the concepts she has organized, each one extending across her multiple corticolimbic levels, with the result that the reader encounters her ideas with an empathic resonance. Even the most abstract of her ideas must engage not only the surface sensorimotor patterns (of the specific words she writes) but also the nested patterns of meaning that extend to the holistic and syncretic representations of value and meaning at the limbic core. The result must be that the reader of her paper will understand not only her motivation but also the detailed articulation of her most abstract ideas.

Abstract Structures of Experience

For each dimension of brain organization, a central theoretical challenge is to understand the formation of general, abstract representations. The brain's networks evolved for mediating between visceral and sensorimotor functions, and yet they have progressed beyond these primitive biological constraints to yield abstract structures of mind.

Abstraction sounds like (and can be) a fancy, academic, or intellectual quality of mind, but it may be useful to think back to Chapter 2, when we considered real-world forms of intelligence, such as the heavy equipment operator's knowledge of running a bulldozer over uneven terrain or a gardener's appreciation of the qualities of soil. In those examples, we were also looking at the abstract concepts of the expert mind. In a formal sense, the expert's automatization of motor or perceptual skills into fast and efficient packages provides a hierarchic, abstract, organization of conceptual patterns.

Intelligence From Structures of Acting, Sensing, and Feeling

Think about the organization of expert motor skills. The novice has to consciously pay attention to the relation between moving the levers and having the bulldozer's tracks go forward or backward. In contrast, the expert experiences the process at a more abstract level (for example, of moving forward or climbing left while positioning the blade to dig in). The subordinate actions of hands on individual levers are incorporated into this hierarchic conceptual level; as a result, the expert's abstract representations allow working memory and attention to be available for strategic evaluations and plans at a level that is clearly more abstract than is available to the novice.

In the perceptual domain, the expert gardener sees the qualities of soil not just because of sensory training in the visual system but also because this expert has a base in experience—an implicit conceptual library of abstract expectations for how to sense the qualities of soil. These expectations guide the visual process. Rather than look at all of the properties of the soil and try to enumerate the possible significance as the novice might do, the expert is able to expect certain abstract categories of perceptual features from years of experience with different kinds of soils. This abstract, expectant representation generates efficient visual inspection.

Abstract representations also generate more elaborate feelings in order to guide expert intelligence. A soldier who is a veteran of urban warfare moves in to clear a room of suspected terrorists with more abstract and effective emotional controls than the novice. He may appear unemotional, but his feelings—of fear, anger, hope—are essential motive forces that optimize his mental function in the critical seconds of evaluations, decisions, and actions. Experience has taught him that anxiety is valuable: If he is too relaxed or too hopeful, he could miss a lethal threat. Nevertheless, the corebrain motive representation must be balanced with its effective control of his mental resources. If the anxiety is unstructured by his confidence and sense of volitional control, it could escalate to panic, and his entire matrix of mental representations may fail to differentiate and instead collapse into the concrete, primitive patterns of motive excess, unbridled arousal, tunnel vision, and flailing actions of the fight-flight reflex. Without the abstract patterns of feeling gained by experience, the novice's own emotions are as much of a threat as the enemy. The more abstract organization of the expert's feelings lets them serve as supports and guides for mental alertness, not distractions from it.

Cognitive abstraction must arise from the same patterns of neural organization that yield these efficient patterns of motor, sensory, and emotional intelligence. The challenge is to say *how*. We need to articulate a theory of how abstract, automated, and hierarchic patterns of mental organization are formed from the roots of visceral and somatic constraints. We have, of course, begun this theory in the preceding material, imagining the general outlines of the psychological processes that arise from visceral and somatic roots. Now we will take a closer look at theoretical articulation, examining the formation of abstract concepts in explicit structural terms. We will see how abstract psychological structure is formed by the dialectical interaction between the specification and differentiation of meaning in the sensorimotor cortices and the holistic and syncretic representations of significance at the limbic core.

A Connectional Architecture for Semantic Complexity

We can review the alignment of psychological with neural structure by looking again at the differing patterns of network architecture across the four levels from core to shell. Figure 2.12 is a schematic of the connection patterns that vary across these levels, from a densely interconnected pattern at the core to the sparsely connected one at the shell. The direct functional interpretation of connection density suggests that sensorimotor representations are differentiated into discrete and separate patterns. These sensorimotor patterns are linked with specific receptors (the specific senses and regions of the skin) and effectors (muscle groups). In contrast, the dense interconnection of corebrain limbic networks across regions suggests that representations of evaluative and motive significance are syncretic, fusing a variety of elements into a holistic amalgam of felt meaning and motive urge.

Given these differences in representational structure, as patterns of activity move across the corticolimbic levels (e.g., from shell to core) a shift in the pattern of representation (e.g., from differentiated and specified to more holistic and diffuse) must occur. There is reason to believe that the dominant network activity does traverse levels. For example, we saw earlier that the recordings of neural activity from a monkey's brain showed a progression of activity in the core-to-shell (limbifugal) direction when the monkey was remembering which stimulus to respond to (as if it were imagining the perceptual qualities from memory). In contrast, the progression of activity was in the shell-to-core (limbipetal) direction when it was perceiving the stimulus to guide its action. The shift in representational structure with these network-level translations must imply a corresponding processing of the representation (i.e., differentiation in the limbifugal [outward from the limbic core] direction, compared to more holistic integration in the limbipetal [inward to the limbic core] direction).

Because memory consolidation seems to require some kind of processing across all four levels, we can infer that some cross-level representations are formed. These would be particularly important in linking the limbic (evaluative and motive) influences with the specifications of perceptions and actions required to interface behavior with environmental events. The cross-level connections are very likely required in forming skilled motor patterns, efficient perceptual patterns, and the well-practiced motivational and emotional patterns that guide adaptive behavior. Networks of neurons that are activated together, both within and across levels, would form efficient cell assemblies, concepts that are readily activated and thus prepared to become building blocks for hierarchic representations (concepts that bind subordinate concepts).

To explain how this structural assembly of visceral and somatic constraints could give rise to abstract concepts, we can formulate what might be called the *hypothesis of encephalized conceptual structure:* Concepts are organized by the same corticolimbic mechanisms that differentiate specific sensations and actions from the integrative motive core. With the extended organization of memory in the mammalian brain and with the corresponding sophistication of perceptual evaluations and action plans that this allowed, the network mechanisms became encephalized. This means they became useful not just for integrating motive control processes in relation to sensorimotor specifics and not just for differentiating specific actions and percepts from general adaptive intentions and expectancies. Rather, they also became useful for the formation of abstract concepts.

These abstract concepts are defined by having more complex referents than the elementary concepts at the core and the shell that are achieved for all mammals. The elementary visceral concepts refer to internal bodily states. As they become somewhat more complex, these concepts refer to emotional categories and motive drives, but they are still emergent from the visceral function. The encephalization of the visceral function allows not only the content of the control of motivation to become more sophisticated but the structure as well. Thus the syncretic patterns at the core would gain a new purpose, that of forming more integrated patterns of meaning (such as Kim's understanding of regional isolation in the United States during the 19th century).

Similarly, the differentiation of specific elements at the somatic shell evolved to articulate sensorimotor patterns. However, with the encephalization of the somatic function, a specialization of the structural operation of somatic differentiation may have taken place, with the result that it became the engine of conceptual differentiation.

Within this framework we can speculate that cognitive representations are organized primarily as intuitive hunches or urges at the syncretic corebrain level. To become more fully differentiated, they may require linking across heteromodal, unimodal, and even primary sensorimotor networks. In this process of increasing differentiation, concepts gain the progressive specification of referents (what they refer to) that is consistent with the differentiated architecture of the shell.

Under the principle that brain function is explained by its connections, these structural operations must be integral to all information processing that proceeds across the core-to-shell architecture of mammalian brains. Motive controls are inherently holistic, whereas perceptual and motor operations require the specification of individual elements of sensation and action.

Even though the fundamental architecture is mammalian, something different seems to have happened in the human brain. This something may have evolved along with our extended developmental period and with the radical plasticity in self-organization that this neoteny allows. Functionally, the human brain gained the capacity for abstract thought, in which mental representations have more complex relations to their referents. Linguistic forms may be particularly important to this process, differentiating elements of meaning in specific ways, yet providing the capacity for generalized referents to classes of sensorimotor events. Working with the assumption that mind emerges from brain in its explicit connectional forms, we must question how more complex patterns of reference—and thus more complex representations of meaning—could be formed from the structural operations we find in human neural networks. These are the structural operations inherent to the somatic and visceral functions.

Rhythms of Abstraction

To address this question, we could extend the hypothesis of encephalized conceptual structure to the case of cyclic or recursive processing of representations across the linked corticolimbic networks. Somehow the structural process became liberated from the concretizing constraints of visceral need or somatic reality interface. One way to explain this is in terms of recursive processing, in which the more general constraints of the visceral (holistic) and somatic (differentiated) networks are combined in rhythms or cycles of interaction, with each cycle organizing hierarchic forms that are linked across all levels.

Thus a concept may begin with a syncretic, evaluative, semantic base at the core and then become differentiated in terms of more specific (sensorimotor) referents toward the shell. Importantly, in the recursive processing back toward the core again, a practiced and balanced brain—with effective memory consolidation—may allow the differentiated structure at the shell to remain intact as an effective structural constraint. The effect of keeping links to this differentiated shell structure intact would be to limit the fusion of elements in the information traverse back toward the core. Rather than the meaning becoming syncretic (fused and undifferentiated) again, it would now remain linked to the differentiated structure at the shell. The result would be a more complex, multileveled representational structure—a new form of semantic organization—that achieves hierarchic integration.

The semantic integration in this account is different from syncretic fusion because it now remains constrained by differentiation. In terms of referents, the differentiation of meaning in relation to specific referents that was achieved through the initial (limbifugal) traverse is not entirely lost, so now it must be arbitrated with the integrating influence of the limbipetal traverse. In order to create a new coming together while staying true to the structural differentiation, the integration must be at a higher level that then subordinates but does not nullify the differentiations at the somatic level.

This is a neat trick. We know something like this must happen functionally because we do create abstract concepts out of the primitive roots of experience.

A young child's concept of "bird" is syncretic, a vague prototype formed from immediate experience with a few birds, whether in real life or in picture books. Most likely, an older child's concept of "bird" becomes differentiated through experience with a variety of birds. Learning of the different lives led by ducks and hawks breaks down the holding of a concrete, syncretic prototype in mind as a concept of "bird." Semantic architecture—structure of meaning—becomes differentiated.

At a somewhat more intellectual level, the older child also learns of the shared features of avians—how they differ from bats and maybe even how they evolved from dinosaurs. By this time, the "bird" concept has already become differentiated, so it will never again take the childlike syncretic form. The result is that the concept may become *hierarchically integrated;* as a result, the concept of "avian" weaves together the essential features of the type while at the same time maintaining the differentiation of individual species.

This is abstraction. If our speculation above is correct, this mental product is forged out of the structural dialectics of the mammalian corticolimbic architecture. The consolidation process across the linked networks from shell to core is dialectical in that an inherent opposition of structural forms—fused versus separated—exists between the core and shell. Inherent conflict occurs in the differing structural forms, and as a result, the primary constraints pull toward either specified differentiation or holistic generality but not both. Each wave in the cycle of abstraction traverses this conflict in some way. In those rare optimal instances of the human mind, the dialectic is extended, recursive, and progressive. A hierarchic and complex pattern of reference then creates an abstract structure in the link of mind to world.

Kim's first sense of an insight into life in the antebellum South was a vague intuition that her fantasy was somehow important in this context of anxiety over her term paper. Then she got the idea that her fantasy about the South might somehow have meaning for her questions about life in those times. The process that followed was recursive and probably chaotic, in that neural activity related to her eventual understanding engaged millions of synapses across the core-shell networks every second of Kim's thoughts. However, there was order to this process, and in a general way we can infer what must have occurred as the primary structural operations.

First, Kim was able to form an abstract question. This in itself was a complex mental operation, of which only intelligent people are capable. It must have involved not only the corebrain representation of uncertainty, anxiety, and an incomplete evaluative concept but also a degree of differentiation across Kim's semantic networks that allowed her mind to resonate to the evidence relevant to her question.

Next, Kim was able to link her question to various classes of evidence that were gathered through sensorimotor channels and organized into subordinate concepts. Her holistic question framed a syncretic template for an idea, and this root of an idea resonated to the relevant patterns of evidence.

At this point, continual processing (out toward differentiation and back toward integration) must have occurred to allow the formation of a mental structure that could select the elements of evidence that would fit both visceral and somatic constraints. These constraints included both the visceral resonance that signaled that the information was relevant to her uncertainty and also the somatic articulation of her emerging concept in relation to the specific facts of her history research.

During the early stages of her experience of this problem—once she had formed the question—Kim was largely unaware of the process. She had no control over the process in any conscious sense, but at a certain point she had a hunch. Her visceral representations produced an awareness that her fantasy was relevant to her intellectual question. This vague hunch was all she had to work with, at a conscious level at least.

Then, as she thought more about the problem, assembling the results of her multiple conceptual products during these few minutes of fantasy and reflection, Kim formed a conscious hypothesis for what might explain the difficulty that people now had in understanding life in those times.

In this story, Kim's understanding was complex. If we look in on the brain to trace the development of a real idea in the mind of a real girl, we might find it seems chaotic. A diverse and dynamic maze of neural networks and psychological constructs would have to be woven together to achieve any given instance of understanding. Even so, the structure of Kim's emerging abstract understanding in this story may illustrate the multileveled structure of abstract ideas. In the iconic microgenetic process, ideas build on a syncretic corebrain base. They differentiate to match the specific data of reality as articulated in the somatic sensorimotor networks. Finally, they integrate more fully at the corebrain level while retaining the full set of connections to the differentiated representation. The result is meaning, organized across multiple levels and yielding a hierarchic conceptual organization distributed across the linked set of corticolimbic networks.

Abstract Information and Relational Experience

An interesting result of applying an explicit structural analysis to the neural organization of psychological representations is that every concept—to the extent that it becomes fully organized—must include both visceral and somatic components. That is, the syncretic evaluative base must form an integral complement to the differentiating evidential base. In this way, the brain's memory system appears to have evolved to capture information, the intersection of need and reality.

As we saw in Chapter 3, this capture seems to reflect a process of forming expectancies, some of which work to automatically incorporate new contextual updates. Others require more vigorous effort to integrate discrepant events to resolve semantic uncertainty. The result, taking differing forms at different stages of development, is a dialectical learning process that is capable of transforming the representation of experience.

To the extent that the learning process provides both the differentiation and integration of reference that leads to abstraction, it may be that information itself takes on a complex form as a result of incorporation within an abstract structure. Every concept would include an evaluative basis and an evidential base. An abstract concept is a hierarchically structured pattern whose form—determined as much by the process of somatic differentiation as by the process of visceral integration—becomes multileveled.

These structural operations of memory consolidation may be the building blocks of intelligence. In any domain of understanding, the expert creates abstract structures that bind a base of motive and evaluative direction with the differentiated evidence of the domain. The result is information packages that are formed through extended consolidation and able to be efficiently addressed when the appropriate context is engaged. The expert's consciousness now comprises a more abstract form of experience. The task at hand can be grasped at a hierarchic level whose broad perspective is supported by the subordinated mental representations of highly organized domain-specific information packages. The expert can then direct conscious attention at the higher level, with the result that personal experience is now energized by a highly efficient and productive intelligence.

By including evaluative and motive influences with the data representations, concepts may achieve the representation not only of information but also of a relation with the world. The concept is not just an organization of knowledge about the data of reality; it is also fully integrated with an evaluative judgment of these data. That means that a well-organized concept, once activated, may not require decisions about its relation to the immediate real-world context. This is because it *is* a decision about the significance as much as it is a categorization of the evidence. Information implies relation.

The Structure of Subjective Intelligence

This theoretical model—integrating contributions from both visceral and somatic structures to the process of abstraction—results in a novel portrayal of the components of the mind, at least compared to what is found in classical psychological theory. For example, we have proposed that primary process cognition, as engaged in daydreams and fantasies, reflects the dominance of visceral, motive constraints in cognition. In contrast, the reality-testing, secondary-process constraints are theorized to be articulated at the somatic boundary. This seems at least generally consistent with Freud's primary-secondary process distinction. However, because the visceral and somatic networks are seen as integral organizing influences on abstract thought, the present way of thinking is quite discordant with Freud's ideas about intellectual achievements and how they arise from basic mental processes. Freud considered primary process cognition to be a primitive mental mechanism, not a productive component of intelligence. He saw the secondary process constraint of testing mental representations against reality as the essential process of the ego, the rational adult mind. In contrast, we have seen that both core and shell processes are equally important to the organization of complex and hierarchically integrated patterns of thought.

To be sure, psychoanalysts who elaborated on Freud's ideas recognized the importance of primary process cognition in acts of creative thought. This emphasis led to the concept of "regression in service of the ego," in which the use of childlike ("regressed") primary process cognition could support more effective reality testing. Yet, even in these approaches, the visceral, emotional mechanisms of the mind were never considered integral to abstract thought, nor were they seen as applying a structural bias to how the mind is organized. If information is indeed created by the intersection of the visceral and somatic frames of experience, then we may need to recognize that all complex mental operations require evaluative as well as evidential components of experience and that each component has its inherent structural bias. Evaluative mechanisms are dynamic and holistic; evidential mechanisms are specified and differentiated.

This formulation of abstraction in relation to the required components of information—need as well as data—also runs counter to traditional notions of objectivity. Certainly most empirical scientists (those who gather evidence in experiments or observation as their major emphasis) would agree that objectivity is to be measured by the closeness of an idea to its basis in the evidence of reality and that objectivity is lost to the extent that an idea is formed in relation to personal feelings of significance. Yet here we have a theory that says that abstract ideas, which may achieve an illuminating quality of objectivity in relation to a whole class of evidence, require the intercession of the visceral evaluative function for the very organization of their referential structure.

We can gain another perspective on this issue by considering what theoretical scientists might say. In fields such as physics or mathematics, in which theory becomes highly abstract, theoreticians may be more inclined to recognize the important role of holistic, evaluative conceptualizations in forming a complete theory of the domain of study. There is often a subjective quality to the evaluation of a new way of thinking, in which the beauty or compelling nature of an idea is the initial guide to its importance.

Certainly validation with objective evidence is required for any scientific theory to be accepted, and the articulation of logic and evidence at the somatic boundary must define the critical operations of objectivity. Furthermore, there is little doubt that the visceral, evaluative function is very close to subjectivity because this is where the person's emotional needs are fused with mental representations, within holistic, syncretic, and fully interconnected patterns of mind.

The corticolimbic dialectic may indeed pose a tension between subjective and objective constraints, just as is commonly assumed and just as Freud understood the tension between primary and secondary processes. Yet if we are correct in this analysis of the organization of abstract concepts, then the complexity of abstract referents cannot be found in the objective domain alone. It can be achieved only by the full engagement of the subjective base of the structural dialectic. An outcome that defaults toward either boundary—whether drawn to the apparent precision and objective certainty of somatic specification or the comforting and oceanic certainty of felt subjective meaning—is a dialectical failure.

This is a failure that—constrained excessively by one boundary—is guaranteed not to meet the structural requirements of abstraction. Just as the combined structural features of abstraction (those of differentiation and hierarchic integration) require the balanced and complete operations of corticolimbic influences, so, too, must the dialectical process reach to the syncretic organizing influence of subjective meaning to provide dialectical balance to the specification of objective, evidential reality.

As a result of the dialectical process, objectivity may be acquired only at a late stage of abstraction. There must be a period in the process of understanding in which the holistic apprehension of the information is accomplished subjectively, at the core of the self. There is no lack of bias here. At this primitive, syncretic evaluative base, we find all of the historical constraints of selfhood set loose.

Certainly in any field it is easy to find a measure of narcissism in the creative process. In the fields I know best, psychology and brain research, I can point to many examples of creative scientists who seem to be so impressed with the ideas in certain papers they read that they take the ideas as their own. Through no apparent malicious intent, they soon forget that the ideas once belonged to someone else. Often I find this very thing happens to me, too. My best ideas turn out to be fragmented memories of the good ideas of others. Of course, this is the task of scholarship, to find the links to the previous literature and to cite them carefully and extensively. Yet the essential narcissism of creative thought is the reason the history of ideas is so interesting. Historians know that once a new insight appears in a field, everyone is irrevocably changed by it even if they are unable to acknowledge its source.

It should, of course, be no surprise that egocentrism inserts itself patiently in any human enterprise. What may be novel is the realization of the locus of mental organization at which subjectivity is most intense. The corebrain representations of general information integration are the same networks that evaluate information for its personal significance.

In the routine adaptive challenges of mammals, this integral direction by visceral representations allows a coherent influence of evaluative and motive direction on information processing. In humans, with the encephalized use of corebrain networks for more general cognitive integration, are the motive functions still inherent to the syncretic neural matrix? We can hope that the recruitment of visceral networks for more complex cognition allows greater distance from primitive egocentrism, yet it may be unavoidable that holistic thought remains sensitive to the motive drives and personal agendas that are latent within corebrain representational networks.

At the first level of organization, then, it must be the natural order of things that objectivity is intrinsic to the somatic networks that are linked to the environment's sensory data and action affordances. With the specialization of networks toward the somatic interface for cognitive differentiation and specification, we can expect that cognitive objectivity similarly accrues to the highly specified semantics in these networks. In a parallel fashion, greater integration at the limbic core—even for cognitive and semantic rather than emotional and personal integration—must invite an embeddedness of that conceptual level with the subjective perspective.

From these alternate frames of experience we can theorize that when the consolidation process yields hierarchic organization of meaning at a higher level, the result is not just semantic complexity but also a reframing of the subjective-objective polarity at a higher, more abstract, level. Information implies relation.

Thus, when a concept has become both differentiated within the shell networks and specified in relation to external referents to a reasonable degree, the subsequent integration in the corebrain networks must be constrained to some degree by that somatic objectivity. Rather than becoming wholly engulfed by the dynamic emotional processes of syncretic visceral representations (with their inevitable personal biases), the hierarchic representations may achieve a more global integration that is semantically consistent with their differentiated referents. This structural result of the dialectical interplay of visceral and somatic networks may be integral to the encephalization of both visceral and somatic functions. It forms a basis for complex, abstract concepts that are motivated by intellectual rather than strictly egocentric concerns.

In this process of abstract conceptualization, the structural outcome may have interesting implications for the representation of the self. Rather than simply framing the understanding of events in relation to visceral urges, the abstract concepts formed by the recursive weaving of core and shell structures may allow the evaluation of events to be based at a more abstract level. The result may be an appreciation of the beauty of an idea or a sense of fairness in the resolution of a social problem. In such abstract emotions we may find not only the encephalization of the visceral function but also a more abstract realization of the subjective sense of self in the process of integrating higher-order patterns of experience.

Certainly any current scientific analysis of the psychological process must recognize that the mind's structure is often implicit, constrained by evaluative and semantic domains whose presence can be inferred only by observing their effects. Nonetheless, the structural analysis of abstract thought results in implications not only for the cognitive results but for the experiential process as well. We can see how, with increased facility in organizing the hierarchic patterns of mind, a more abstract form of conscious experience must emerge.

6
Subjective Intelligence

The most complex and abstract constructions of the mind must emerge from the confluence of personal motivation with the data of the world. These constructions arise directly from the mind's bodily structures to make up each episode of intelligence. They allow us to solve problems flexibly, to recognize the significant patterns in experience, and to grasp essential principles from the flow of events.

Of course, if the product of intelligence is to be more than fantasy, the dialectical process of concept formation must be firmly attached to the constraints of reality. Otherwise we fail to proceed beyond primary process cognition, and subjective experience is then satisfied with fantasies of personal grandiosity. Or, the mind is attracted to the salient myths of the popular culture, whether these run to past lives or alien abduction. Only when cognition is grounded in objective evidence can intelligence achieve validity.

And yet, the fundamental fact is that the mind's control processes evolved from elementary motive mechanisms. As a result, the generative mechanisms of intelligence engage an unavoidable process of subjective participation. The individual's unique personality provides the evaluative and motive roots of each conceptual process. Recognizing the personal, neuroanatomical structures of the mind leads to an appreciation for the evolved, biological context of experience. After appreciating these adaptive, personal roots of the experience, we face the question of how objectivity itself could arise from such a motivated substrate.

In this chapter we return to the structures of intelligence discussed in Chapter 2. The first goal is scientific theory. I suggest that the new theory I have outlined here (of the visceral and somatic structures of mind) could lead to new perspectives on the other functional dimensions of the brain, including left and right hemisphere specialization and the specialization of anterior and posterior networks. As with any puzzle, if a new idea truly fits, it will yield new insights into how other pieces fit together. The general idea is that the dimensions of brain network organization (lateral, anterior/posterior, and core/shell) are not independent but have evolved together. Recognizing how cognition emerges from the visceral and somatic constraints allows us to look at lateralization and anterior/posterior specialization in a new way.

Another goal of this chapter is philosophical: to learn how to take hold of the subjective base of the mental process. The promise of the rapid developments in brain science is a new perspective on conscious participation in the process of intelligence. However, just as we have seen that complex constructions of mind weave patterns that are not reducible to either somatic specifics or visceral needs, the perspective on subjectivity in the present framework turns out to be a complex, even dialectical one. A fully developed intelligence may lead to balanced states of consciousness and abstract constructions of mind that incorporate objective as well as subjective perspectives.

Balancing subjective and objective perspectives is challenging in many domains of intelligence. In the intellectual domain, the difficulty of this balance may be the essential problem underlying the schism between humanism and science. This balance may also be a fundamental problem of daily affairs, as we attempt to organize the subjective experience of self within the context of interpersonal relations.

If this analysis could be developed beyond the preliminary form presented here, we could envision new ways of understanding the subjective control of consciousness. When we study the control of representations in the mammalian brain, we find each that representation is regulated by motive processes (Chapter 2). As research on animal learning and the motivational control of human cognition demonstrates, these processes have the remarkable effect of organizing a kind of predictive control of experience. This is a control in which the motive representations of expected rewards provide a frame for the evaluation of ongoing events (Chapter 3). The limbic core of each hemisphere holds the (subjective) motive expectancy. This then negotiates with the (objective) specification of events in reality that is articulated in the sensory and motor networks of the neocortical shell. We discover that consciousness—forged at this confluence of core and shell—is not simply an immediate memory. Rather, it turns out to be a kind of predictive expectancy.

Even as motives operate unconsciously to shape experience and behavior in the naïve mind, we may learn from a scientific analysis to draw more explicit inferences on where consciousness must arise. This generative origin is the sphere of influence—at the motive base of subjectivity—where consciousness must be expanded if we are to overcome the motive blindness of the naïve view. Even as we know we must validate intelligence in relation to its objective products—its success in fitting the environmental demands—we can see that we must also understand its development in relation to the regulatory operations of its subjective and yet largely unconscious motive base.

As we clarify these facts in some future scientific development, it will be natural and obvious to train young people in both subjective and objective exercises of mind. Intelligence is controlled at its subjective, motivational basis; this is the topic of the current chapter. Yet it is only through objective constructions that the mind becomes capable of approaching the world as more than a projection of the self; this is the subject of Chapter 7.

For the moment we will consider the subjective process of cognition as Jared's friend Andrew faces a decision that places considerable demands on his intelligence. This is the intelligence of real life, and it requires control.

Andrew leaves the party soon after Jared and Kim. He gets in his car and drives toward a neighborhood on the east end of town. Along the way, he drives through a fast food place and picks up a coffee. Unlike Jared, Andrew was careful with his beer drinking tonight. He knew he would need a clear head for his late-night surveillance, looking out for cars to steal.

Even though other kids were still at the party, Andrew didn't hang around. He wanted some time to think. He realizes that Jared is a solid friend, someone who cares what happens to him, but Jared doesn't know what's up with boosting cars. As long as they don't get greedy, Andrew knows that he and his crew can make a lot of money on parts and never go to jail. Only the "juvies," kids under 18, actually lift the cars, so if they are caught, the worst they get is a month or two in "juvie central."

Now that Andrew is 18, he no longer does the lifting. Instead, he scopes for targets. Tonight he drives through a street in this typical neighborhood with a couple of targets at the curb. Hondas are hot now. He circles once and parks at the end of the block, scanning the windows of the houses to see whether anyone is watching. He settles in to spend a couple of hours on stakeout. He'll watch and listen to the traffic here to see who is up late and who comes home from a night shift. If they set up a boost here, Andrew wants to be sure they avoid traffic.

In addition, tonight he wants some time to think. Even though Jared doesn't really know what the dangerous part of boosting cars is, Andrew does. Tonight at the party Jared said Andrew was too smart to waste his life in jail. Andrew knows that jail is not the danger here now. One of the guys in his crew got his cousin to clear not only parts but also whole cars out of state. This cousin is in a gang, and that could mean real trouble. Andrew knows of a kid in another city who disappeared when he crossed this gang.

Andrew thinks it's funny that Jared said he was smart. Jared is the one with the date with the smart girl, yet Jared was so dumb he got drunk and had to be driven home. But now that Jared had said this about him being smart, Andrew has to ponder whether it is really true. He knows he could have done better in school, but it was always so boring. He knows there are jobs he would like, but none of them pay much. Stealing cars is fun and profit. Still, this would be the time to walk away.

In some arbitrary sense, it is clear that Andrew has the intelligence required to understand his life choices, and it is true that he is motivated in his own self-interest, so he should think as intelligently as possible about his choices. Yet there is more to it. If we wanted to teach Andrew how to frame a key life decision in his mind, we would need to appreciate that this is a subjective process. It must be motived in relation to his personal experience and his sense of self. Subjectively, Andrew is like the rest of us; he has a limited knowledge of the intrinsic mechanisms of his intelligence. If we wanted to teach him how to take control, maybe we could teach him about the visceral and somatic structures and how to use the properties of these alternate modes of conceptualization to frame his choice in abstract terms.

This is not a choice that can be made on a concrete basis, in terms of money, thrills, jail, or even the risk of gang violence. Those are indeed essential realities, but the concrete facts do not add up to a decision. Andrew must organize an evaluation of his life now, one that builds from his subjective appraisal of himself and his values. To apply his intelligence, Andrew must think in abstract terms about who he is, where he wants to be, and what it takes to get there. Without a more abstract conceptual understanding to guide his actions, Andrew's choices in the coming days will be directed by the concrete default states of mind, the easy affordances that are familiar habits, the urges that arise spontaneously, and the assumptions that shape experience at its unexamined roots.

We can return to the basics of Chapter 2 to examine the structures of Andrew's intelligence that help shape the process of his decision.

Hemispheric Mechanisms of Intelligence

In right-handers, at least, we have seen that the left hemisphere is especially skilled in analytic cognition. Furthermore, by examining the nature of distributed representation in the left hemisphere, we have seen that this skill in cognitive structure seems to be associated with a unique network structure, a focal pattern of cortical organization. In Josephine Semmes's instructive model, this focal pattern involves *similar* informational elements grouped together at any given spot in the cortex. This pattern can be contrasted with the right hemisphere's more diffuse organizational pattern, in which, at any given spot in the cortical network, a *variety* of information is represented (Figure 2.6).

Roots of Hemispheric Structure

Considering what we have learned about the corticolimbic architecture within each hemisphere, this pattern of lateral specialization suggests that the left hemisphere is more fully developed at the neocortical shell, where specific sensory and motor networks are differentiated. In contrast, the right hemisphere's diffuse and syncretic organization seems to emphasize the network pattern at the limbic core, where a greater connection density integrates multiple network influences in a more holistic pattern.

Of course, limbic core networks occur in *both* hemispheres of the cerebrum, as well as fully developed neocortical networks for sensory and motor representations on both sides. However, the hemispheres' differing functional abilities and the associated patterns of network organization imply a relative balance toward elaborating the shell (on the left) and the core (on the right). In fact, this would be consistent with the proposal on anatomical grounds that the left hemisphere becomes specialized in child development for primary sensory and motor networks (at the shell). In contrast, the right hemisphere seems to become specialized for more fully developed association networks of the cortex (toward the core).

Other evidence from research in child neuropsychology has suggested that the right hemisphere's skill in integrating information, such as across multiple sensory modalities, gives it a particular advantage when a person encounters a new situation. With a new context, few defined patterns of thought are available to serve as guidelines; consequently, the right hemisphere's holistic concepts allow a diverse range of information to be apprehended even if the grasp is diffuse and syncretic. In contrast, as thought becomes increasingly differentiated into specific patterns, the left hemisphere's better-routinized cognitive operations allow fast and efficient handling of the now more highly organized information.

The developmental basis for this right-to-left shift of hemispheric contributions may be seen in the first years of life. Babies appear to start organizing intelligence within the right hemisphere, as their dominant mode of processing, in the first year of life. In the second year, as motor and language patterns become more routinized and practiced, the left hemisphere takes on a greater role, possibly even earning the title of "dominant hemisphere."

We can see that this progression—from syncretic on the right toward differentiated on the left—is the same one that we have deduced from examining the core-to-shell progression of network organization within each hemisphere. At the visceral core, the fully distributed pattern of network organization leads to syncretic representations, within which all of the elements are fused in dynamic interactions. With organization of concepts that integrate the differentiating structure of the somatic shell, the greater specification of referents provides a more definite separation of information elements. The result of this differentiation is a more routinized, articulated cognitive process that is well suited to automatized information operations such as language.

The theoretical implication is that the evolution of the human brain has allowed each architectural strategy (the syncretic core or the differentiated shell) to become elaborated as the primary organizational theme within an entire hemisphere. If this is correct, then the progression of ideas in many new situations would involve dialectical balance not only between core and shell within each hemisphere but also between right and left hemisphere contributions to the constructions of mind.

Developing Intuition Into Consciousness

In early attempts to understand a novel situation, therefore, the first patterns to be formed may be holistic, visceral representations at the limbic network level. Reflecting the residuals of personal history, these representations take the form of what psychoanalysts called *transferences,* inherent expectancies for what should happen. Such visceral concepts are formed at the core of each hemisphere. However, because the right hemisphere is specialized for elaborating this holistic form of network integration, it is particularly suited to maintaining memory structures (consolidated throughout its architecture) that reflect the patterns of limbic resonance. The result is a fast and somewhat abstract—if syncretic—comprehension of the novel situation that is organized within the linked network architecture of the right hemisphere.

In contrast, the left hemisphere appears specialized for differentiating the shell networks out of the holistic limbic base. An essential mechanism for this differentiation, which we have seen in the studies of meaning, may be inhibitory specification. At least in the domain of language, the left hemisphere seems uniquely able to select one focus of meaning while inhibiting related meanings that are outside this focus. Within the corticolimbic architecture, this appears to reflect a kind of memory consolidation that isolates the somatic representations at the shell from the embedding visceral base of syncretic meaning. The result is a capacity for differentiating, inhibiting, and specifying. This is a formation of meaning in left hemisphere networks that is in many ways the dialectical complement to the right hemisphere's holistic apprehension.

The result of these structural semantic forms is a lateral dialectical struggle that starts with the fast, holistic, intuitive, and viscerally rooted grasp of the right hemisphere. It is then specified and differentiated by the somatic articulation skills of the more constrained, modular, and explicit cognition of the left hemisphere. Certainly this lateral tension must be rooted in the core-shell dialectics within each hemisphere. However, within our study of the neural structures of mind, we can see how this productive organizational tension of the consolidation process takes on a new and more abstract form when it is embodied in the larger frame of hemispheric contributions to intelligence.

In the car Andrew sits quietly and sips his coffee. With Jared's encouragement, he *does* feel like he is smart. He reasons carefully about his future in car theft, and he knows about the real dangers. Every time he thinks about what could happen, his anxiety cranks up another notch. Instead of pushing the anxiety away, Andrew lets it motivate and differentiate his cognition. He thinks about the risks, and he reasons. He articulates some of his thoughts in words, as if he were still trying to explain himself to Jared. Although Andrew is of course unaware of the neuropsychological mechanism, the structure of his linguistic cognition provides a grammar for his thoughts. This is a sequential order, a structure of mind that helps to organize and specify the logical relations among the concepts that are now active in his working memory. He can reason actively and systematically to differentiate the important facts and their logical relations.

After letting his thoughts settle for a few minutes, knowing that he will remember them clearly now, Andrew quits worrying, relaxes completely, and takes in the night scene. His experience is now holistic; he looks around to form a comprehensive perception of the space here and now. He listens through the partly open window to the sounds of the night. In this quiet reflection, it is as if he can also listen to the feelings that guide his sense of himself and his life now.

The Visceral Basis of Holistic Experience

Andrew's focused reasoning is not without an emotional basis; his anxiety is important to framing his expectancy for the dangers of his lifestyle. Even more so, his intuitive reflection has a strong emotional quality. This example could illustrate an important feature of the structure of intelligence. This is a scientific question that might be seen in a new light through the theory of lateralized core-shell organization: hemispheric specialization for emotion.

As the scientific evidence for right hemisphere specialization for emotional communication became increasingly strong in the last several decades, researchers tried to understand what could explain this asymmetry. Both hemispheres contain limbic core networks, and those networks on both sides would be closely interdependent with the visceral regulatory functions (emotions and motivations) of the hypothalamic centers. A theory of right hemisphere elaboration of the corebrain networks could allow us to revisit this question with new insight. If the right hemisphere has evolved its entire architecture in line with the densely interconnected organization of corebrain networks, the result would be not only a more holistic conceptual structure but also a greater influence of limbic visceral functions within each mental representation.

The right hemisphere is important to both expressing and interpreting emotional communication through tone of voice and facial expression. We saw this in an earlier example of nonverbal communication, as Jared's right hemisphere skills allowed him to reflect on Andrew's angry reply when Andrew stormed out of the diner. In that example, the holistic, nonverbal conceptual skills of the right hemisphere were important to allow Jared to reflect on the analogical qualities of Andrew's communication: his tone of voice, facial expression, posture, and body movements.

As Andrew sits in his car and reflects on his conversation with Jared at the party, he also relies on holistic, nonverbal concepts to interpret the emotional significance of the conversation. As he replays the conversation in his memory, his memory of Jared's communication is evaluated for its emotional meaning. Initially, at the party, Andrew's emotional responses had been egocentric. He was still mad at Jared and did not understand why Jared was making such a big deal about stealing cars.

Later, however, as Andrew recalled this incident, his emotional response became more *allocentric* (other centered) as he considered how Jared must have felt in the situation. Andrew can now develop an empathic emotional response by mirroring the feelings that would naturally be associated with Jared's emotional expressions. When Andrew situated these feelings in the context of his knowledge of Jared, he was able to achieve a new insight into the meaning of Jared's refusal of the transmission.

Andrew also uses his emotional empathy in interpreting Jared's concern for him. Jared was not just being righteous; he *cares* about Andrew. These insights are abstract constructions, integrating at a hierarchic level not only an array of evidence about Jared and the life context of this interaction but also Andrew's emotional evaluation of the evidence. The emotional evaluation is an essential foundation for integrating the evidence into a coherent semantic structure. Integrating personal values into the semantic context, Andrew's mind creates meaning.

The interesting conclusion suggested by the core-shell theory of abstract thought is that holistic integration of information may always have the quality of emotional evaluation. This is because it is achieved within corebrain networks that, although they can form concepts with abstract, hierarchic integration, are fundamentally concerned with the visceral, emotional evaluation of ongoing events. When meaning cannot be fully grasped through concepts differentiated with specific referents in the world, then it may need to be approached with more intuitive concepts. These are more diffuse, inarticulate ideas that represent the world through feelings, hunches, and qualities of the common sense. This is the visceral basis of mind—gut-level experience.

Andrew's quiet, unfocused reflections on the evening and on his life generally have an important emotional quality. Both the holism and the emotional richness of the right hemisphere's representations figure importantly in the mind's consciousness of the self. When someone thinks about important personal values and choices, the mix of relevant information is so complex that it is difficult to form a single, logical line of thought. Furthermore, even in its diffuse intuitive form, the information is subjectively significant, with a perceptible emotional quality. This emotional quality may become the defining handle that allows a realization to be grasped and then pulled out of the preconscious void.

A corresponding and opposite effect appears to occur within the left hemisphere. The more differentiated conceptual processes—those anchored toward the sensorimotor shell—seem to be relatively distant from limbic emotional constraints. Andrew called upon his anxiety to motivate his reasoning about dangers. His formulation of the reasons in a logical, sequential order then provided a memorable specificity that he can return to in future thoughts. This persistence of ideas is essential to the continuity of self.

Just as we can assume that evolutionary advantages for elaborating the corebrain connectivity—with its intrinsic emotionality—must have arisen within the right hemisphere, we can also think that there must be an advantage to separating the left hemisphere's network processes from limbic constraints. The result may be to allow the left hemisphere's verbal constructions to be formed with more explicit and specified symbolic referents. This may be the basis for rational cognition. Rational thought is specific, isolated, and extracted from the connotative constraints that may be inherent in the right hemisphere's densely interconnected corebrain architecture.

Lateral Dialectics of Conceptual Structure

Although this alignment seems reasonable enough, it results in a somewhat surprising model of human brain organization. Using the simple reasoning we derived from studying computational models of distributed representation, we have adopted the simple hypothesis that what you see is what you get. Mind has only the properties of brain. Connection implies function. The right hemisphere's network organization is responsible for a more holistic pattern of conceptual structure. At the same time, the right hemisphere seems to integrate emotional influences within its cognitive operations. Because representations are not easily separated from each other within distributed networks, holistic thinking may inherently engage the emotional and motivational qualities associated with the adaptive networks of the visceral brain.

We do not have to assume these are always *primitive* visceral qualities, particularly if we recognize the capacity for abstract concepts that can become hierarchically linked to both core and shell constraints. However, at a basic level, the holistic creations of the right hemisphere may be products of the visceral domain. Very likely they bring along the intrinsic affective qualities of this domain (gut-level experience) as they form each of their contributions to the daily phenomena of mind.

Thus, in the early stages of understanding, knowledge may be inherently syncretic and subjective. A broad apprehension of meaning takes place, elaborated primarily in the right hemisphere. Yet because of the diffuse cognitive structure and memory representation of this primitive conceptual level and because of the pervasive nature of motive blindness, the conscious awareness is limited. This form of knowledge has the diffuse quality of preconscious impression, as well as the affective coloring of an intuition or a hunch.

This frame of mind contrasts with the explicit, more objective relationship among differentiated elements that can be examined with the articulated structures (typically in verbal form) within the left hemisphere. For those whose intelligence makes effective use of the right hemisphere's syncretic form of experience, the concepts that take more articulated form within the left hemisphere will retain the holism of the right hemisphere core base. With elaborate roots in the hemispheric core, right hemisphere conceptual organization may bring a sensitivity to preconscious and quasi-conscious intuitions, a tolerance of unspecific and inarticulate ideas and a capacity to engage affective, intuitive qualities of thought. The result is a strong motive base, giving both direction and meaning to the unfolding subjective process.

Within the core-shell theory of abstract thought, the development of ideas can be seen as a process of what Yakovlev called *exteriorization*. The base structures are internal, the adaptive visceral core. The process of organizing ideas requires the articulation of these structures in relation to the world, as the world is grasped by the somatic sensory and motor networks. At the internal core, the diffuse syncretic constructions of mind are infused with motive significance, yet shrouded in motive blindness. We have only vague impressions and feelings of this domain of mind. Forming ideas and becoming conscious of them requires exteriorization—developing and organizing the structures of mind so that they take form against the world and are then tested against the concrete forms of images or the discipline of real action. As we recognize hemispheric specialization for core and shell networks, we can see that the lateralized structures of intelligence, the holistic and analytic conceptual patterns, have each been elaborated toward one pole of the dialectic, the holistic core for the right hemisphere and the differentiated shell for the left.

The result of hemispheric specialization is that the dialectic tension within either hemisphere between core and shell is no longer balanced. The right has been captured by the core, and the left by the shell. The result is that the dialectical organization of mind is now achieved through opposing structures on two sides of the brain. The exteriorization of consciousness is rooted in the syncretic, implicit right hemisphere and must be articulated within the more differentiated—and possibly more explicitly conscious—constructions of the left hemisphere.

Hemispheric Contents: Symbolic Logic and Analogical Mirrors

In addition to providing differing forms of conceptual structure, the left and right hemispheres (as we saw in Chapter 2) bring different forms of mental contents to the operations of intelligence. In the great majority of people (including many left-handers), it is the left hemisphere that is uniquely able to form ideas based on words. This is a form of mental representation that can be described as *symbolic*. The words are arbitrary symbols for their referents in reality.

The remarkable properties of symbolic representation can best be understood by contrasting it with the *analogical* representation, which is the more basic form of concept in mammalian brains and constitutes the primary mental contents of the right hemisphere in humans. An analogical representation is an analog or a concrete model of the event in reality. If you are sitting alone in the forest by a campfire and you hear a twig snap, the qualities of the sound may resonate in your mind—as an analog representation of the auditory event in the world—as you attempt to understand the causal significance of this event. What resonates is a mental representation of the sound itself, an image of the sensory data.

If, on the other hand, you were sitting with a friend, you might say, "Did you hear that snap?" In this case, your friend would need to decode your words—within the current context—to understand your question, operating on the symbolic representations to derive meaning from them.

Certain important properties of analogical representation are integral to human intelligence. If, for example, your friend did not hear anything, he or she could still infer the meaning of your question and then begin to wonder whether something might be moving nearby. But your friend would not be able to reflect on the sound itself. In contrast, with the analogical representation reverberating in your echoic memory, you could reflect on the sound. You might decide that it was just the crackling of the fire, or, by replaying the memory in your mind, you might decide it could only have been caused by a significant weight, at least the size of a cougar, on a large dry twig a few feet to your right.

Analogical representation is thus a continuous mirror of the sensory experience. If the snap started with a squeaking, splintering noise before the sharp crack, your representation of the percept could hold that progression of sound in mind.

In contrast, a symbolic concept provides a substitute for the experience in the form of an arbitrary token, like a word. Some words—like "snap"—sound like what they stand for. However, most words are arbitrary, substitutive codes that could take other forms if we just learned their referents correctly.

Once experience is organized in words, we are able to structure it with grammar. Grammar is a remarkable automatism, a brief cultural ritual, a synchronizing habit of the people who share a language. It allows the other minds of the culture to interpret meaning easily and unconsciously. Through these small preconscious rituals, grammar gives us fast, automatic, and reprogrammable skills in creating semantic progressions. The capacity for linear and propositional structuring of meaning is unique to an adequate grammar and is simply not possible with analogical concepts. Given the social context in which we learn to use grammar, it should not be surprising that when we reason systematically, we often do so by engaging in a covert dialog, as Andrew did in reasoning through his decision tonight. As if to gain order to his thoughts, he imagined he was still talking with Jared.

It has long been apparent that the left hemisphere's capacity for symbolic representation is a fundamental building block of human intelligence. This was perhaps the key evolutionary change that allowed intelligence to become extracorporeal—not restricted to a given body. Stories could be told, and eventually they could be written.

Language, captured in the words of a culture, became an effective, disembodied vehicle of mind. However, it remains a product of bodily structures in each generative act. By recognizing the visceral and somatic frames of corticolimbic network organization, we may approach the bodily context of the mind's language in a fresh way. We can see the left hemisphere's specialization framed closely by the properties of inhibitory specification and differentiation within the networks of the sensorimotor shell. Importantly, as I have already proposed, this characteristic of corticolimbic structure may allow symbolic representations to be extricated from the semantic fusion that is intrinsic to the distributed network representations formed at the limbic core.

As a child's brain learns a language, the left hemisphere becomes highly skilled in specifying and routinizing the sensorimotor components of speech. Once captured in symbolic codes, meaning can become modularized, separated from the embedding visceral matrix. The result is a capacity for logical rather than analogical computational operations. Structured and repeatable operations (language) can be performed on the words, rather than on visceral representation of the raw meaning.

In an even more structured form, the units are numeric or algebraic, and the operations are mathematics. Both language and mathematics are fundamental abstraction layers of computation, and each offers a degree of public semantic transparency. The regularization of ideas within these frames provides a corresponding objective separation from the syncretic matrix of the subjective self.

As a child's left hemisphere becomes specialized for the structured network operations close to the somatic shell, the right hemisphere appears to retain the primitive analogical representation that is inherent in the fully distributed network patterns of the visceral hemispheric core. In fact, right hemisphere specialization appears to have allowed it to elaborate both core and shell networks as variations on the theme of this fully connected corebrain network pattern. The result seems to be that the right hemisphere gains advantages not only of holistic structure but also of analogical content. The right hemisphere's concepts are continuous mirrors of both environmental events and the internal evaluative, visceral response to those events. In this intuitive domain of mind, the analogical concepts are powerful organizers of information. Yet they remain fused in context and only vaguely articulated, interlaced within the unwieldy matrix of fully distributed representations.

Neural Modes of Perception and Action

By clarifying the way in which motivational influences are organized across the core-to-shell scaffolding of corticolimbic networks, we may also gain insight into the organization of intelligence and experience along another major dimension of brain organization: the anterior/posterior division of the brain that constitutes the action/perception division of the mind.

We saw in Chapter 2 how the differing architectures of the perception and action networks suggest different structures of intelligence. Because multiple sensory channels can be processed in parallel, the perceptual organization of experience in the posterior association cortices can be holistic and integrative. Because actions must be sequenced in a linear fashion, their organization in the anterior brain requires a more focal mode of attention, in which only a single action gains dominance at any given time. We have also seen how the memory system interposes a delay between perception and action, providing grounds for a more complex organization of experience and behavior. By understanding the organization of motive control across the hierarchy of core-shell networks within the sensory and motor pathways of each hemisphere, we may gain new insight into the elaboration of more complex forms of mind from more elementary patterns of perceptual and motor representation.

For both perception and action networks, the core-shell processing is recursive; as a result, a two-way arbitration takes place between somatic specificity and visceral significance. Furthermore, abstract representations are formed only by extensive consolidation, creating linked, hierarchic patterns to satisfy the opponent requirements for differentiation at the somatic boundary and holistic integration at the visceral core. The input and output functions of information processing operate in different—and in fact opposite—directions within these networks, causing the consolidation process to take on different forms in the anterior and posterior networks. These are different vectors in the process of consolidation, and they suggest different qualities in the process of experience.

Aesthetic Experience

In perception, the primary direction of information flow is from the input channels, creating limbipetal traffic from shell to core. Since the nature of consolidation is that it is recursive, perception requires limbifugal, core-to-shell processing as well. Expectancies create a base for perception as they reach out to structure the sensory data on the basis of past experience.

However, the perceptual information is rooted in the sensory networks; consequently, abstraction takes this specific vectoral (directional) form in the posterior brain. The direction of information is essentially a shell-to-core function, such that abstract concepts that are formed primarily in the perceptual networks have the purpose of representing the perceptual referents. In line with this, the expectancies reaching out *(exteriorizing)* from corebrain concepts have the purpose of linking the motivational base to the perceptual representational skills of the somasthetic, auditory, and visual sensory systems.

These features of the perceptual networks lead to a quality of abstraction in which the visceral evaluative function serves to support the quality of perception in its own right. The theoretical analysis of abstraction in the perceptual corticolimbic networks may actually help to explain Pribram's intuition that the mind's embodiment in the posterior brain tends toward an *aesthetic* mode of experience.

The result of abstract representations in this aesthetic mode may be understood by contrasting them with concrete ones. The most concrete evaluation of perception is in its relation to personal biological needs. The expectancy born of need reaches out to select the sensory data that provides the affordance for this need. A hungry child wants to be fed and does not perceive other qualities of the world that are not constrained by this need. The needs of adults—for belonging, for mastery, for respect—are somewhat more complex, but they apply concrete biases on perception as well. A needy, egocentric person not only seeks out self-affirming situations but also fails to experience qualities of life scenes that are not construed as self-relevant. Data, when unconstrained by need, fails to become information.

When abstraction is achieved in perceptual networks through the hierarchic organization of links to the core, the visceral function may itself become abstract rather than constrained by needs. It can then provide evaluative controls that are suited to optimizing the information quality of the posterior brain both generally and abstractly. Within the sensory and perceptual networks, the qualities are those of representation. In the language of the Gestalt psychologists, these are qualities of optimizing the isomorphism of mind to world. With artistic training and discipline, the aesthetic sense may become highly sophisticated, with the result that the experienced artist or critic makes subtle distinctions that are lost on the novice. In other examples, the aesthetic experience may be very simple, such as Andrew's enjoyment of the sensory qualities of his quiet moments in the car. Need becomes encephalized to support the properties of the representation itself.

In each case, the simple sensory qualities are paramount, even though they become organized in higher-order percepts that give them abstract meaning. Still, even though their defining roots are in sensory networks, aesthetic representations have a structure that requires the involvement of the entire corticolimbic hierarchy. The visceral evaluative function operating at the base of this hierarchy is essential to provide the quality of emotional significance that energizes and motivates the perceptual apparatus. Yet aesthetic experience is clearly an encephalized visceral function that does not constrain or degrade perception out of bodily need. Rather, the consolidation process is subordinated to the representational (informational) function itself.

Andrew's sense of the beauty of the night scene and his feeling of the significance of this perception are not forced by biological drives or ego needs. Rather, his feelings have as their essential purpose the support and appreciation of the intrinsic properties of the aesthetic experience. When the aesthetic mode of experience is fully developed, the properties of mind then span both the elemental sensory qualities and their incorporation in the hierarchically integrated form of abstract perception.

Receptive Cognition in Self-Monitoring

A variation of the receptive cognition of the posterior brain is *critical appraisal*. Criticism is of course an integral component of aesthetics, when evaluations are formed at the base of the corticolimbic hierarchy in the process of perception. In addition to critical evaluation of the events in the world, the receptive cognition of the posterior brain is essential to self-criticism as well. What we commonly call "self-consciousness" not only is a normal property of mind but also may be necessary for ongoing critical monitoring of behavior.

An example of impaired self-monitoring is seen in the language deficits of people with damage to posterior regions of the left hemisphere. Such damage produces Wernicke's aphasia, in which one's understanding of speech is impaired. Of course, this impaired comprehension would be expected with damage to the receptive faculties of the left hemisphere, but striking problems also arise in language expression. A person with Wernicke's aphasia speaks fluently but with unusual language constructions, rambling semantic themes, or even nonsensical jargon. What these people seem to have lost is the ability for critical self-monitoring of speech.

An analysis of the visceral and somatic boundaries of consolidation can provide insight into the mechanisms of self-monitoring in speech and perhaps in self-consciousness more generally. In speech and in general self-control of behavior, perceptual channels for monitoring one's own actions must exist. In speech, this monitoring is primarily auditory, and the damage that causes Wernicke's aphasia is in the association cortex (typically heteromodal), which is important for integrating auditory and other information. However, hearing the sounds of speech is not the issue; patients with Wernicke's are not deaf. Rather, what is missing is the ongoing *semantic* evaluation of speech; this is what seems to require effective integration across the linked set of corticolimbic networks in the posterior left hemisphere.

As an astute observer of the neurology of language, Jason Brown has pointed out that when a semantic deficit in aphasia occurs, the lesion invariably involves damage to limbic cortex. A semantic impairment means a problem in understanding or expressing the *meaning* of language, rather than more purely sensorimotor deficits in comprehension or expression. We can surmise that two things are missing in people with Wernicke's aphasia: (1) the ongoing semantic interpretation of one's own speech and (2) the critical evaluation of the semantics (meaning) in the context of the interpersonal interaction. Because limbic cortices provide both the dense connectivity for global representations and the motive response to personally significant information, self-monitoring may reflect the close integration of motives and meaning at the core of the hemispheric network hierarchy. This architecture and these mechanisms of self-monitoring appear to be integral to self-consciousness more generally.

This piece of neuropsychological evidence could provide another small insight into the multiple levels of cognition that are opaque to naïve introspection. We speak, and we have some awareness of the process. What we can learn from neuropsychology is that the mechanisms of this awareness exhibit structure.

Of course, we all know that, in speaking, some people are more self-conscious than others. For some people, words just roll off their tongues. It is the fortunate extrovert who speaks skillfully. Others are just loose blabbers. For fortunate introverts, self-consciousness provides critical monitoring, so that words may be carefully chosen. But for others, self-monitoring leads to a pathology of consciousness, so that trying to speak in public leads to a self-reflective gridlock. Neuropsychology offers us a new way of understanding the mechanisms of monitoring speech, with multiple network levels representing not only the auditory data stream but the linguistic interpretation and the deeper levels of personal semantics as well. By studying this evidence, we can better appreciate both the individual styles of motivational control of people around us and the component mechanisms that are integral to the subjective process of receptive communication.

The Limbipetal Vector in Cognition

The limbipetal direction is inward, from shell to core. In the posterior brain, this is the direction that sensory data must travel to be integrated within meaningful experience. Of course, we now know that perception requires backflow of some sort in the highly ordered architecture of Figure 4.6. In addition, consistent with the critical role of memory in perception, the act of perception can be considered as another form of exteriorization, as interior expectancies meet the exterior world at the sensory boundary and are either confirmed or refuted. Fundamentally, however, the direction for information control in the perceptual networks is eventually inward, as the brain samples the properties of the sensory surround.

The present theory of achieving abstract cognition through corticolimbic consolidation of visceral and somatic representations takes on a unique form in the architecture of the posterior brain. The limbipetal vector of control means that the goal is for the mind to adequately mirror the world. This is a certain form of cognition, an intelligence of understanding reality, not forcing preconceptions upon it. In the formal context of scientific inquiry, this mode of cognition begins with *observation*. Interpretations must be suspended so that the mind can apprehend reality in its unexpected guises.

Of course, the cognition of the posterior brain is not just a sensory operation. The study of the brain has taught us that higher cognition is embodied in the literal form of arising from the body's operations of sensation, feeling, and action. So the task of observing reality with judgment suspended is a cognitive task and is required of some of the most complex and abstract instances of intelligence—in science, art, and daily life.

If abstract thought consolidates both visceral and somatic domains, then we can approach a new, more specific analysis of the perceptual cognition of the posterior brain. What we have learned in building our theory is that expectancies are the base of perception, that these are motivated by internal needs, and that information itself is formed only when data resonates with personal significance. Does this mean that we cannot be impartial observers? That the subjective base of the mind effects a patient and inexorable distortion of experience? The answer is both yes and no. The subjective basis of cognition, with its roots in personal motives, is clearly implied by the considerations discussed in Chapters 4 and 5, but the nature of abstract cognitive representation is that it extends beyond the concrete constraints at either the shell or the core. By interlocking the constraints of visceral syncretic motives and somatic specified realities, abstract cognition achieves a new level of representation, with structure in which neither boundary prevails. In the posterior brain, this means that concepts may be formed in which meaning may be informed by the sensory data.

Of course, by themselves, the data of the senses never tell much of a story. Information emerges only in a relational context, and interpretations require motivation. Yet by understanding the bodily structures of cognition, we are able to reformulate our understanding of the way in which perception can achieve observation and the mind can be informed in spite of its preconceptions.

Moreover, through a more explicit neuropsychological analysis, the nature of subjective bias might itself be specified and deliberately handled. Interpretations of the world are by their very nature motive-memories, infused with egocentric needs and presumptions. Pretending objectivity without understanding the fundamental subjective basis of the mind is to invite bias to operate freely and fully cloaked in the preconscious domain of motive blindness.

As he relaxes quietly in the dark, for the first few minutes Andrew is preoccupied with the sensory qualities of the night scene: the subtle colors of the houses and yards under the streetlights, the sound of an occasional car on the next block, a muffled TV down the street, and then a train in the distance. Soon his mind seems to wander, as the day's images and Jared's remarks appear in chaotic and even surprising sequences. Earlier, Andrew was actively thinking about himself and his life choices. At this moment, the structures of mind he formed in those thoughts (now largely unconscious) are available for a more extended consolidation, organized by the architecture of his posterior brain in its current mode of reflection. His typical pretenses about his life choices and his rationalizations about stealing cars are for a short time unnecessary.

Many of Andrew's mental operations are running in the background, as his consciousness is passive and receptive, yet the result of his reflection is a subjective realization. He is able to take in a different perspective of himself, one that started with Jared's comments. His own talents and the dangers of his life circumstances are now aligned in a new, somehow more obvious pattern.

Ethical Experience

The vectoral balance—the directional quality of information—takes an opposite form in the motor system of the anterior brain. It remains fundamentally subjective, but this is a different opportunity for elaboration of the visceral function. In the anterior brain, thought motivates action. As Karl Pribram has suggested, the psychological result of this elaboration in the frontal lobes may be an *ethical* form of experience. By studying the corticolimbic architecture of the anterior brain and theorizing on the structure of abstract thought, we may see how values are organized upon a primitive, visceral base. Developing their motive engines from this base, values can then provide more complex and reasoned direction for both extended intentions and immediate actions.

As we saw in Chapter 2, the motor system must organize a single course of action, serially and sequentially, from the diverse and heterogeneous causal influences operating in the paralimbic core. This is a "fan-in" architecture, from the many (diffuse limbic connections) to the few (focused neocortical modules), and must be achieved with the flow of information proceeding from core to shell. This form of distributed architecture presents a particularly challenging operation for the cognitive process in the anterior brain. Only certain information elements can dominate, and many connections must be inhibited in order to achieve the necessary serial order of behavior.

This organizational demand for focusing and sequencing the action plan is embedded within a corticolimbic architecture that has evolved to support the motivational function of corebrain limbic networks. When Pribram and his associates stimulated regions of the frontal cortex in the mid-20th century, they were surprised to find that the result was autonomic (visceral) responses. This was supposed to be the seat of the intellect, and yet here were gross bodily responses.

What we now realize is that the motivation of action from visceral controls was key to understanding the frontal lobe. This insight remains relevant for today's understanding of the executive functions of the personality. As we consider the inhibitory sculpting of behavior required to select effective actions from an array of impulses and opportunities, we must keep in mind that both the motor acts and the corresponding motives must be selected. Furthermore, this selection is developed largely through inhibitory differentiation within the information architecture. The multiple options for action must be narrowed to the most important choice. There is a semantic differentiation of motive self-control. The result may be Pribram's ethical mode of experience and behavior.

Just as perceptual operations are usually constrained by biological and egocentric motives, the organization of actions is typically linked with concrete motives. These motive vectors, if not concretely visceral, are familiar constraints under the umbrella of egocentric concerns and form what psychoanalysts describe as the "ego halo," the narcissistic surroundings of the self.

From this egocentric base, we can theorize that, through the dialectical organization of memory consolidation, complex and abstract mental structures can escape the concretizing influences encountered at both the core and the shell. The process of consolidation must be understood in relation to not only the expectant nature of perception but also the motive base of action prediction and control. Motives in their concrete form are instigators of actions in a narrow and restricted sense. Yet they may also take on more complex, abstract forms as the information structures they guide become differentiated and hierarchically organized episodes of action consolidation in the anterior brain. In this way, visceral directives are themselves revised recursively in complementary balance with the action plans that are assembled closer to the motor shell. As with Jared and his refusal of the transmission, our values become sophisticated and consciously known by the experience of exercising them.

Even complex motives have roots in the primitive controls on arousal and drive emerging from subcortical circuits. Such controls are essential to activate the brain and direct attention adaptively. But more complex motive structures develop within the child's increasing intelligence. As they do, they become able to support a broad range of action classes rather than specific behaviors. The increasingly abstract evaluative support in the developing mind is able to engage thought and behavior that can be described not just as motivated but also as reasoned in relation to deliberate values—as *ethical*.

Although concepts can be activated in milliseconds, the primary time frame for developing abstract motives may be longer: that of the person's developmental history. When motives are differentiated, they do so through operations of the self, as action choices are not only urged by desires but also buffered by the child's increasing knowledge of reality constraints. This is the arbitration of primary and secondary process cognition that Freud saw as integral to the development of the ego.

Expressive Cognition and the Production of Meaning

The study of the brain's production of language also provides insight into the implicit levels of mind, specifically those in the frontolimbic networks. As we speak or write, consciousness becomes crystallized with the production itself as we listen to or read the results. We may be aware of elements of the intentions to create meaning (although even these all too often evaporate). But the generative process is automatic and unconscious. The words appear.

At the level close to motor patterns, language production is a good example of automatization. This is habit formation, making something work so well that it no longer requires attention or awareness. Attention can then be directed at the higher level of semantic intention. Only when attention is required at the elementary production level, as in speaking a foreign language, do we recognize the effort of what is normally automatic. Automatization inserts certain practiced neocortical cognitive operations into the unconscious domain. Motive blindness shows the inaccessibility of the more visceral cognitive operations at the limbic domain. It is a wonder we can think at all.

For most of us, of course, language production supports some of the most explicit creation of meaning that we experience. Though we may not be fully conscious of the process, by saying things we learn the contents of our own minds. Certain people with damage to the frontal lobes (particularly on the left side) exhibit striking deficits in the ability to generate language, deficits that are highly frustrating for the process of self-expression.

As with people with Wernicke's (receptive) aphasia, the syndrome of Broca's (production) aphasia is most remarkable for the unexpected problems. It is not just that damage to the inferior frontal lobe (at base of the motor system) leads to a deficit of inaction. Rather, the person is unable to produce language in the presence of what seems to be intact knowledge of the meaning that should be produced. People with Broca's aphasia may find only one or two words, but these are invariably appropriate to the communication context. It is as if the semantic constraints are intact, just disconnected from the intermediate generative process.

Having studied the problems of people with Wernicke's aphasia in the preceding material, we have seen the importance of self-monitoring of meaning in the process of accurate language production. That is what those people have lost. With this self-monitoring capacity intact, we can imagine that people with Broca's aphasia struggle with a damaged production system that is now overwhelmed by critical constraints. Yet, even without considering the reciprocity of anterior (expressive) and posterior (receptive) networks, the syndrome of production aphasia illustrates the multiple levels of cognition, this time within the motor control networks of the anterior brain. This is not a motor disorder of making language sounds; many people with Broca's aphasia can repeat sentences adequately. It is also not a semantic disorder; given help, these patients can demonstrate that they understand the meaning of what they want to communicate. Rather, this form of aphasia is a deficit of a middle level of cognition caused by damage to a middle network of the frontolimbic hierarchy. Recognizing this, we can better appreciate the multiple components of neurocognitive representation, the multiple representations of mind, that must be assembled if we are to communicate normally. In the naïve mind, speech appears as a unitary, nearly automatic process. However, its embodiment incorporates multiple linked representations, each constrained by the unique demands of its level of the corticolimbic hierarchy.

The Limbifugal Vector in Cognition

The limbifugal direction is outward from core to shell. Organizing actions requires the activation of motives within the limbic base of the anterior brain and then the translation of those motives into the expectancies for goals that guide the unfolding action plan. This is exteriorization. Hedonic expectancies are formed at the visceral level to direct actions that actualize them. At the motor interface with the world, these expectancies meet the reality of the context of this environment. Compared to the less intricate cognition of simpler creatures, the increasing temporal span of human cognition allows increasing complexity in the exteriorization strategy. When urges arise, they cannot manifest in action but rather must motivate extended planning. Nonetheless, the architecture of the planning process remains the corticolimbic (core-shell) dialectics, only now in representational form. The primary process cognition allows plans to serve as free vehicles for motives and as fantasy explorations of possible valued scenarios. The secondary process cognition requires plans to be tested against the constraints of the world.

In testing plans at the external interface, the experienced planner packages the motor operations of the plan within both familiar forms and a set of action habits that have captured the regularities of the world in reusable, modular form. These packages of action habits and the environmental constraints they fit are what James Gibson called "affordances." Drawing from the ethologists' observations of animal behavior in natural conditions, Gibson emphasized that behavior is usually not a response to a stimulus but an opportunistic realization of an affordance. The perching bird finds a perch because this is the environmental constraint that fits its action habits and their associated needs. In Yakovlev's terms, the affordance effectively exteriorizes the animal's current need state, manifesting it in the world as an action plan for satisfaction. A person looks for a path to follow up a hillside because these are the affordances that occur to a brain that has a motive to go up the hill and the motor experience in walking.

Motives and affordances make for natural behavioral scenarios, and between them we fill up most of our days. But these are default positions. Abstract thought in the planning process often requires a constructive tension that disrupts familiar behavioral patterns. At the core, a motive becomes an intention, a somewhat more complex representation that incorporates a valued action. Or the motive-intention incorporates an environmental goal that is consistent with the motive and the associated actions that might achieve it. At the shell, the affordance organizes an action pattern that instantiates a talent, a capacity of the person.

These are then more complex elements in the limbifugal translation from need to action in the anterior brain. The construction of effective plans requires a recursive arbitration of these constraints, so that the plan is an abstract concept with hierarchic integration of the motive vector and its possible affordance vehicles. Awareness monitors this unfolding development through whatever limited channels are available, such as a sense of need, a felt intention, or a familiar confidence in the affordance vehicles.

To be achieved through these elements, the cognition of action must be organized in a kind of mirror image of the cognition of perception. The pathways of the frontal lobe run from limbic networks toward motor cortex, following the basic corticolimbic architecture shown in Figure 4.6. Now, however, the direction is left to right rather than right to left. As a result, the processing pattern in the corticolimbic architecture must be turned inside out. Action begins within the corebrain limbic motive base and progresses toward articulation at the neocortical shell.

The inferior-to-superior connections in Figure 4.6 provide the back projections in perception, modulating the forward projections of sensory data. Now they must provide the primary traffic from the visceral motive-semantic base to be articulated within the somatic articulation of neocortical action sequences. Because cognition is embodied at every level, the architecture of the limbic-motor pathways shows us a major dimension of mind. In the anterior brain, thinking is a motive-motor process. An idea is a developmental process, beginning with personal significance, organized as a vectoral intention by the attraction of possible affordances, and then finally differentiated into discrete form—words or actions—that are then crystallized into conscious form as the palpable products of the experiential process.

As Andrew mulled over his choices tonight, his concepts were organized within the differing structures of his subjective experience. These provided order to the consolidation process, which continues to organize his thoughts into lasting constructions of the self. His reasoning—about the realities of his life of crime—was structured in the analytic, sequential, linguistic structure of the left hemisphere. Instantiated in the active cognition of the anterior brain, this reasoning provided Andrew with a rational basis for a clear, conscious realization of the implications of his choice.

In contrast, when he relaxed his focused attention, adopting a more holistic conceptualization of himself in this scene and this time of his life, Andrew engaged a different frame of mind. He could then build a novel, intuitive construct of his life now, one that was responsive to the many influences within his ongoing cognitive consolidation. The result of his reflection was an organized impression of how things should go from here.

Andrew's cognition is now an operation of his subjectivity, but in important ways it is also an intersubjective process. Andrew understands himself and his situation tonight in a somewhat more abstract way because of his talk with Jared at the party. He sees himself in a fresh perspective, framed in more abstract relief by Jared's point of view. In the way of many social contributions to the organization of the self, Andrew's neural structures of mind are more complex tonight because of their intersubjective perspective. They have been expanded in viewpoint by the careful attention of someone else.

7
Objective Experience

The idea that intelligence emerges from a subjective basis will not rest easy with many people—particularly scientists. As we saw in Chapter 1, subjectivity has been the aspect of the mind that has proven intractable to scientific analysis. Yet through the reasoning in these pages, we have concluded that subjectivity forms the very foundation of meaning by organizing the personal, visceral motive base of the mind's embodied neural hierarchy. For those hard-nosed scientists making up one of the two cultures of the academy, this characterization of mind contradicts the cherished notion of intelligence as an objective capacity.

However, the literary intellectuals should not rejoice too soon. Subjectivity is a fact of the mind, not a justification for its efforts. If it is correct that abstract thought is a hierarchic construction linking corticolimbic networks, then the process of constructing thought has an inevitable effect on the subjective base of personal motives, which become incorporated in more complex concepts. And, through the reentrant coupling with complex representations, the mind's motive base is itself transformed. From the natural childish egocentrism of the subjective perspective, the transformation yields adult forms of mind. Experience itself becomes more objective.

I admit this story may be optimistic. In a basic analysis of neurocognitive structure, the subjective-objective dimension aligns with the core-shell dimension. At the core, personal motives constrain the subjective perspective, so the natural mode of mental function is primary process cognition (fantasy). At the shell, environmental realities constrain objectivity, so the natural mode is secondary process cognition (validation). I have developed the present theory to consider ways that these constraints could be arbitrated into hierarchic constructions of complex thought. Yet the majority of the evidence around us (in universities at least) suggests that the tension is uncomfortable. So, as with sexual orientation, people may claim to explore the interesting middle zones, but in the end they go either one way or the other. Some embrace the felt significance of the subjective view and actualize their visceral projections as literary humanists. Others are drawn to the security of the objective somatic boundary and restrict the mind's operations to the definite certainty of empirical science.

Objectivity Through Intersubjectivity

For those rare episodes in which core-shell dialectics do produce abstract ideas and objective experience, the most important preparation may come from intersubjective reasoning. The sharing of perspectives in the process of interaction is a complex cognitive process. Understanding another's subjective view is cognitively challenging, and it is then transformational for the default subjective view. Practice in shifting perspectives in this way tempers subjectivity, and yet it allows each child and adolescent to organize abstract concepts that remain strongly rooted in the limbic base of personal meaning.

I have organized the theory in these pages as a brain theory, with the mind emergent from linked neural networks, so it may seem as though abstract thought arises wholly within the body, not between people. In addition, we know that the minds of those who are flatly deficient in understanding social interactions may become capable of abstraction in many technical domains. Nonetheless, I am convinced that it is in the process of understanding social interactions, through intersubjective reasoning, that abstractions are formed with the strongest integration of personal semantics with a perspective that challenges the default subjectivity of the naïve mind.

In this chapter I balance the analysis of bodily mechanisms by considering the social context of both visceral and somatic representations. This allows me to introduce some of the ongoing research in the areas of social and affective neuroscience, including that in my own laboratory. We are gaining important new evidence on the neural structures of both subjective and objective cognition. Many studies are now directed specifically at the neural mechanisms of intersubjective cognition that represent and interpret the experience of others. In surprising ways, this research may yield insight into the mechanisms through which the mind achieves objectivity, not as an assumption or a pretense but as a product of differentiating the cognitive representational structures of mind beyond the default motive constraints of naïve subjectivity.

There is an emotional as well as a cognitive basis for Andrew's reflections on his talk with Jared. As sometimes happens in adolescence, the friendship between Jared and Andrew is formative for both of them. Most adults can now remember each friendship of adolescence even if, at the time, it soon seemed fleeting and inconsequential as they moved on to other things. One reason friendships are so formative is that they provide a context in which people gain an objective perspective on how they appear to others. This objectivity is often most understandable and influential when provided by a friend, someone who cares.

As we consider how the brain's structures organize intelligence, even a simple theory is incomplete if we consider the mind in isolation. We all recognize the importance of social emotions—love, bonding, and rejection—to subjective experience, but there is also an important lesson on objectivity. Both the motive and the cognitive structures of social perspective taking allow more abstract and objective structures of mind. For most people, the greatest achievements of intelligence are made in developing accurate understandings of social interactions, and these achievements remain embodied. Practice in understanding others' perspectives yields sophisticated forms of experience, and yet this understanding has fundamental roots in bodily mechanisms.

Bodily Integrity and Social Bonding

One of the most important realizations in modern neuroscience is that the evolution of social attachment has occurred through elaboration (and I would say encephalization) of the neural circuits that mediate the response to bodily pain. Pain is a primitive, essential signal that alerts an organism to a threat to bodily integrity. There is an immediate neural response. In all vertebrates, the response to pain may be reflexive and immediate, such as the withdrawal of a limb from a painful stimulus. Limb withdrawal requires no cognition, not even a brain, because it is mediated directly by a spinal reflex. In more complex mammals, pain may require an extended coping effort, such as seeking social support. Regardless of the specific context, an intense pain will command attentional resources until it is resolved. The loss of bodily integrity is the primordial threat to organisms and requires a persistent and effective motivational response.

An important clue to the physical basis of pain has been known for many years. The neural circuits mediating pain perception are divided between the *visceral* and the *somatic* domains of the central nervous system. Consistent with the present core-shell theory, the visceral representation of pain, elaborated considerably within limbic (core) networks, is essential to the psychological evaluation of pain. In contrast, the somatic circuits, which project to the somatosensory neocortex on the lateral surface (shell) of the hemisphere, are less important to the subjective aspect and serve primarily to localize pain to the particular body part. Even when the pain does not come from the internal organs but is limited entirely to the somatosensory domain (e.g., in the skin), it is not only the somatic representation that is important to the functional experience of pain. The visceral networks of the limbic system are also integral to evaluating the emotional significance (and thus understanding the meaning) of somatosensory pain.

Striking evidence of this limbic role in the emotional evaluation of pain has come from a medical practice that is no longer common but was used in years past for the treatment of chronic pain. This was psychosurgery. The neurosurgeon would open the patient's head and cut out a section of the limbic networks, particularly the anterior cingulate cortex. The effect of this operation on the patients' subjective responses was consistent with the functional division between visceral and somatic representations. The patients would report that they still felt the pain as intensely as before and in the same place (because the somatic networks of body surface localization were intact), but now, because of damage to the corebrain visceral networks, they just did not care.

It makes sense that the subjective monitoring of bodily integrity is organized within the core-shell corticolimbic architecture. Converging evidence comes from the localization of opiate receptors in the brain. Opiate drugs (opium, morphine, heroin) have long been known as effective painkillers. Opiate receptors (the chemical ports on neurons that react to the drugs) are particularly dense in the corebrain limbic cortex (anterior cingulate and insular cortex). Importantly, the effect of opiate drugs is similar to that of cingulate lesions: The person still recognizes pain and can localize it but no longer cares.

We know from this research that opiate drugs have their effect because they act like the natural (endogenous) opiates in the brain. The endogenous opiates have integral roles in modulating the subjective and motivational response to painful stimuli. For example, if you are in a life-threatening emergency, such as a car accident, the release of endogenous opiates is a natural response that provides a necessary analgesia to allow you—even if you are injured—to take necessary actions such as seeking help. Many people have been surprised by the seriousness of an injury because they continued to function during a crisis with minimal distraction from the pain. Over time, of course, the natural opiate analgesia wears off, and the chronic pain demands that the injury be attended to. The density of opiate receptors in limbic cortex points to a key role of corebrain visceral networks in evaluating and thus experiencing bodily integrity.

Although it may be unnoticed in an emergency reaction, the pleasure of opiate release is a powerful motivator, as seen in the powerfully addicting effects of opiate drugs. The opiate abuser soon loses interest in basic human values and particularly in the values of sexual and family relationships. This specific substitution of opiate drugs for affection and social attachment turns out to have a fundamental explanation: The normal motivation for personal relationships and social contact appears to be, at least in part, controlled by the release of endogenous opiates within the limbic pain evaluation system.

A number of years ago, research by Jaak Panksepp provided remarkable evidence of this evolutionary adaptation of the pain system to support social bonding, which is observable not only in mammals but in birds as well. A young chick cries out when separated from its mother, so that the mother can find it. A small dose of an opiate, a dose too small to be sedating, stops this vocalization. It also stops the chick's search for the mother. A similar effect is readily observed with mammals, such as dogs. It is as if (not unlike the pain patients after cingulotomy) the puppy is abandoned but no longer cares.

The implication of these findings is that the evolution of social bonding in mammals has elaborated the necessary motivational control out of the very primitive motivational systems, including both visceral and somatic forms, that modulate the response to pain. Some evidence suggests that the human fetus is exposed to high levels of endogenous opiates in the womb, providing not only a tranquil state but also a suppression of respiration. Birth is then like withdrawal. Welcome to life.

The only anodyne is then the mother. Certainly the comfort of feeding is fundamental, but research with monkeys has shown that food alone is inadequate to provide for the emotional needs of infants, at least of primate infants. For the baby monkey and almost certainly the baby person, physical contact with the mother is a primary need. The opiate release associated with being held by the mother then provides the necessary antidote to what is now the emotional pain of separation.

The Pain of Rejection

In later social relationships, humans bring not only elementary social needs but also considerable cognitive capacities (which allow them to understand the intentions of others) to frame their own interpretations and to exert at least some voluntary control over their emotional reactions. However, just as the consolidation of cognition requires an extended organization across the linked set of corticolimbic networks, so too does social cognition require an integration of visceral evaluative representations with the somatic patterns of the sensorimotor interface with the social interaction. Recent research using magnetic resonance imaging of brain activity has shown that the networks responsible for physical pain appear to form a basis for the adult's response to the psychological pain caused by social rejection.

As their heads were being scanned, the volunteer subjects in this research played a game with other subjects that required cooperation. As a secret conspiracy of the experiment, the other subjects banded together against the naïve subjects and deliberately rejected them. Even though the naïve subjects were in the unusual context of a neuroimaging experiment, they showed an emotional response to this rejection, marked by increased activity in limbic regions of the cortex. These were the same regions that are responsible for the evaluation of physical pain. Rejection hurts. This is pain in literal, bodily form.

Psychological study for many years has shown that the quality of the infant's early social attachment will be influential in all later relationships. Even though the earliest bonding experiences soon recede into the long-forgotten unconscious, they appear to remain as transferences, templates for what to expect from relationships. These implicit expectancies shape each new interaction. Although humans retain some capacity for emotional learning and for recovery from emotional stress, psychotherapists know this capacity is limited.

Even in the popular culture, of course, we recognize that social rejection is painful. Love hurts. However, until the recent research, most people would have thought that the pain of rejection is just a metaphor rather than a physical process. Yet Panksepp pointed out years ago out that the opiate mediation of close relationships would explain a number of features of the emotional qualities of these relationships in a literal rather than metaphoric way.

For example, after the euphoric response to the initial experience with an opiate drug, the addict soon habituates to the dose and adapts to the presence of the drug as the neutral hedonic state. This sounds a lot like the love relationships that many people are euphoric to find but then soon take for granted. For the addict, the need for the drug becomes apparent only upon withdrawal, when the deficit state leads to intense (and once again fully conscious) cravings. Many of us find that relationships are valued only when they are lost.

The psychological pain of social rejection does not have the same subjective experience as physical pain, of course, even if it shares the same physiological substrate. It is more cognitively elaborated, and its emotional quality is more representational and implicit. Yet, in the language of neuropharmacology, there is evidence of direct "cross tolerance" of physical and psychological pain.

A striking behavioral disorder of adolescent girls is "cutting," in which a girl repeatedly lacerates the skin on her arm or leg. This is often done in private on a region of the body that does not show, so it is not just a cry for help. In the naïve view of the mind, this makes no sense. Furthermore, conventional psychological theory, whether cognitive or behaviorist, cannot explain why an organism inflicts pain on itself. This behavior seems to violate the basic motive principle of psychological theories of approaching pleasure and avoiding pain.

Yet, once we learn to recognize the evolved structure of the human emotional system, this behavior can be seen in a new light that reflects on the inherent intersubjective sensitivity of adolescence. Most people would think that physical pain and bodily integrity would take precedence over social emotions. However, adolescence is a developmental transition that brings strong needs for social attachment, needs that are integral to the function of the self. Childhood is a time when people take care of you. Adolescence is when you have to reject your family of origin so you can go out into the world and create a new family. In the transition, a big gap opens up. It is an interpersonal vacuum in which the attachments that are integral to the self are missing, and subjective experience is readily overwhelmed by the pain of abandonment.

Thus, for a young person who experiences intense emotional pain, whether from a past trauma or a current deprivation, the opiate release associated with self-injury may dull the intense experience of pathological loneliness. The experience of relief from this subjective pain is powerful and even overshadows the physical pain caused by sacrificing bodily integrity.

Pain, Hope, and Information

The control exerted by limbic networks is not limited to social experience; it can be seen in many situations in which self-regulation is required. For example, the visceromotor networks of the anterior cingulate cortex are engaged not only in evaluating pain but also in monitoring the significance of many perceptual and motor events. Research with brain waves has shown that when people make an error, a strong electrical response can be recorded from the surface of the head, right over the anterior cingulate cortex (Figure 2.9)—that is, as long as they know they made an error. If an error is made but the person is unaware of it (as in the early stages of learning a task), then there is no error-related negativity. The implication is that the visceral networks responsible for evaluating emotional significance provide integral controls in the ongoing self-monitoring of actions.

Although self-monitoring and self-regulation are often considered to be among the higher, executive functions of the brain, there is certainly an emotional quality when the subjects are trying to do their best but they know they just pressed the wrong button. In fact, when Phan Luu conducted one of the early studies of the error-related negativity in our laboratory at the University of Oregon, he often knew immediately when the undergraduate subjects had made an error. Their frustrated vocalizations could be heard through the walls of the next room. In fact, the students working in the lab (who all had been subjects themselves) proposed a "scientific" term for this brain wave response: the "Oh shit!" wave.

One straightforward interpretation of this limbic mechanism of self-monitoring is that there is a kind of painful experience in detecting one's own errors. However, a similar engagement of limbic cortex is seen in many examples of traditionally cognitive evaluation, such as in evaluating the outcomes of economic decision tasks. The science of economics is of course built on the assumption that people make rational choices to maximize pleasure (and should thus never do things like cut themselves). When subjects view an outcome of a decision that is less favorable than they expect, even if it is still a good outcome, they emit a robust medial frontal negativity. The visceral basis of economic value. Just like the rats.

In fact, this brain wave evidence suggests that the subjects in these decision tasks maintain a positive expectancy, just like the one that creates the learning effects in Chapter 3. In this hopeful state, violations of the hedonic expectancy seem to create a brief, even painful limbic modulation of the attention process.

This is an exciting time in research on the neural mechanisms of self-regulation. Our new technologies for imaging brain activity are showing the dynamic nature of the emotional control of cognition. It is striking how much of what we have considered rational thought requires ongoing motive regulation, yet the emotional nature of the limbic influence is apparent in many studies as well. Led by Phan Luu's theoretical work, the studies in our lab have recently focused on the role of the limbic pain networks in directing attention. What we have seen is that brief engagements of these emotional networks are essential in the dynamic control of the mental process, even for the most rational of cognitive evaluations.

For example, we wanted to see whether informative feedback would engage a self-monitoring response similar to the error-related negativity even if no adjustment of the correct response was required. For help with this research, we went to experts in experimental studies of emotional influences on attention at Oregon State University, Douglas Derryberry and Marjorie Reed. Derryberry and Reed had developed a method to delay feedback on performance, so that it was still significant emotionally but no longer required corrections of the response. It was the equivalent of getting course grades two weeks after taking the final exam.

In fact, Derryberry and Reed had used letter grades as feedback on how fast subjects were able to provide a correct response. The task is simple: Press the right arrow key if the arrow on the screen points right and the left arrow key if the arrow points left. However, to make it interesting, the arrows on the screen can be either in the left visual field (left side of the screen) or the right visual field (right side). There is then an inherent conflict between the pointing direction of the arrow and the side of the body it is on. This conflict makes it challenging to push the correct button, especially when the subjects are pressed to perform as quickly as possible. One way to pressure the subjects is to arrange the task so that, if they do not keep up, they get bad grades on their performance.

Like all college professors, Derryberry and Reed knew from long experience and the occasional explicit vocalizations, that university students hate to get F or even C grades. When Luu and I applied our dense-array brain-wave measures to the Derryberry and Reed experiment, we predicted that the negative feedback would engage an error negativity over the limbic frontal lobe (anterior cingulate cortex) even though the informative feedback was not linked to the response. As predicted, when the students saw a C, they showed an increased medial frontal negativity. When they saw an F, it got even bigger.

The feedback provided the students with information and reduced their uncertainty about their performance. It became information because it was significant for evaluating their success in achieving a goal. The negative feedback, even in this cognitive task, was associated with activity in the same visceral networks that carry out the more elementary function of evaluating pain.

For many people, psychological pain becomes chronic, and they become clinically depressed. The connection to social rejection or a loss of a significant relationship is all too common. Depression is so often associated with social deprivation that I teach my doctoral psychology students that depression is a social disease. This reminds them to look for the social causes whenever they see depression. As noted earlier, the state of depression is associated with a pervasive negative shift in the evaluation of life events, even as the person is unaware of the motive bias. With no introspection to the emotional basis of the negative cognitive bias, it just seems like the world has gone bad.

In an effort to use our neuropsychological research tools to help understand clinical depression, our research team recruited depressed people from the local community and asked them to perform the same speeded task that we used with the undergraduates. We found that, compared to normal community subjects, most of the depressed subjects showed highly exaggerated frontolimbic responses to the negative feedback, as if this self-relevant information was psychologically painful.

This research is providing us with clues to the mechanisms of self-regulation. In control theory, negative feedback is highly informative. Positive feedback simply reinforces the current state, but negative feedback signals that a change is needed. As Luu and I studied these issues, we consulted with Steve Keele, an expert in the psychology of attention at the University of Oregon. Although we were trying to reason through these mechanisms in terms of neural control theory, Keele pointed out that similar issues of self-regulation were already well known in animal learning theory. As we saw in Chapter 3, animal learning is not a simple association of stimulus and response but rather a cognitive process in which animals appear to hold a valued expectancy for what will happen, and they use that expectancy as the standard for evaluating each life event.

With Keele's theoretical guidance, we were able to see that the mechanisms of expectant self-regulation in our brain wave studies of depressed people were fully predicted by modern animal learning theory. Hope is the measure of the current lot. In other words, we gauge our evaluation of events by the hope that we hold. Negative feedback disrupts hedonic expectancy, and it demands change. Specifically, it demands learning.

Cybernetic theory says that negative feedback is the most informative signal. Evolution has created mammalian brains that operate in an ongoing mode of positive expectancy; as a result, negative feedback disrupts the ongoing process of hedonic self-control, demanding a shift to a new mode. This is painful. In fact, the depressed people we studied seemed to be prepared for pain of this kind. When they saw negative feedback, their frontolimbic response was exaggerated compared to that of the nondepressed subjects.

That is, most of them. We were surprised to find that the more severely depressed people in this sample did *not* show increased medial frontal responses to the negative feedback. In fact, each of these more severely depressed people showed a virtual lack of a medial frontal response to negative feedback. Although we are still trying to understand this effect (and in fact to replicate it to be sure it is real), we are hypothesizing that this effect may provide a clue to the shift in self-regulation from moderate to severe depression. We think this effect may be a second mechanism of depression. Whereas aversive events may lead us initially to be sensitized to negative feedback, a more chronic loss of the valued events in life causes us to lose hope altogether.

This reasoning might be consistent with classical observations of the progression of increasingly severe depression in children. A number of years ago, one of Freud's students, René Spitz, observed children whose mothers were taken away (and who were then virtually abandoned in a prison nursery). These young kids showed an initial response described as *protest*. They cried out for their mothers and showed an active, anxious depressive response. After time passed and the mothers still did not appear, the children became withdrawn and listless—a response described as *despair*.

We think that adult depression may be similar to the response of these children. The limbic evaluative response (indexed by the medial frontal negativity) *increases* in mild to moderate depression because at this stage people are more sensitive to the painful information of life. However, in more severe depression, perhaps a different adaptive response sets in, one that causes the depressed person to become less sensitive to pain and less viscerally responsive to negative information. The normal positive expectancy is withdrawn, so that failures are less painful after hope fades.

In this brief review I have touched on only a few of the findings in current research on the neural basis of the social and emotional mechanisms of self-regulation, specifically the studies from my own laboratory. Nonetheless, it should be clear that we now have measures of brain activity that can reveal the component mechanisms of the mind. We can measure brain electrical activity (or blood flow) from the primitive limbic regions of the hemisphere, and, to our surprise, these are engaged not only in overt emotional response but also in processes of evaluation and decision making that are integral to rational cognition. In fact, one of the fastest-growing areas of brain research is neuroeconomics, the application of brain systems analysis to economic decision making. As with other areas of human endeavor, the traditional assumption has been that economic decisions are made objectively and rationally. Many forms of evidence now show this assumption is wrong. As we analyze the motive basis of evaluation and decision, we are gaining new explanatory power in business and economics.

The Evolved Roots of Family Values

Pain is only one of the primitive building blocks of neural self-control. In the normal state of hopeful expectancy, for example, the elementary mechanisms of arousal, including both anxiety and elation, influence the brain's ongoing processes of attention and memory. When a person becomes depressed, a shift occurs not only in the emotional control of cognition (as expectancy is biased negatively) but also in the multiple neural mechanisms associated with positive hedonic tone. Anxiety appears to increase at certain points, and yet at other points the critical process seems to be the loss of hedonic arousal (a negative shift on the depression-elation dimension). The result is that the more severely depressed person loses interest in normally pleasant events, withdraws from social interaction, and eventually shuts down activity altogether in a state called *psychomotor retardation.*

Although pain is just one component, it is a critical one. Furthermore, it serves as an instructive example because it illustrates an elementary motive that has been elaborated into more complex motives of social attachment through the process of evolution. In today's research, we are learning how to reason from such elementary mechanisms of self-regulation to understand significant psychological processes, including monitoring behavior, evaluating feedback, and making economic decisions. We are coming to understand how the lateral specialization of the brain allows not only the precision of linguistic communication but the embedded richness of emotional comprehension of social interaction in the right hemisphere as well. We are finding that the structures of consolidation provide the motive basis for the brain's networks of comprehension and expression, thereby affording aesthetic sensitivity to perception and ethical direction to action. In this framework, we are beginning to comprehend both the inherent subjectivity of the self and the objectivity that is achieved by integrating the perspectives of others.

We are thus building a knowledge of elementary human value systems, the emotional and motivational mechanisms that allow us to evaluate the significance of events. The new neuroimaging technologies are enabling us to see these mechanisms in action—how they shape cognition generally, wherever value judgments are required.

In the close relationships of friendships and families, the mediation of feelings and values by elaborated pain mechanisms teaches us about the vertical integration of subjective experience. The bonds of close relationships are complex psychological processes, yet they can be broken by opiate drugs because the drugs substitute for their motive basis. These are not just effects in depraved people who become addicts; similar effects have been observed in experimental subjects given opiates for two weeks. We may not want to think that important personal and cultural values depend on maintaining the balance in our biological substrates, but this is what the evidence on opiates implies.

The vertical integration of pain and social attachment is evident across multiple levels of the nervous system. The circuits for representing and responding to pain have been maintained across each advance of vertebrate neural evolution (mesencephalic, diencephalic, and telencephalic), and at each level the pain circuits are linked with the circuits controlling vocalization. For the higher vertebrates (birds and mammals), care of the young has required not only communication (often through vocalization) but also motives for social attachment (imprinting and bonding) for both parents and juveniles.

If we take stock of the core values of subjective intelligence, social bonds must be among the most significant, for we experience these bonds as essential emotional qualities of ethical attitudes and choices. For any particular person, the experience of social values is a complex product of personal history and of the relationships in a particular family, community, and society. For humans generally, the evidence on pain mechanisms implies that the experience of social values draws on a common set of neural mechanisms that are forged through multiple transformations of bodily structure over several hundred million years of vertebrate neural evolution. These are the bodily roots of family values.

Organizing Love

Do the multiple neural levels of pain and vocalization imply that subjective experience can be fractionated into qualities at each evolutionary level? The available evidence suggests not. Human conscious experience appears emergent only from the more recently evolved levels, even though *all* levels are required to maintain the functioning mind, which includes implicit unconscious mechanisms, as well as conscious experience. Although our current state of knowledge is limited, examining the evidence may be instructive as it probes the roots of experience.

Behaviorally, the control of vocalization has been subjected to careful scientific analysis at each evolved level of neural representation. Research by Uwe Jürgens, Detlev Ploog, and their associates, using both lesions and electrical stimulation, has clarified the behavioral characteristics of nonhuman primate (monkey) vocalization at each level of the neuraxis. At the brain stem level they found discrete motor regions for respiration and vocal control. At the midbrain level are patterns for coordinated vocalizations that represent the discrete calls of that species (such as the cackling and growling in the aggressive vocalizations of the squirrel monkey). Only at the limbic level of control does the vocalization appear to have an emotional quality. This suggests that, at least for mammals as complex as primates, the limbic networks have taken on the elaboration of vocalization within complete emotional responses. Then, at the neocortical level, the control of vocalization appears to allow a deliberate, intentional use of vocalization. This voluntary control appears to reflect a more recently evolved form of cybernetics, now modulating the spontaneous emotional control at the limbic level.

Although we must be cautious in generalizing this vertical control hierarchy of monkey brains to human brains, clinical observations of emotional vocalization in humans appear consistent with the basic primate roadmap outlined by Jürgens and Ploog. In a condition that is quaintly termed "pseudobulbar palsy," the patient exhibits spontaneous episodes of crying. The neural damage leading to this condition typically involves disconnection of higher brain networks from the brain stem (a region traditionally called the bulbar region). Therefore, the disorder appears to reflect a cortical disinhibition of a crying response that is fully organized at the brain stem level.

However, the crying response, although behaviorally complete, appears not to be subjectively complete. The report of patients is that they do not *feel* like crying. They just feel surprised and embarrassed by these inappropriate outbursts.

As with the monkey vocalizations, the human emotional display of crying involves functional centers at multiple levels of the neuraxis. Yet the integration with emotional experience and behavior requires engagement of the higher endbrain systems of the telencephalon, including both limbic and neocortical networks.

The brain stem networks are essential to the behavior, and we know that we often learn of our own emotional responses by observing our actions. So if you get pseudobulbar palsy, the first time you burst out crying, you would have to wonder whether you felt sad. But you would soon figure out that you did not and that the disorder had fractionated your experience and behavior. Evolution has apportioned complementary processes across the multiple levels of the neuraxis, and the limbic and neocortical networks of the telencephalon appear most directly active in structuring subjective experience. The conscious mind has arisen in the terminal additions.

In the brains in which these multiple levels have been systematically studied (those of monkeys), the limbic networks integrate vocalizations with spontaneous, ongoing emotional processes. In contrast, the neocortical networks allow greater articulation of discrete actions, such as in voluntary vocalizations intended to achieve a specific goal. If we extend this analysis to the colocalization of pain and vocalization across the neuraxis, we would conclude that the human *experience* of social needs and values arises most importantly from the telencephalic levels, the linked networks of the cerebral hemispheres. Certainly we depend in each waking instant on vertical integration. The brain stem circuits must continue to play essential roles, including maintaining those qualities of arousal (such as anxiety and elation) that are critical to emotional experience. However, through encephalization the brain stem controls have coevolved to provide what are now largely support roles, allowing the massive networks of the endbrain to organize the multileveled conceptual representations. In humans, these representations of mind appear to be necessary for integrated subjective experience.

In the hemispheric construction of subjective experience, we see the functions of both limbic and neocortical networks that are consistent with core-shell differentiation: spontaneous emotional control at the core and more deliberate specification of actions at the shell. We saw earlier that human emotional communications may be organized with different patterns at both the core and shell levels. Even in their roles in representing the complex concepts of interpersonal relations, the somatic and visceral constraints are inherently concrete. Mirroring in the sensorimotor networks is primitive and reflexive. If someone has a particle of food stuck in their teeth, you almost cannot prevent yourself from reaching up to your own teeth, as if you were looking in the mirror, even though you know this is an obviously inappropriate thing to do. In fact, people with frontal lobe damage—or with an obsessive-compulsive personality disorder—may not be able to inhibit this action as they exhibit an obligatory mirroring response.

In a similar manner, the visceral, emotional responses to observing another's emotion are not necessarily productive of real empathy. These are primarily responses of emotional contagion, primitive feelings that mirror perception with little integration of more complex facets of experience. Cows show emotional contagion, but we would not consider them capable of deep empathy.

Emotional contagion certainly has an adaptive purpose in that it excites and coordinates the emotions of social animals (for example, a herd of cows as they panic in response to a predator, or a department meeting of university faculty as they panic in response to a message from the dean). However, even as they are mirrored in both sensorimotor and visceral levels, social as well as egocentric representations form what is fundamentally a concrete basis for the mind. If we are to understand the neural basis of more complex intellectual functions, we require a theory that explains the formation of abstract representations within the network space bounded by the bodily structures of mind.

So it is with the experience of affection in social relationships, whether mediated by encephalized pain or more complex mixes of elementary emotions. Real empathy must be more than just mirroring another's actions and responding with emotional contagion. It must be organized through a similar developmental process as other forms of abstract cognition. The emotional roots of this process must engage some degree of emotional contagion within the syncretic, visceral networks of the limbic core. They also must engage some kind of affection, so that there is concern for the other and not just irritation at the emotional cost. Together, these patterns at the limbic core energize an extended process of consolidation that engages the more specified representations of the other's actions, which are mediated by the somatic sensorimotor networks. Through the process of hierarchic integration, weaving the motivation at the core with the veridical observations at the shell, empathy may be organized in an abstract form, a conceptual representation of the other with both significance and accuracy.

If Jared falls in love with Kim, he will have an opportunity to progress from his initial infatuation toward a more fully organized compassionate bond. As Freud recognized years ago, adolescent infatuation is narcissistic. Jared initially projects his own egocentric fantasies upon Kim, imagining her as a kind of reflection of his ego halo. Mature love is allocentric, integrating an actual knowledge of the other person and incorporating that person's perspective on life events. In every young romance there comes a time when the people wake up to find they are entangled with a strange person, someone who has real needs and a unique personality that they have not yet understood. Infatuation quickly fades, and the opportunity for mature love arises.

The psychological structures of the family bond, which capture knowledge of the self and the other, are essential to the developmental construction of love. They are conceptual structures, and we can speculate about how they could be organized out of the dialectical interplay between the syncretic emotional resonance of the limbic core and the specification of differentiated conceptual form at the neocortical shell. As with any consolidation of hierarchic meaning, the abstract experience of love is an organized structure that must span the diffuse responsiveness of visceral self-regulation and the specified demands of reality.

Even when they respond through emotional contagion, the visceral needs are egocentric, providing an evaluative base that grounds the organizational process in the subjective vantage. Even when they capture the intimate experience of the other person, the sensorimotor networks of the shell are objective and match the constraints of the outside context even if that context is the other's body and mind. The shell and core thus provide consistent boundaries of the mind even as the mind spans the interpersonal relation. Just as all abstraction mixes personal, visceral need with the data of reality to create meaningful information, the organization of love creates abstract qualities of experience that are fundamentally subjective. Yet in the constraints of the relationship, the love attachment progressively incorporates objective qualities of the other person within the conceptual structure of the experience.

The Objective Self

Mature love is thus a conceptual achievement, a sophisticated and abstract form of experience, but we have seen that it is also powerfully motivated by primitive roots in the most elementary of motive processes, those monitoring the bodily integrity of the self. Pain is a companion of love, not just in deprivation but also in the ongoing subjective quality of the attachment experience.

Through the process of loving, the self thus incorporates valued objective information, the realities of the loved object. We have considered this process in the context of adolescent romantic relationships, where it is intense, supported by new abstract intellectual capacity, and formative for the adult mind. However, the basic process is intrinsic to the organization of the self throughout infancy and childhood as well. The child learns by attending to the parent's attention. Words gain meaning because the child understands the mother's intention. The objective constraint is integral to the interpersonal matrix and offers a perspective on reality that is antithetical to the fundamental default of narcissistic subjectivity. When the child learns to effectively negotiate perspectives with the parent, then the self becomes increasingly abstract, organized as a dialectic construction around the opposing but complementary perspectives of narcissistic need and social reality.

Latent Structures of Awareness

Neuropsychological research is necessary because the principles of psychological theory cannot be comprehended through naked introspection. Otherwise, smart philosophers would have deduced an accurate structure of mind centuries ago. Rather, the necessary principles must be grasped through careful and rational interpretation of the scientific evidence. In the exciting state of science in psychology and neuroscience today, we have both classical insights and the burgeoning evidence of modern technologies to draw upon. To the extent that we have entertained a meaningful neuropsychological hypothesis in these pages—assembling selected evidence into a reasonable conceptual model—it is a hypothesis not of the explicit but of the implicit structure of mind. These are the patterns that we cannot quite grasp through introspection. They are implicit structures at the base of consciousness.

Yet as we come to understand the neurocybernetic principles we can evaluate the scientific evidence and infer that certain patterns must be operative. These are structures of mind. They may be latent within unconscious consolidation, but they operate to organize the information of experience.

It should be a priority of scientific research to bring a more explicit understanding to the internal, subjective mechanisms of mind. The recent advances in understanding the nature of intersubjectivity show important progress in this direction. The capacity for intersubjective reasoning, which draws inferences about the perspectives of others, has been called having a "theory of mind." Children develop a theory of mind as they learn to reason about the psychological perspectives of other people and gain abstract conceptual skills in interpersonal relations. As we have seen, these interpersonal conceptual skills provide exercise with perspectives that contrast with the default subjective view of the mind, allowing greater objectivity in thought and experience. It should not be too much to hope that further advances in science may bring a similar objective perspective to understanding our own minds, thereby offering us ways of participating more deliberately in the process of experience.

Expanding the Context of Consciousness

It is not obvious that consciousness can be expanded. Perhaps we must be content with scientific knowledge of the mind as an inferential process. At first, this seems to suggest a hard curtain dropping at the subjective-objective boundary. We can infer, for example, that we evaluate success through a change in mood level as we become more elated (just like our hopeful rodent cousins). But there is little subjective participation in this mechanism. Subjectively, when we are successful, the world just looks better, and we feel more powerful.

The knowledge of emotional constraints is then something to be held objectively by a scientist or therapist, not by the subjective owner of the mind. The mind's owner has only the distorted subjective vantage, the naïve view, which is continually shaped by the visceral evaluative frame.

However, this is perhaps too narrow a view of subjectivity. In fact, applying our very hypothesis of visceral and somatic frames of mind, the traditional philosopher's view of subjectivity may stretch the fabric of experience too tightly on the conscious aspect—the articulated somatic frame. Even though consciousness is explicit and must often be articulated toward the sensorimotor boundary, perhaps some forms of consciousness may shade into more abstract (conceptual and evaluative) qualities as well. Certainly the visceral manifestations of feelings and moods often take form in the gray areas of consciousness. We then learn to evaluate events in relation to subtle—and only partially conscious—changes in subjective feeling states.

Even if the inferences we develop must be applied to domains of experience that are unconscious, the result of this application could perhaps still produce a richer and more powerful quality of experience. A useful scientific theory of mental process could give us a knowledge of what to expect from the workings of the unconscious. If we cannot plumb the depths, then at least we might be more sensitive to those processes at the fringes. Perhaps we could even make a distinction between articulate consciousness and this more implicit knowledge of experience and describe them together as "awareness."

Of course, sometimes we are surprised by what emerges from the consolidation process. Like Jared in understanding his reactions to Andrew's outburst, we may need to comprehend ourselves by observing our own actions and drawing the appropriate inferences at a more or less conscious level. Still, it is an intriguing possibility that greater knowledge of the unconscious mechanisms of the mind could allow us to participate in those mechanisms more deliberately. This knowledge might begin in a general, objective form, such as understanding how memories are organized between the boundaries of visceral and somatic frames. Within a person's process of experience, the knowledge could then become specific, reflecting the structures of the self that have formed to resonate to certain patterns of meaning or to extract certain event configurations from the flow of complex life information.

It sometimes seems easier to formulate general theoretical principles than to explain how they would help any real person understand any real life event. With the stories of Jared, Kim, and Andrew, I have illustrated how neural mechanisms might operate in real minds. It is, of course, one thing to tell stories and another to apply a knowledge of neural mechanisms to actually help young people take charge of their psychological capacities.

The remarkable thing is that no one has ever tried this, at least not scientifically. Because science has ignored subjective experience, no one has tried to teach people how their minds work and certainly not from the subjective view. (Perhaps the exception was psychoanalysis, but the scientific process seemed to be abandoned too early, and the ideas became enshrined in the cult of Freudianism rather than revised through new evidence and critical reasoning.) It *might* work to apply a modern, scientific analysis to clarify the workings of subjective experience. It would be interesting to try.

Lessons of Encephalization

Even as the products of a useful neuropsychology must be tested in personal experience, our scientific analysis shows that the latent structures of awareness, including the visceral and somatic forms, are fundamentally those that evolved with the vertebrate neuraxis. Of course, any theory of the biological roots of the mind could be seen as reducing complex "mental" properties to "mere physical" quantities. There is no need for such concrete thinking in contemplating either the nature of the mind generally or the qualities of one's own experience specifically. Rather, the acceptance of the mind's physical embodiment must include some appreciation of how the basic vertebrate architecture, extended through the unique, self-organizing human telencephalon, can be recruited to support all of the complex and subtle properties of human psychological function that appear before us. Otherwise, the theory is incomplete.

We can begin by understanding first the continuity of human nature with our vertebrate ancestry and then the new functions that emerge from the old forms. It does not seem so difficult to understand encephalization—not so difficult, that is, once you become a disciplined student of the brain. Simply examine how basic functions, such as sensation, are elaborated by more recently evolved levels of the brain; then learn how the more primitive levels are subordinated to the more recent ones.

For example, for a frog, the midbrain visual area is a fully functional visual system. It does quite well in informing the frog of important properties of the world, such as bugs flying by. In fact, the frog's midbrain has specialized neurons that seem to act as bug detectors, coding the perceptual object of the bug in order to direct adaptive actions, such as thrusting a ballistic tongue toward the bug's exact spatial coordinates. This is, of course, the guidance problem that remains intractable for today's U.S. missile defense efforts. In contrast, with the most elemental neural circuits, the frog cranks out direct hits.

For us, there is a midbrain visual system as well, but it is fully subordinated to the higher levels of the visual system. This is probably fortunate, at least in the case of bugs flying by. Our midbrain system is in fact subordinated to the diencephalic (thalamic) visual system, which in turn is subordinated to the cortical visual system. The encephalization of vision is illustrated clearly when damage to the cortical visual system in humans occurs, making the person effectively blind. The fully featured sensory function of vision in humans has become encephalized so that it is now primarily a cortical function and cannot be carried out by subcortical systems alone.

Yet these subcortical visual systems continue to be important in carrying out support functions for vision, such as orienting attention and eye movements reflexively and efficiently when an object (such as a bug) traverses the visual scene. This is not to say that there is some residual amphibian ancestral memory in our midbrains that would make that bug even remotely attractive, but, in humans, the qualities of orienting vision and attention are still supported by the midbrain. In fact, if you become cortically blind and are forced to guess where an object has just appeared in your visual field, you will guess fairly accurately, using your subcortical visual systems, even though your conscious experience (as indicated by your verbal report) is that you can see nothing.

Here again, as with the release of pathological crying in pseudobulbar palsy, consciousness seems restricted to the corticolimbic networks (where consolidation does its work) even as the ongoing function of the mind is fully dependent on the effective operation of the brain stem substrate of the neural hierarchy. It almost seems as if those old frog routines are still in there, organizing primitive patterns within the latent structure of attention.

How do we understand these multiple structures that are coordinated below the surface of experience? We need a principle of coencephalization. The vertical integration of the brain has required higher levels to be supported in their encephalized capacities by effective controls from the now subordinated levels. Our capacity to examine a complex and dynamic visual scene, such as when driving on the freeway, is largely supported by the highly developed primate visual cortex and the association cortices within the visual corticolimbic architecture. But the midbrain visual system still has an integral role to play in helping to maintain elementary spatial attention and eye movement control. We know this because a midbrain neurological disorder seriously impairs visual capacities. Although evolution has favored terminal additions, making the largest changes in the more recently evolved (and thus encephalized) systems, the entire neuraxis is the effective unit of natural selection. Moreover complex telencephalic systems, such as the primate visual system, can achieve their specialized capacities only when they are supported by what is often a coordinated dedifferentiation of function within the lower brain centers.

Hierarchic encephalization and its complement—the dedifferentiation of subordinate structures—are general principles of neurophysiological evolution, yet they may suggest lessons for psychological processes as well.

First, understanding the principles of neurocybernetics means we do not have to see the presence of the mind's evolved structures as implying a sort of reductionism. The subcortical levels of the human brain have evolved to carry out finely tuned support functions, such as maintaining arousal, regulating emotions, and directing spatial attention. Although these functions are fully opaque to introspection and fully inaccessible to volitional control, they are as essential to the mind's complex operations as any cortical process.

Second, in order to achieve the effective participation of the entire neuraxis in intellectual functions, the neural mechanisms that *are* under volitional control may need to *recruit* the contributions of the more primitive brain areas that carry out essential support functions.

Thus, as we have seen, it is not the case that intellectual function is without emotion. Indeed, emotional and motive mechanisms may be essential to the most subtle and abstract concepts we can achieve. In this process, the emotional influences from corebrain and subcortical mechanisms must not run wild but must be tightly subordinated to the intellectual function. Without this full integration of the neuraxis in support of the higher functions, experience and thought degrade to a primitive psychological level. Intelligence may be fundamentally subjective, but through abstract organization it gains objectivity and freedom from egoistic constraints.

Although I have formulated this idea in neuroanatomical terms, it may be comprehended (and in fact it may be familiar to many people) in psychological terms. Effective intelligence requires that our resources be dedicated to the intellectual process. When motives degrade to baser forms, the capacity for abstract experience degrades as well. When we look on the structure of the unconscious mind, the evolutionary organization is not an indication of primitive forms per se but of primitive neuropsychological processes that may be recruited to play key roles in the subtlest and most precise operations of intelligence.

In each of the stories that illustrate real-world examples, we have seen how the mind's operations require contributions from all of the levels of the evolved neuraxis. When Kim organized her strategy to get Jared's car keys, she did so (implicitly, of course) by calling on a linked set of self-regulatory mechanisms. She visualized the goal through a combination of conceptual skills that engaged not only visual but motor and somasthetic imagery as well. This imagery was recruited by the more abstract goal representations (get keys, have fun) that were held on-line in her frontal lobe. Her midline frontal structures progressively elaborated the incipient urges and motives of the limbic networks (move ahead, take charge) within the overall frame of these goals. At the same time, the limbic representations of the more elemental motives were able to resonate with the requirements for global arousal (pay attention, stay alert) supported by the brain stem projection systems. And all the while, Kim's limbic networks were able to maintain a motive balance with multiple motivational and emotional processes currently active in the extended limbic circuitry (sexual arousal, fear of embarrassment). These processes took shape across various levels of the neuraxis—telencephalic, diencephalic, mesencephalic—composing the products of eons of neural evolution. Yet these were undeniably the mechanisms of Kim's mind. *Their* products—the results of these processes—were the psychophysiological manifestations impinging on Kim's conscious experience.

The hypothesis of this book is that the interaction between integrating and differentiating mechanisms within the visceral and somatic neural structures yields the conceptual organization of abstract intelligence. In the fictional example of Kim's extraction of Jared's keys, her intelligence was clearly abstract. She was able to recruit the necessary neuropsychological processes in service of her abstract concept of a complex and challenging goal. The abstractness of her concept can best be understood perhaps by contrast with the concretizing, degrading influences that can arise from either the visceral or the somatic boundaries of her neural apparatus.

Kim was briefly angry about her predicament. The visceral processes associated with her anger were energizing and motivated her to take control of the situation. In the form it took in this story, this brief anger was a productive control process that supported Kim's capacity to think clearly and to organize more complex and abstract motives and concepts.

In contrast, when many people get angry, they stay angry, and the effect is concretizing, causing immediate and dense stupidity. Kim could have gotten resentful with Jared and, feeling righteous, vocalized her anger reflexively. The effect of such a concrete manifestation of a visceral impulse would have been to make a scene and very likely to cause Jared to drive away drunk and angry.

Similarly, Kim's abstract reasoning allowed her to assemble a complex action plan (getting the keys) that was only minimally concretized by the somatic constraints of the shell (how she would manage to get the keys out of Jared's pocket). If her consciousness had been focused by a narrow, concrete concept of the required physical movements, she would have walked over to Jared, perhaps stated her intentions, and tried to reach in his pocket. Seeing her intention, he probably would have easily blocked her actions reflexively. Instead, her action plan was organized at a higher, more abstract level (in reference to multiple motives) in a way Freud would describe as *overdetermined.*

One motive was to move up against Jared quickly enough that he would not turn around before she pressed against him. Another compelling motive/emotional complex was the anticipation and then the realization of the sexual response—both hers and her empathic sense of his—caused by this contact. These control processes were then subordinated to the goal of getting both hands into both pockets and finding and extracting the keys, all the while anticipating the flow of her actions into her turn toward the door and invitation for him to follow.

Just as the visceral mechanisms of her self-regulation in these minutes could not be allowed to degrade her mental process to a reflexive expression of her feelings, the somatic practicalities of her desired skeletal/muscular movements could not be allowed to concretize her action plan, which needed to be kept as a tentative, dynamic, loosely regulated structure. The effective embodiment that Kim recruited allowed her to instantiate her complex motive determinants within a coherent and abstract action plan.

Specification of Awareness

Productive thought and action have roots in the vertically integrated neuraxis, but they emerge primarily at the somatic shell. Only when the minds contents are articulated to a moderate degree of specificity through actions (including covert ones, such as internal verbalizations) or perceptions (including images) can we forge them with enough clarity to support introspection. It may be that greater discipline in articulating a more differentiated consciousness will occur as we more fully understand the mechanisms of the mind articulated at the somatic frame. Perhaps in some future training program we may instruct our children in the differentiating processes of abstraction that are gained through discipline in the skills of somatic articulation.

Certainly every intelligent person is familiar with the specifying disciplines of motor skills, of linguistic expression, or of an active imagination, but these somatic realizations are often considered to be ends in themselves rather than mechanisms of forging the elements of awareness. Then, too, because both visceral significance and somatic specification are required for abstraction, we may find that abstract knowledge cannot be derived solely from conscious focus. Consciousness may illuminate only one side of the dialectic, the articulation of a specific somatic form. If so, then productive abstractions may require the capacity to incubate more implicit conceptual foundations at the visceral base. Because, in our conscious reflections, we seem simply to "receive" these insights rather than construct them through effortful reasoning, the conceptual elements emerging from the visceral base of the dialectic often take the guise of an external gift, as from a dream or a vision.

Yet some thinkers learn to actively participate in the preconscious developmental process. As mentioned earlier, Einstein worked out ideas through bodily sensations (visual and motorific images as he called them) before he could articulate an explicit theoretical (and eventually mathematical) formulation. For another example, the structuralist philosopher and pioneering anthropologist Claude Lévi-Strauss described himself as thinking in impulses and feelings, in tensions and pulls, rather than in words.

Through a scientific analysis of the neural basis of experience, we are now gaining insight into the formative processes at the visceral as well as somatic bounds of abstraction. Consciousness requires some persistence in memory that requires not only motive resonance—to energize the consolidation process—but also an adequate somatic articulation of form. We have a degree of awareness of visceral processes, to be sure. Feelings and urges are powerful forces in experience. Yet even in those rare people with fairly developed emotional self-knowledge, we have seen that the operations of the visceral core are often implicit and invisible, achieving their effects on experience before there is consciousness that experience has been changed.

These visceral roots of experience are the engines of the mind. They excite qualities of mind known to every observant thinker: the excitement of curiosity, the confidence that comes from reasoning successfully applied, the frustration from effort that cannot achieve a goal. Scientific research is providing important clarification of neuropsychological processes, such as the mechanisms of focused attention and the control of working memory. It may be that understanding the motivational basis of these processes will lead to more effective subjective control of them. At the implicit, visceral level, control may be gained not through direct volitional action but by recruiting and nurturing the feelings and intuitions that generate new conceptual forms.

Given the properties of these bodily foundations of mind, the opportunities for apprehending experience may require recognizing states of mind that are quasi-conscious. There are clearly domains of mind that are unconscious. Yet, as Freud proposed, we can work to bring the unconscious into the light of awareness. The result may be an expanded consciousness. As science offers us clearer inferences into the mind's mechanisms, we may learn to reach into the unconscious and to take hold of processes that then become, in some new way, volitional.

Training and discipline in organized structures of mind, such as language and mathematics, lead us to powerful structures of intelligence. So, too, could explicit training in subjective neuropsychology—in the control of mental operations at their base—lead to a new, more deliberate quality of intellectual capacity and perhaps an expanded scope of conscious experience.

Developmental Knowledge

Subjectivity may become more abstract, then, by dual disciplines. On the one hand, we must acquire the discipline to more clearly articulate specific informatic forms close to the somatic surface. On the other hand, we must cultivate the patience to recruit more holistic and intuitive feelings that provide the exciting, generative base for mental operations. But how do we negotiate such opposing disciplines? The implication of the theory of bodily abstraction is that, taken alone, neither of these frames of mind are adequate as modes of experience. Rather, they must be set in dialectical opposition and then maintained through some degree of reentrant interaction in order to achieve conceptual abstraction. Well, this may sound good in theory, but can you try it at home?

One helpful realization may be that experience is a developmental process. The mind achieves its remarkable structures by growing them. To participate in this process of the recursive consolidation of experience, we need to learn how to grow.

The 19th century witnessed two major theoretical insights into the nature of life on this planet: embryology and evolution. These are both developmental processes of growth through transformation. Some of the most influential theoretical psychologists in the early 20th century, including Sigmund Freud, Jean Piaget, and Heinz Werner, recognized that the mind is best approached through a similar developmental reasoning. Werner in particular adopted the view that development for all levels of biology—whether a species, an individual, or a particular thought—follows a general progression. This begins with a holistic, diffuse form and then progressively becomes differentiated and hierarchically integrated.

Much like Freud and Piaget, Werner proposed that the developmental reasoning from evolution and embryology should be applied to understanding psychology. He coined the term "microgenesis" to describe the growth process of an idea. The idea begins in a diffuse, holistic form and gradually becomes differentiated and more fully organized until it achieves the complex form of a mature concept. Throughout the 20th century, Werner's reasoning was largely ignored.

An unusual exception was the neuropsychological theory of Jason Brown. As both a neurologist and a philosopher, Brown studied the brain and drew a conclusion that was widely dismissed as audacious. He proposed that the microgenesis of ideas recapitulates the phylogenetic hierarchy of the brain's evolved architecture in important ways. The early impetus for a thought may form in the primitive arousal mechanisms of the brain stem. The motive that sustains its development (over the following seconds or days) may be recruited in the hypothalamus and limbic circuits. The global pattern of personal meaning may be represented within limbic networks. Only in a final stage, after elaboration in the intermediate, association networks of the cortex, is the specific differentiation of form instantiated in the sensory or motor networks that will articulate the thought into actions, words, or images.

You can see that in the preceding pages, in my speculations on the neural structures of experience, I have adopted Brown's daring, literal interpretation of cognitive function emerging from the multiple levels of the evolved neuraxis. Moreover, I believe that a developmental interpretation—explaining how ideas grow—may be necessary for a practical discipline of the mind. Experience may be organized volitionally by taking charge of the growth not only of the conscious contents—that which is above ground, as it were—but also of the roots that anchor these contents in the deep unconscious. We must grasp these roots at the formative stages of experience. Otherwise, we may be limited to participating in the preconscious life of the mind only after the fact, by inferring implicit motives and feelings.

Thus, when Jared first sat on the curb and reflected on his interaction at lunch with Andrew, something did not seem right. It was an intuitive sense of incompleteness, a kind of anxiety brought on by uncertainty, that caused Jared to search for a more complete understanding. This consolidation operation then went underground in Jared's mind while he was preoccupied with other tasks, but the uncertainty was a motive engine in his limbic networks, energizing consolidation. Later he realized that Andrew had intended the transmission as a gift.

When Kim tried to understand her history lesson, she drew a blank. She reasoned through the issues but could not understand how people could have been so insular in different parts of the country. Then, through a fantasy that seemed completely unrelated, she saw the incongruity between her assumptions about life both now and then. This realization came out of unconscious rumination, but it announced itself through certain intuitive feelings—of incongruence and possible significance. Following those self-regulatory feelings, Kim guided her rumination into form and worked it into a conscious realization.

As he sat in his car, Andrew wanted to reflect on what Jared had said at the party. It was gratifying that someone thought he was smart and cared about the danger he faces. But there was more to his reflection than just feeling good that someone had paid attention to him. It was time to rethink some of his rationalizations about his life and to take stock in his current lot. He sensed the significance of a new understanding of his life situation, even though he was unsure exactly what it was.

Most of us are familiar with these kinds of things: the intuition that a certain direction of thought is significant; a dream that reminds us of a past event that is somehow relevant to current experience; the nagging anxiety that a decision will lead—in a way not yet understood—to unseen dangers. These are hidden roots. Yet from them, we grow ideas. Just as the skilled gardener understands the discipline required to cultivate the growth of plants from embryonic to mature and productive forms, so the student of the neural mechanisms of experience may come to appreciate the discipline required to bring ideas from their nascent, barely intuited forms into the light of conscious realization.

A Rhythmic Epistemology

Within the theory of the body dialectic, the growth of mental structures is not linear. Rather, it is recursive and reentrant, weaving the tapestry of experience from the opponent processes of visceral generality and somatic specification. The mind is framed by its visceral core and somatic shell. Although experience must to some extent span these boundaries, this cannot be a static span. Growth is dynamic. As we saw in Chapter 5, if we are to create abstract concepts that are both differentiated and integrated, this growth must shift between the bodily frames of mind, oscillating in a tentative balance between the recruitment of visceral significance and the articulation of somatic evidence. At the core, the new idea is diffuse and incipient. At the shell it is focused and crystallized. The organization of experience is then a negotiation of these forms, which develop increasing complexity through each translation from inspiration to expiration.

Epistemology is the study of the basis of knowledge. Jean Piaget began his observations of the growth of intelligence in infants and children as a way to understand the developmental basis of knowledge, which he described as a "genetic epistemology." "Genetic" in this sense means a developmental process rather than being controlled by genes. Although knowledge is sometimes held up as an ideal, it may have its basis in the concrete biological mechanisms of growth that Piaget suspected, even when the end state is an abstract form of knowledge.

If abstractions are indeed formed through the corticolimbic mechanisms of consolidation as a well-organized structural outcome of the mechanisms that create cognition generally, then an effective epistemology may need to be oscillatory, as well as developmental. We do not grow a complex idea in a linear trajectory. The discipline of mental growth may require a shifting of perspectives, carving the patterns in the structures of abstraction through a rhythm shifting between the core and shell boundaries of the consolidation process. In each swing of the beat, we exercise a new pole of dialectical constraint. At the core, the nascent concept engages a certain resonance of personal significance, which is then energized and then differentiated with specific reference to instantiation in the world as the traverse extends to the shell. Retaining its roots in the holism of the core, an opportunity arises to form a hierarchic construction (an abstract form) that links both internal (significant uncertainty) and external (environmental data) constraints to create information.

If we imagine that it is possible to gain experiential access to multiple levels of the consolidation structure, then full participation in each phase of dialectical growth—deep core or surface shell—may require corresponding shifts in the quality of awareness. A conscious focusing of attention draws us to the articulated surface, where the content of mind is composed of definite, palpable, and thus memorable specifics. The grasp of deep implications may require a more inarticulate, intuitive state, perhaps even a dream state. It is a loose grasp and at the same time general.

The dialectics of abstraction are thus charged by the inherent opposition of the fundamental forms of mental structure. The dialectic process is exactly the same as the process of memory consolidation, by which organized patterns of mind are linked across the corticolimbic architecture. This is a spontaneous process that allows even the youngest child to hold and build qualities of mind.

Yet we can hope that scientific analysis will soon offer ways of making it deliberate. Rather than operating from the default state—the simple, naïve assumption of mind—science may point toward a more conscious articulation of the cognitive process.

On Monday after the party, Kim calls Jared before class to see whether he wants to have lunch at the diner. They walk over from school at lunchtime, not saying much, and find a booth.

"Thanks for driving me home," Jared says.

"You can't drink so much when you're driving."

Jared nods, looking Kim in the eyes. She doesn't pull punches, but still, she invited him to lunch. She seems to like him.

Kim is looking out the window and waves at someone to come in. Jared follows her gaze. "It's Andrew," Kim says.

"Do you know him?" asks Jared.

"You introduced us at the party," she reminds him.

Andrew comes in and walks to their booth. "So you made it home okay?" he says to both of them with mock concern.

"I drove Jared's Honda," Kim says. "It goes pretty well, but it kept popping out of 5th gear."

Andrew and Jared look at each other and laugh. Kim does not see why this is funny, but she smiles anyway. She knows there must be some story here. She motions for Andrew to sit down next to her and then tells him, "When I get a job this summer, I want to get a car. What do you think I should get?"

Jared is impressed that Kim has not only reached out to him today but has included Andrew in their interaction. She knows his friendship with Andrew is important.

Kim likes Jared and tries to let him know in the way she looks at him, even as she talks with Andrew.

Andrew is glad to sit with Jared and Kim today because they are both important to the new view of himself that he is trying out.

The three friends share an important few minutes together, each working to be attentive to the others. It is a time of change—the last weeks of high school—and a large uncertain future is opening up after that. Each of them senses that these interactions are defining moments for their lives. Through the more or less conscious processes of personal and social experience, they are each working out their identities through the actions they chose and the attention they share with their friends.

8
Information in the Age of Mind

What evolved as control systems for elementary bodily processes must now be recruited to support abstract qualities of intelligence. In the remarkable developmental logic of evolution, each brain is formed by retracing the phyletic path. This fact—of morphogenesis (growth of the bodily form) through phyletic recapitulation (building on the outline of species evolution)—shows that the scientific explanation of life must be developmental. We need a developmental analysis to explain how we got here, whether as a species or as individuals.

A developmental analysis may also point to where we go from here. Only historians have the perspective to understand the future. We may need to become biological historians—evolutionary theorists—if we are to understand the continuing evolution of mind in the Information Age. We can even consider the possibility that the Information Age will itself evolve into an Age of Mind.

The developmental momentum of embryogenesis (morphogenesis before birth) is the essential context for understanding the child's development of intelligence. The embryo's own spontaneous, spasmodic actions become the motive vehicle for the early stages of neural self-organization. The brain's infantile architecture is woven from an overly dense pattern of neural connectivity, and the functional networks are sculpted through Darwinian subtractive elimination—the death of unused connections. In the womb, where sensation and action are functional only as self-organizing vectors for neural morphogenesis, the vehicle for this process—the "use" that avoids connective retraction and synaptic death—is the embryo's spontaneous actions. As with the structural differentiation of the nervous system (the rhombencephalic, mesencephalic, diencephalic, and telencephalic tissues), the embryo's actions serve to organize the brain functionally, not just in a diffuse sense but also in relation to the literal pattern of the vertical integration of the evolved neuraxis.

In this process, what may seem like vestigial foundations of the neuraxis, such as the diencephalic roots of the telencephalon, are never literal reenactments of earlier evolutionary forms. Even many scientists get confused on this point. These are just the scaffolding of morphogenesis, nature's way to make a vertebrate brain in the only way available—by building on the cumulative variations on the phyletic DNA theme. The genes code the blueprint progressively, like methodical architects, building the primordial foundations first and then the more recently mutated extensions. None of the archaic structures we find in the developing brain is literally an archaic brain but rather an evolved developmental mechanism—an old form put to new use in the context of hierarchic encephalization. The nascent mind is then prepared to emerge—through the extension of embryological operations—into the postnatal experiment of this child's life.

For simpler creatures, birth signals a transition to a mature stage at which embryogenesis is over, the neural algorithms are mostly formed, and the brain's circuits operate more or less as wired. The flexibility in mammalian behavioral evolution was brought on by genetic mutations of the *rate* of development, as well as the development of specific structural features. Certain switches that mark the transition to the stable neural patterns of the adult seem to have been delayed.

The amazing result is that the mammalian juvenile period continues the self-organizing process begun during embryogenesis, but now the sensory and motor systems—and their motive substrates—have working contact with the world as they mature into their functional forms. From this contact emerged *learning*, an articulation of the brain's architecture to match the information demands of the world. For humans, certain mutations of the hominid line seem to have produced an extreme protraction of this juvenile state and created an extended neoteny of neurophysiological self-definition that extends over decades. In a certain irony, it is literally a kind of mental retardation, a protraction of the morphogenetic process, through which human intelligence emerges. This is a radical neoteny, an almost unnaturally extended maintenance of the brain's fetal form. It creates an opportunity for cognitive growth throughout life.

Welcome to the Evolved Mind

Regardless of whether the theory in these pages captures the essential mechanisms of experience, we can be confident that, in some eventual scientific analysis, the mind's structural vessel must be what we find in the anatomy. This is the vertically integrated hierarchy of the evolved levels of the mammalian neuraxis.

The requirement for some form of evolutionary analysis seems scientifically undeniable, but are we ready for it philosophically? Is it this acceptable? To study the evidence on the generative forms of mind (lurking at the fringes of awareness) and find (at this most intimate experiential distance) the residuals of animal brains? Not just one animal, such as a suitably dignified great ape, but a pattern that is nested within the implicit unconscious mind, embedding—in explicit anatomical detail—the full (rhombencephalic, mesencephalic, diencephalic, telencephalic) phyletic order?

Although Charles Darwin gave voice to evolutionary theory in the 19th century, naturalists and philosophers had approached the idea of evolution for many years. What made Darwin's writing so important to scientists—and then the public at large—was that he gave evolutionary theory a mechanism, an explanation for how it might happen. This was natural selection, the variation of traits among individuals and then survival of adapted forms. It was an account of a mechanism for evolutionary change and progression that allowed the articulation of a theory of evolution whose validity could be evaluated in detail and then bid for a place in the body of scientific knowledge.

For the public, of course, it was the simple assertion of our animal origins that was intolerable. Darwin was well aware of the societal distress and revulsion that would meet his ideas. In his first presentation of his theory, *Origin of Species*, he carefully avoided any reference to human evolution. However, this was obviously the interesting question. In *The Descent of Man*, he took it on directly. Although the specifics of genetic mutation and variation would not be understood for many years, scientists quickly understood that this story of evolution explains the natural order. It is our history. However, for many people, this was a history that was at odds with their concept of themselves and their family tree.

The outrage that Darwin's ideas triggered appears to have been painful for his sensitive temperament. For the most part he left it to others to defend the theory against the many attacks it received. In his later book, *The Expression of the Emotions in Man and Animals*, Darwin used his skills as a naturalist to catalog emotional expressions and to discuss the functions they appear to have in coordinating the lives of social animals. To many people, this seemed like a retirement project for Darwin, perhaps along the lines of his earlier work, *The Formation of Vegetable Mould Through the Actions of Worms With Reflections on Their Habits*. Apparently in the same vein, *The Expression of Emotions* seemed like an uncontroversial study, a pastime for a sensitive, reclusive scientist. It avoided the spite of a vengeful public.

But did meek old Darwin have a subversive agenda in this work? When we examine the social functions of our emotions, we see the parallels with the adaptations of other animals, and we are—at the same time—dealing with the most intimate elements of subjective experience. The nasty expression of a spiteful coworker is not exactly the snarl of a wolf whose status in the pack has been threatened, but we can understand the parallel to animal status motives in both form and function. It is then in our responses to this emotional communication—in our most intimate emotional experience—that we must recognize the powerful adaptive residuals of our evolved origins. Evolution gains subjective realization.

Whatever Darwin's intention, I believe he would appreciate the importance of evolutionary analysis in understanding the brain's mechanisms of emotion in modern research. The intellectual challenge is to explain how all of the levels of the neuraxis work together in each of the species' variants of the vertebrate neural plan, including our own. This is an evolved plan, so the necessary understanding must include not only the recent variations but also the fundamental themes that have maintained themselves over countless speciation events.

Darwin would also appreciate, I think, the theoretical challenge of explaining the fundamental mechanism. In this case, it is the neural system of psychological function—a physical mechanism. For the question of emotions, the hypothesis that we have entertained is that, by motivating the very means of memory consolidation, the emotional processes at the limbic core are responsible not only for the subjective properties of emotional experience but also for the motive and evaluative roots of cognition. Through dialectical interchange with the specification of neuropsychological forms in the sensorimotor neocortical networks, the corebrain emotional responses generate the fundamental conceptual processes of abstraction and concept formation.

In this way, the intellectual capacities that we hold as the most unique of human traits cannot be isolated from the matrix of adapted processes in vertebrate evolution. Rather, they emerge directly from this matrix. The mind has evolved, and its integral motive engines are the emotional controls that we share with animals. As I believe Darwin recognized, to understand the psychological significance of motivation and emotion may be to appreciate the very presence of evolved forms in personal experience.

Thus, as we look to take hold of the unconscious substrate of the mind's bodily structures, we find the evolutionary continuity of the human neuraxis. As science continues to clarify the functional importance of neural organization, we may appreciate the continuity of the human neuraxis both through objective, academic theory and personal insight into the structure and mechanisms of experience. Science is offering us the opportunity for an Age of Mind.

An Evolving Informatic Interface

Evolution has tweaked and retweaked the vertebrate developmental plan, capitalizing on successive waves of mutation to achieve differentiation at the top of the neuraxis. As a result, the most important insight into the deep structure of awareness may require a Darwinian analysis. In addition, the rules have recently changed. At some point in the hominid line, the volatile plasticity of radical neoteny produced a novel process for the cultural transmission of information. This was, of course, language. Language evolved to transcend evolution.

To be sure, language itself emerges from the neural structures of cognition. Although certain psychologists, linguists, and philosophers may attempt to argue that language is a mental operation that cannot be reduced to neural mechanisms, this argument is no longer compelling in light of today's scientific understanding of the neurophysiology of language. We now have a growing understanding of the neural mechanisms not only of language production and comprehension but of its motivational and semantic foundations as well.

Yet, with the evolution of linguistic capacity, humans gained the capacity to internalize cultural information structures and operations to an unprecedented extent. The relation of the brain to intelligence then changed. Rather than being just a vehicle for genetic causation, the brain became a vehicle for the cultural elaboration of the genetic potential. The shared communication rituals of the culture became internalized as mechanisms of mind.

Certainly we humans are not alone in cultural transmission. The social transmission of information, made possible by the mechanisms of extended neural self-organization in juveniles, is ubiquitous in mammalian development. The result is culture, even if it is constrained by the limited plasticity of the particular mammal. A juvenile cougar learns the world of its mother and adapts to her hunting skills, regardless of whether they target rabbits or to deer. However, the animal's cognitive repertoire remains limited. Its perception, attention, and hunting skills remain restricted to the primordial feline pattern.

With the neotenic capacity for language, humans crossed a threshold of abstraction plasticity. We can now entertain an arbitrary informatic code with few limits of variation and with an internal structure (grammar) that allows efficient hierarchic forms of sentences and stories. With this new capacity for extracorporeal structure—structure outside the individual—the developmental process changed.

Changed, too, was the embodiment of mind. The capacity for language and eventually for mathematics transformed human intelligence as they became internalized. These informatic codes of the culture are now integral structures of the mind, giving it voice. We routinize these cultural structures at the somatic shell into versatile affordances by means of the same mechanisms of automatization that we use to form other motor habits. The only difference is that now the flexible and abstract representational qualities of the symbol habits of language allow an efficient specification of differentiated concepts never before possible.

The symbolic artifacts of culture have become as essential to the physical embodiment of mind as the dynamic, plastic neural architectures in which they are embedded. The self-organizing nature of brain anatomy causes a blurring of the boundary between genetic and epigenetic mechanisms. For those of us fortunate enough to encounter effective cultural transmission (that is, good enough teachers), we internalized the artifacts themselves. Rather than translating thoughts into language and mathematics, we learned to think in words and numbers. The medium is the message, and now so are we.

As with other patterns of motor affordance, the symbolic structures of language and mathematics quickly become automatized to the point of unconsciousness. As experts, we do not think about saying words and then produce them through conscious, detailed intention. Instead, we have vague and global semantic intentions, and language automatically provides concrete (somatic) actualization. This specified actualization appears upon our lips, or, as I write now, within our fingers. For the skilled user, the symbolic embodiments of mind are soon absorbed into the opaque domain of preconscious mental process, with the result that conscious attention can implicitly assume their presence. Then, at a higher level, the mind takes up a consciousness of intentions, of signs of progress toward objectives, and of the abstract, executive control of the mental process.

Machine Informatics at the Shell

The linguistic code is thus incorporated—embodied—at the somatic shell, creating automatic structures of specification that support complex cultural artifacts of mind. Captured in writing and the mathematical communications of science, these structures enable intelligence to be cumulative. Each learner has access to the literature, not just the personal skills of individual parents and teachers. Although the specification of the data code is at the sensorimotor shell (and those representations immediately abstracted from it), this specification creates a corresponding demand for control—energized by uncertainty—within the embedding visceral representations. The result is a deep level of cognition and of significant, personal representations of meaning.

This transformation of biological evolution through language and cultural evolution may be the essential historical context within which we can craft an understanding of the next steps in the evolution of intelligence. For the children of the early third millennium, cybernetic appendages are now redefining the shell. Computer skills are becoming as important to participation in the culture as language itself. These skills are defining a new zone of the shell, the data interfaces of the mind. Internet informatics are the daily companions of our information hunger. Cell phones and instant messaging are weaving the close and intimate tapestry of our relationships. And we crave bandwidth.

Because the shell operations are automatized, whether qwerty keyboards or spoken English, they are unconscious symbolic interfaces. We are then loath to change them because that would require painful conscious effort. So I am now typing with a (qwerty) keyboard layout that was developed to alternate frequent letters between hands, thus minimizing the mechanical key jam of the first primitive typewriters. The rigidity of change is problematic only for adults. With instant messaging, for example, children are picking up the available informatic appliances and then defining a new language. We can see the evolving somatic interface appliances of the next generation.

The craving for information bandwidth will lead to rapid progress at the interface layer, and the result may be a new kind of machine language that we internalize, not unlike speech capacity. Or, because machine intelligence is progressing so rapidly, we may soon find sensor systems that interpret brain activity patterns. Energy direction systems will soon be available that will gently modulate brain activity, allowing efficient and rapid two-way brain-machine communication. Although direct brain-machine interface systems may sound like science fiction, they are now an active focus of study in many industry and university laboratories. Progress in this area will be rapid because it is strongly motivated. For today's young people, the limit on informatics bandwidth is not just a limit on practical efficiency. It is a limit on the quality of experience.

Internalizing the Informatic Interfaces

Of course, significant progress in intelligence may have nothing to do with humans or their brains but may emerge in our machines. Yet, unless we make the mistake of allowing machines autonomous motivational capacity, human motivation will drive the investment in technological change. The lessons of history have convinced me that a defining motive will be information hunger. The expansion of the informatic functions of the shell in some form will provide the information bandwidth to feed that hunger—and, of course, to stimulate it further.

To be sure, the nature of progress in technology is that it is difficult to anticipate. What we can do is pay attention to those technologies that offer the potential for rapid and profound development. Although the medium of digital electronics has defined our recent informatic progress, it does not match the intrinsic communication channels of brain networks particularly well. Major advances in biotechnologies may soon offer capacities for neurotechnology— the direct measurement and manipulation of neural populations. For example, we are discovering fundamental mechanisms of neural plasticity, the self-organization of neural networks, that allow learning and change. As we have seen, an important insight has been that these mechanisms of plasticity reflect extensions of embryonic systems that allow the nascent brain to articulate its own architecture by means of a bootstrap operation of functional self-organization. As we gain increasing knowledge of the genetic mechanisms of the cell, with the remarkable flexibility in their embryonic (stem) forms, we will find new opportunities for manipulating bodily tissues, including those of brain networks.

To reason with any credibility in order to anticipate the trajectory of information technology, we may find it useful to adopt the historian's perspective. To forecast farther into the future than the immediately obvious, we can reach farther into history and consider the structural momentum of evolution. Differentiation of increasing neural complexity in living forms has required elaboration of the fundamental—in our case, vertebrate—neural plan. Each generation could tolerate only minor mutated elaborations on the basic theme because it had to maintain the survival skills of the previous generation. Mutations were least fatal when engaging the most recently evolved level of neural elaboration. The result was progress through terminal additions: through rhombencephalic, then mesencephalic, then diencephalic, and finally the extensive variations of the telencephalic pattern that have produced the planet's extensive varieties of vertebrate intelligence.

In each elaboration, the operational function of the brain was encephalized through that terminal addition. The result was a more general, flexible form of those capacities of behavioral self-control that had evolved—in a more rigid and reflexive cybernetic pattern—in the previous level. Neuroanatomists understandably assumed that the cerebral cortex of the human brain is the endpoint of evolution, so they named it the endbrain, the telencephalon.

But perhaps this is not the end. We can imagine an exoencephalon, an informatic interface to extend cybernetic capacity to silicon or other computational substrates in the evolving embodiment of mind. At least one view of the trajectory of evolutionary history sees the evolution of intelligence achieved by a direct interface of computational devices with the human brain, modeling the brain's own neural networks precisely enough to communicate directly with them. Through the principle of terminal additions and the theoretical model of increasing evolutionary differentiation at the somatic sensory and motor interface level, we would expect this interface to be attained at the somatic shell, where it would be linked to the neocortical networks for sensory and motor interface with the world.

Among the many neurotechnologies now in development, several are obvious candidates for direct neural interfaces, including silicon grids that support electrical contacts with neurons. Of course, with many possible technologies, we can now only guess at what will prove effective, and there is no hard and fast requirement that only the sensorimotor networks of the somatic shell must support the direct neural interfaces. However, I think this is the prediction from evolutionary principles. It would be the first tendency of continued evolution to elaborate an exoencephalic interface through the same developmental progression—elaboration primarily through terminal additions—that has obtained for hundreds of millions of years of increasing cybernetic complexity within the vertebrate line.

A New Kind of Science

What properties of mind would support the conceptual differentiation that would operate such a transformational mechanism? Certainly there is strong demand at present for technology that engages us intimately at the sensorimotor interface. In movies now, sensory reality is free to take on any form with computer-generated graphics. In video games we can participate in fantasy worlds created with remarkable realism. As the informatic media become still more interactive, we will find increasing ways for motoric participation in these cybernetic fantasies.

Although it is important to realize that the new entertainment technologies are transforming the sensorimotor experience itself, these may not be the routes to the most profound intellectual transformation. From this reasoning, I maintain that the best guide to what can happen as we internalize the informatic interface is the mechanism of the first major cultural transformation of biological intelligence, human language. In its basic form, language evolved as speech when reflexive, emotional vocalizations appear to have become volitional, directed, and then structured within the cultural rituals of grammar. With the development of reading and writing, language became more deliberate and perhaps more conscious as both the writer and the reader could reflect on the communication intent in relation to the vehicle. As you read these words, your understanding must begin with knowledge of the language in which they are written. This is a somewhat arbitrary mental process, but it is so highly structured and routinized that, with adequate training and the appropriate translation, all skilled readers have access to the information. The informatic vehicle (the printed word) is fast, automatic, and more or less reliable in producing consistent meaning, at least at the first level of interpretation. With that first level accessed almost unconsciously, you can then use your mental capacity to reflect on possible deeper, more abstract levels of meaning.

In the course of cultural evolution, soon both language and mathematics developed, allowing human intelligence to become highly structured and automatized, with the most routinized operations dependent strongly on the neocortical networks at the cortex's somatic shell. It must be a similarly structured cognitive capacity that will allow a closer, more effective interface with machine intelligence.

In this context, it may be important to recognize the unique qualities of computational analysis and simulation, particularly in recent scientific advances. Stephan Wolfram, a mathematician and developer of scientific software, has argued that the capacity for computation in modern scientific research has led to a revolution in our ability to gain insights into nature. Wolfram calls this "a new kind of science." In my own research experience, I have seen important advances from computational simulations that are unique from—and in many applications superior to—what can be captured by mathematical equations. Maybe Wolfram is right. Machine intelligence allows us to capture whole domains of nature with such fidelity that we can run the simulation so as to reveal the systematic patterning of the underlying mechanisms. From this, the principles of the mechanisms emerge. In many areas of science, simulations are providing a depth of insight that is very difficult, if not impossible, to gain either by analyzing equations or conducting experiments.

Of course, this process of computational simulation is just a more concrete form of the perception and conceptualization we have always done in science and more or less in daily life. In perception, we grasp elements of the world, and through conceptual reasoning we conduct operations on those elements, simulating the world within the mind's operations. In classical science, we begin with observation and then develop a theoretical model of the world that we can conceptually play with to explore its properties. We then test this model by means of experiments. With computational simulations, we can now follow the principles of natural phenomena that are recursive, modifying themselves and progressing in ways that cannot be modeled in equations. Even experienced theoreticians who are able to model reality through their powerful skills in conceptual modeling are often unable to discern the principles that can now be gained through recursive, self-modifying simulations in high-performance computing. With evolving computation, we may indeed find a new kind of science.

With the intelligence that will emerge from the continued expansion of computational capacity, the evolution of language and the subsequent achievements of mathematics and scientific computation may thus serve as historical guides to the objective intellectual capacity that will soon become possible in the Information Age. These are also clues to the emergent capacities of mind that structure the increasing bandwidth as we internalize increasingly effective informatic appliances.

Understanding Embodiment

With the mutation of each more complex terminal addition in brain evolution, a corresponding dedifferentiation of the subordinate, more primitive brain levels occurred; as a result, these primitive levels became effective support systems for the new "higher" centers. As support systems, the more primitive brain regions provided functions such as arousal, neural modulation, and motivation that allowed more effective realization of the flexible representational capacities of higher centers such as the cerebral cortex. Perhaps this is the fate of our biological brains.

We have already learned about the differential rates of ontogenesis (individual development) in the core and shell networks of the mammalian cortex. The effect is an early maturation, specification, and fixity of the neocortical sensorimotor shell networks, in contrast to the continued plasticity and modifiability of the visceral networks at the limbic core. Of course, even with today's primitive neurotechnology, we have many clues to how neural plasticity itself can be altered. Yet the asymmetry of the plasticity between the core and shell networks may be an important principle in anticipating continued brain evolution. As the first interfaces with the exoencephalon are created, they are likely to be limited and rigid; thus the fixity at the shell would be appropriate. The flexibility of the integrative semantic networks at the visceral hemispheric core would then be required to organize new conceptual capacities to match the expanding power of high-performance computing at the somatic shell.

Thus, as we project the developmental momentum of evolutionary history to forecast our immediate future, it seems that the next few decades will bring expanded informatic appliances to interface with the brain's (increasingly dedifferentiated) sensorimotor shell. The critical theoretical question then becomes the complementary development of the visceral function. We can imagine that the powerful cybernetic capacities of the expanding exoencephalic shell will need to be controlled by the motivational and emotional capacities of the corebrain visceral networks. In fact, the progression may be one in which the existing human telencephalon—in its entirety—becomes adapted to the evaluation, control, and support functions of what are currently the visceral limbic networks.

It is difficult to imagine the minds that will emerge within these uncharted avenues of cybernetic structure. To my way of thinking, the best approach to imagining them may be by means of the classical theoretical insights gained from studying vertebrate neural evolution. Variation through mutation led to evolution through terminal additions because the root structures of the brain could not be changed—at least not much—and still ensure survival and reproduction. Therefore the new evolutionary experiments that gave rise to more complex forms of neural organization were relegated to the terminal additions. The continuation of this principle suggests that, as we gain bandwidth through creative adaptations that allow the human brain to interface more effectively with computational resources, it will be through exoencephalic adaptations achieved at the neocortical networks of the brain's somatic shell. We will then face the challenge of taking charge of the increasingly complex emotional and motivational representations within the limbic corebrain networks to serve as integrative motive structures to organize the new informatic resources into abstract representations of meaning.

Signs of an Age of Mind

It is hard enough to accept the residual levels of vertebrate neural evolution within our intimate mental equipment. It may be logical, but it is unintuitive at best to imagine what this progression means for the neurocybernetics of the future. Certainly, few of us expect to take up roles as neuromorphic computational hybrids any time soon. Are there practical implications for the present times? Will the convergence of neuroscience, learning theory, and computational theory, as outlined in these pages, provide useful insights into the psychology of actual people?

Certain signs indicate that our knowledge of the brain, at least, is progressing to the point that it will soon be obvious how to apply it to personal experience. If there is to be an Age of Mind and not just an age of intelligent machines, it will be essential that we take up this new knowledge and use it to extend the volitional base of human consciousness. In fact, in reflecting on these issues for some time, I have come to think that the challenge of developing conscious insight into neuropsychological processes is as important to the future of intelligence as the burgeoning of machine informatics.

In the last few years we have seen increasing excitement over the insights from brain research in many areas of society. Psychology, political science, business, and economics have all adopted research into brain mechanisms to address their traditional questions. A new paradigm of "neuroeconomics" is emerging. If this takes hold, it will of course be advertising that first makes the business case to invest in progress. However, the progress is remarkably widespread, and it is extending our knowledge of brain mechanisms beyond the narrower objective domains of cognitive neuroscience (perception, memory, attention) into traditionally more subjective and humanistic concerns, including the processes of emotional response and social understanding.

In fact, the paradigm of neuroeconomics is strongly dependent on research in emotion and motivation to understand how people assign value to events. Not unlike the traditional theory of animal learning, the traditional theory of economics assumed that people make rational decisions of self-interest through straightforward pleasure-pain discriminations (related, of course, to the value of goods in commercial interactions). The failure of traditional economic theory to explain people's actual choices can be compared quite closely to the failure of traditional learning theory to explain animal behavior. Neither economics nor learning theory got the fundamental mechanisms right. As a result, the basic questions of how people assign value to events is informed quite clearly by the modern animal learning evidence reviewed in Chapter 3.

Neuroeconomics research, using measures of brain activity such as the dense array electroencephalogram (EEG) or functional magnetic resonance imaging (fMRI), is generating excitement in both academic and business circles because it shows that the important predictors of people's choices are often the brain regions that mediate emotional responses. For many scientists this close correspondence between meaningful decisions and the brain mechanisms of emotion is still a mystery. Why are the motivational circuits at the core of the hemisphere active in complex evaluations of meaning? From Chapter 4, of course, we know the answer: The corebrain limbic networks that control memory consolidation provide the intrinsic motivational control. Moreover, the densely connected limbic networks, which by their very nature pull for syncretic, holistic concepts, must support an integrated understanding.

At the time of this writing, most of the research on the social, emotional, and economic implications of the brain's motivational mechanisms has come from academic centers. However, it is now taking broader hold in business circles, where it is influencing society at large. Even more important are the signs that we are gaining philosophical insight from this new realization of the motivated base of the mind's operations. In several disciplines, including neuroscience, linguistics, and even philosophy, modern thinkers are beginning to understand the mind as *embodied*, as literally rooted in bodily processes. One of my colleagues at the University of Oregon, philosopher Mark Johnson, has articulated this line of reasoning particularly well. Johnson points to the continuity of this new approach with the thought of pragmatic philosophers in the United States, as well as psychologist William James. Philosophers, psychologists, and humanists have typically adopted a naïve, introspective view of the mind in which they separate mental processes from the physical events of the world. The fact that neuropsychological mechanisms can be identified for so many fundamental mental processes provides evidence that we can now critically examine all of the properties of mind in relation to specific bodily processes.

The notion of an embodied mind promises to enable new interdisciplinary work in universities because it offers a way of bridging the chasm of the two cultures of the academy. Even though it is couched in brain mechanisms and is thus amenable to scientific validation or refutation, this research addresses basic issues of human experience, which may include the values that shape a decision or the feelings that support empathy for another's pain. Issues such as these are integral to traditional humanistic studies. In the neuroeconomics vein, the goal is to understand and be able to predict economic behavior, something that can be objectified in the traditional manner. However, in philosophy, the goal can be broader: understanding the nature of experience. Science can help us grasp the answers to the fundamental questions of subjectivity.

Thus, whether applied to business decisions, philosophical analysis, or the clinical problems of emotional disorders such as depression, research on the neural mechanisms of the embodied mind is now addressing questions of subjective human experience. In interesting ways, the integrative nature of the theoretical framework required to understand the neural mechanisms, such as we have considered in these pages, is relevant to traditional humanistic methods of study. We have seen from neuroanatomy and neurophysiology that the dense interconnectivity at the limbic core causes the contributions from the corebrain level to be both integrative and syncretic, in that motive influences are fused with a rich amalgam of meaning. From computational principles, we have seen that distributed representations are not easily compartmentalized but are highly interactive; consequently, the history of the brain shapes the new learning through a kind of arbitration. When significant information reaches the matrix of embodied experience, the result is stability or plasticity but not both.

Although these are neurocomputational principles, they have psychological and even humanistic implications. They imply that motivated experience is a function of the self and of the sum total of one's life experience. A new event is understood as it interacts with the rich distributed representations of prior experience. This is not a discrete, logical, ordered comparison but a multifaceted pattern of resonance with the existing network. At the corebrain level we evaluate significance through highly personal, idiographic concepts. This is not due to some defect of objectivity. Rather, it arises from the very nature of subjectivity. To understand the interpretation of meaning at the mind's adaptive base therefore requires a humanistic psychological analysis, a consideration of cognition not as a machinelike information-processing operation but as the manifestation of a unique self.

Because the syncretic level of experience is an integral function of the self, it is also a function of the culture in which each mind is formed. The influences on the self, mediated within the corebrain networks, are cultural, whether through the family, the community, or literature. In this way, the literary intellectuals' traditional assumptions about the mind's inherent cultural embeddedness can be related to a specific level of neuropsychological analysis. This is the level of neural networks that provide the motive engines for memory consolidation.

Perhaps surprisingly, neuropsychological principles then imply that analytic, deconstructionist reasoning—the analysis of people's experience and behavior as a product of their culture and times—is necessary to make sense of the influence of the cultural context that shapes any creative effort, whether in science or the humanities.

The Unbearable Tension of Abstract Thought

In mediating between personal values and the evidence of the world, I have speculated that it may be the tension between holistic significance (at the syncretic limbic core) and specific differentiation (at the articulated neocortical shell) that allows abstract concepts to become hierarchically integrated. Considering this process in practical terms, we should perhaps take heed of a lesson from the many years in which the intellectual work of modern society has been fractured by the chasm of the two cultures: This tension hurts. It is therefore optimistic to think either that scientists are ready to address humanistic questions or that, even if they were, the literary intellectuals would be ready to listen. Buffeted by the tension, braced only by the weakest intellectual stamina, we are naturally attracted to one pole. We are then drawn inexorably toward the comfortable concrete certainty of whichever boundary most closely matches our natural personal biases.

For some people, the greatest attraction is the specification of definite evidence. At this (somatic) boundary, the ambiguity of theory or interpretation can be avoided. In business, this attraction is making decisions by the numbers, where the exquisite certainty of the facts offers the illusion of a safe harbor from the storms of economic complexity. Thus, when the bean counters give evidence that can be captured only by bureaucratic rules, corporations of any size become immobilized and incapable of innovation. In the sciences, this attraction leads researchers to be drawn to empirical evidence. A walk through the scientific posters at the Society for Neuroscience, for example, reveals an impressive array of the brain's molecular details specified with great enthusiasm but precious little interpretation of what each fact means for how the brain works. As a result, because the concepts never develop sufficient breadth to integrate more than the narrowest observations, neuroscience has no real paradigms but only microparadigms at best. At the somatic boundary, the mind is drawn to exact identification with the exquisite detail of specified reality. Experience is then constrained to the specified facts so narrowly that it loses all depth. Yet for the mind thus captured, the intensity of apparent certainty is mesmerizing.

Shifting to the syncretic and holistic concepts at the visceral core, to the frame of mind that dominates more humanistic ways of thinking, we may find an equally strong certainty forged out of the emotional conviction of strongly held beliefs. At the visceral boundary, certainty is a subjective affirmation, a gut-level truth, where evidence (which can never really be trusted to fit your assumptions) then becomes the demon that threatens to destroy the cherished certainty of felt belief.

Of course, even though the dialectics are asymmetric in the two cultures, they are operative in both. An integrative idea (a theory that explains things) is important in scientific as well as humanistic pursuits. In fact, it may be the inertia of felt certainty in theories in the sciences that causes progress to be resisted until the catastrophic upheaval of the scientific revolution comes about.

From either pole, progress toward abstraction requires abandoning certainty and tolerating ambiguity. Because it is the mind's nature to form patterns, we discover that consolidating experience brings anxiety and pain when the patterns (the structures of the self) remain elusive. When abstraction is the goal, the patterns of the self are invariably elusive because they must—at the same time—fit the differentiated specificity of evidence and the holistic integrity of insight. In the vacuum of the arbitration, it hurts too much. So, after a brief struggle, we default to one boundary or the other. We then embrace the comforting certainty of a stable, if biased, conceptualization, taking up residence in one or another of the two cultures of the academy.

If we entertain hope for an Age of Mind, it is fairly straightforward to reason from the present theory of consolidating abstraction to anticipate the structural requirements. We can then anticipate the work required to create more complex minds in the Information Age. We must find increasing abstraction through increasing psychological complexity at both boundaries—the interface shell structures of informatic capacity and the subjective core structures that integrate evaluative significance.

Information in the Age of Mind

Building complexity at both boundaries—by increasing embeddedness within abstract concepts—may change the nature of information. At the sensorimotor interface layer, as it is buffered by more complex cybernetic implements, information will become more and more organized within meaningful contexts. With our senses and actions (direct sensorimotor experience) we encounter the data of reality directly. With ever more powerful computational resources, we come upon selected, organized data that has been transformed into information as it is prepared and organized for its significance. If I want a car now, rather than looking in newspaper ads or at cars on car lots, I search websites for reviews and categorizations of brands that meet relevant features. It is a relatively easy and yet highly organized process of research. In one hour on the internet I conduct a research project on cars to match my needs that would have taken several days just a few years ago.

Complexity must also be developed at the visceral motive layer, again through embeddedness of the visceral mechanisms within abstract representations. As the evaluative and regulatory processes of mind become more transparent to scientific explanation, those that assign significance to certain forms of information will become apparent. The corebrain evaluative mechanisms will then operate as conscious processes rather than as blind egoistic constraints on experience. I can now only speculate on what this will be like. However, to continue this example, I may gain greater insight into my personal motives for appreciating certain car features, recognizing that my personal history applies hedonic biases that shape my experience implicitly, whether these are modern imitations of the hot rods that recall a 1950s adolescence or the street-racing tuners that bring to mind a 1990s adolescence.

At both boundaries, as they become exercised by abstraction, the constraints on the mind will become more flexible and allow in turn more complex and abstract structures of mind to emerge. To understand how this might happen, it is important for us to consider the constraints of certainty that lead to the concretizing tendencies of either somatic or visceral domains. At both boundaries, the primary constraints on information are the attractions of certainty. At the interface with the world, certainty means that concepts are closely aligned with the data of the world. At the visceral core, certainty means that concepts are aligned closely with the needs of the self. In both cases, certainty is a property of the relational basis of information. At its outer boundary, the mind seeks to fit with the world and to achieve a relational alignment with the data of the senses and actions. At its inner boundary, it seeks to fit with the self, which also requires a kind of relational alignment of concepts not only with present needs but also with the consolidated values of the historical self. At each boundary, the demands of the relational fit draw concepts toward achieving certainty. In the logic of information theory, concepts—thus constrained—become informed.

Interestingly, in this way of thinking, information, if it is defined as the reduction of uncertainty, is antithetical to abstract thought. The very nature of the abstract concept is that it represents not just the concrete instance of relation implied by a single perception or a single urge. Rather, it achieves generality. In the language of the theory of visceral and somatic structures of mind, abstract concepts are achieved as they emerge from the syncretic unconscious to first become differentiated by specific referents at the somatic boundary and then—keeping this accurate external relational structure—to become hierarchically integrated by a recursive round of processing at the visceral core. The recursive process very likely requires many cycles of arbitration between core and shell constraints, and the challenge of the arbitration is to avoid the concretizing constraints at either boundary. Concreteness comes from the constraints of certainty, which are offered by the secure alignment with evidence on the one hand and the comforting values of the self on the other. The only way that abstract concepts can be created is to tolerate the ambiguity of first maintaining distance from both constraints, and yet engaging them both in order to achieve a hierarchic, general level of meaning.

At the abstract level of meaning, a new kind of relational alignment then emerges. The concept fits both the world and the self but not in the concrete way of the specific urge or datum of the senses. Rather, in abstract form, the concept (the instantiation of the self-world relation) organizes broad domains of both emotions and experience with accuracy and generality.

Although it may seem that I have framed this process in somewhat arcane theoretical terms, what I have just described is nothing new. It is the process of arbitrating core and shell constraints that we have formulated as the basic mechanism of memory consolidation. Taking the arbitration in this basic form of elemental corticolimbic dynamics, we would have to say that all mammalian memory has as its goal a kind of abstraction. This is because consolidation from limbic core to neocortical shell leads to concepts (memory structures) that create *learning*, the mediation of personal need with sensorimotor competence. At this basic level, the tension of abstract thought is not different in kind from that of basic memory consolidation. The relational alignment with visceral needs is a different set of constraints from the relational alignment with sensorimotor reality. As a result, the work of the mammalian brain is to arbitrate both sets of constraints and maintain the productive tension that drives corticolimbic consolidation throughout the day and into our dreams at night.

What is new in an evolutionary sense is the power of human consolidation of abstract thought made possible by cultural and now technological artifacts. What may be new in the Age of Mind is an expansion in the capacity for abstract thought that is afforded by more complex informatics on the one hand

and increasing conscious control over evaluative and motivational mechanisms on the other.

Because the dialectical relation between certainty and abstraction has always been a property of mammalian brains, the result is not really a new *kind* of information. To the extent that a rodent has greater mediation between stimulus and response than a reptile, information must be more contextual and the reduction of uncertainty less reflexive. But if there is to be Age of Mind, the power of informatic organization at the shell will be matched by expanded insight into the visceral evaluative core.

Complex structures of mind may then emerge. These must still meet the timeless demand to satisfy both the core and the shell, both biological need and environmental reality. However, they may now organize complex concepts more freely because the constraints at both boundaries will be transformed. There the process of relational alignment will itself become more embedded in abstract forms of experience. At the shell the informatic appendages will allow inherently complex forms of sensation and action, interfaces with the world that are accurate links to classes of evidence. Perhaps most important, at the visceral core the increasing knowledge of the mind's regulatory mechanisms will afford a personal understanding of the process of control that allows the process of intelligence to become transformed. As we bring scientific insight to clarify the mind's subjective base, we may find that the motive influence can progress from egoistically asserting the default assumptions of the self to more deliberately organizing the values and choices encountered in the process of experience.

Conscious Self-Actualization

To approach an Age of Mind, we will need to balance the informatics explosion with a personal, practical knowledge of the brain's workings. This will require additional progress in neuroscientific research so that we can understand ourselves with the technologies made possible by computational advances. Given the scientific insight, the task will be to acquire the skills necessary to actualize, first, awareness of the mind's motive mechanisms and, then, more deliberate control of these mechanisms. Considering how we might achieve such a thing, it is perhaps a surprise to find that the only answer is the neglected discipline: education.

Of course, in many ways education has been the purpose of this book, which outlines the skills required for understanding the brain's implicit organizational processes. We need to comprehend first the brain's connectional architecture and then the principles of representation and control that may explain the implications of that architecture. Most important are the integral emotional and motivational mechanisms that regulate the consolidation of memory. We can use these several principles and insights to clarify both our own subjective perspectives and the capacity for objectivity exercised by appreciating the perspectives of others. We can then attempt to formulate a theory to explain how the mind's finest actualizations—abstract concepts—can be derived from the elementary biological mechanisms of consolidating memory.

Regardless of whether the particular ideas in these pages prove useful, I am convinced that any good theory of mind must be anatomically correct. This means it must fit the evolved plan of the human neural architecture. If only he could have known, the fact that we eventually recognized the mind's evolved structure might have brought a smile to old Darwin's face. As we focus the technologies of modern neuroscience to illuminate the mind's implicit recesses, we discover the evolutionary order of neural mechanisms at every turn. The multileveled structure of memories is evident not only in the uniquely human domains of the cortex but rather is distributed across multiple levels of cortical network evolution, with the most critical regulatory influences on semantic integration applied by the emotional controls of the primitive limbic cortices. The operations of the telencephalon as a whole are, in their every waking and sleeping moment, directed by the two domains of the diencephalon, the thalamus and hypothalamus, that were fully formed in an era of our vertebrate neural history before the first hint of a telencephalon had ever been implied by our mutant neural architecture. From even more primitive levels, those of the mesencephalic stage of neural evolution, we continue to receive the ongoing neuromodulation of moods in daily experience, whether it is the transient rush of elation or the crushing despair of depression. The vertical integration across these levels is essential to each operation of mind, whether it is recruiting the effortful anxiety to reason through a hard problem or simply savoring the quietude of the last light of the day.

As we make use of science to illuminate the implicit motives underlying conscious experience, we find our emotional lives to be intimate reminders of the mind's evolved structure, just as Darwin tried to tell us. These are the residuals of hundreds of millions of years of adaptive struggles, handed down by the legacy of our successful ancestry. What Darwin could not envision is the mechanism of embryogenesis, which has evolved to achieve the fine structure of individual experience. Extending the activity-dependent selection of synaptic connectivity of embryogenesis into postnatal life has been the evolutionary strategy for extending the process of self-determination, a process that is begun only by the phyletic recapitulation of the species genome.

As science rolls back the cover of motive blindness, this is perhaps the most important lesson: The personal operations of unconscious motive control are framed only in broad outline on a phyletic scaffold. These evolved neurophysiological mechanisms are just the beginning. Through ongoing embryogenesis we have the capacity and the necessity to self-organize and to form neural connectivity in an ongoing program of self-actualization that captures the contents of personal experience and consolidates it within the fine structure of the neuropil. The self-organization of fine synaptic structure is the continuing task of psychological development, traversing the core-shell boundary in each motive-memory, whether it is a concrete impulse or an abstract reflection. We can now see the task of neuropsychological education: to bring skill and discipline into the effective motive control of the evolved hierarchy and thereby into the ongoing self-organization of each mind's neural form. It is the incredible promise of the Age of Mind that, with increasing scientific insight into the motive mechanisms, this process may soon become fully conscious.

Notes

These notes provide additional reading as background on each chapter. In some cases I point out the scientific papers in which my associates and I have developed ideas related to the chapter. Wherever possible, I cite the work in the scientific literature that either is the original source of the ideas or provides classical insights into key problems of the chapter.

Chapter 1. Mind in the Information Age

Modern efforts to frame the progression of informatics in human society, such as those by Sagan (1989) and Kurzweil (1999), draw on a long history of studies of information theory in relation to both theoretical and practical studies of the mind. Classic studies in information theory were published in the mid-20th century by Wiener (1948) and Shannon and Weaver (1949). The ideas of cybernetics and information theory had wide influence in electronics and engineering, but Gregory Bateson saw their relevance for psychology and anthropology and also applied them to human family communication and mental disorders. An interesting and readable history of the postwar enthusiasm for these ideas is provided by Heims (1991).

Although physiological concepts were fundamental to the ideas of feedback and control, the integration of control theory with brain research was limited, as behaviorist stimulus-response ideas dominated animal behavior studies. The exception was the novel speculations of McCulloch and Pitts (1990/1943), who, by applying one of the first mathematical approaches to neurophysiology, attempted to formulate principles of control in nerve nets.

In Chapter 3 we pick up the more recent resurgence of neural network models in the latter decades of the 20th century. One highly influential work is that of Miller, Galanter, and Pribram (1960). Explicitly incorporating principles of feedback and information, Miller et al. brought these concepts to theorizing about the brain's executive functions, the higher cognitive capacities of everyday planning and self-regulation.

The basic idea of the core-shell organization of the brain has arisen from a long history of research and theory on brain evolution by Dart, Abbie, and others that is now largely forgotten. An excellent review of the early studies that continues to guide my thinking is that by Sanides (1970). Building on his scholarship, Sanides developed his own speculations on cortical evolution that then inspired the interpretations of the anatomical studies of Pandya and Barbas that are discussed in later chapters.

My own efforts to relate corticolimbic organization to memory consolidation and abstraction were first formulated in core-shell terms in my contribution to the *Handbook of Neuropsychology* (2001) and then developed in relation to language in Givon and Malle's *Evolution of Language out of Prelanguage* (2002).

As I explain in several places in this book, a theory of corticolimbic architecture and function must be built on an understanding of vertical integration across multiple levels of the neuraxis. The remarkable, foundational statement of this way of thinking is found in Jason Brown's *Neuropsychology of Cognition* (1977). Derryberry, Luu, and I attempted a more recent description of vertical integration in relation to emotion and self-regulation for Borod's *Handbook of the Neuropsychology of Emotion* (2000).

Brown, J. W. (1977). *Mind, brain, and consciousness: The neuropsychology of cognition.* New York: Academic Press.

Heims, S. J. (1991). *The cybernetics group.* Cambridge, MA: MIT Press.

Kurzweil, R. (1999). *The age of spiritual machines: When computers exceed human intelligence.* New York: Viking.

McCulloch, W. S., & Pitts, W. (1990). A logical calculus of the ideas immanent in nervous activity. *Bull Math Biol,* 52(1–2), 99–115. (Original work published 1943)

Miller, G. A., Galanter, E. H., and Pribram, K. H. (1960). *Plans and the structure of behavior.* New York: Henry Holt.

Sagan, C. (1989). *The dragons of Eden: Speculations on the evolution of human intelligence.* New York: Ballentine.

Sanides, F. (1970). Functional architecture of motor and sensory cortices in primates in the light of a new concept of neocortex evolution. In C. R. Noback & W. Montagna (Eds.), *The primate brain: Advances in primatology: Vol. 1* (pp. 137–208). New York: Appleton-Century-Crofts.

Shannon, C. E., & Weaver, W. (1949). *The mathematical theory of communication.* Champaign-Urbana: University of Illinois Press.

Tucker, D. M. (2001). Motivated anatomy: A core-and-shell model of corticolimbic architecture. In G. Gainotti (Ed.), *Handbook of neuropsychology: Vol. 5. Emotional behavior and its disorders* (2nd ed., pp. 125–160). Amsterdam: Elsevier.

Tucker, D. M. (2002). Embodied meaning: An evolutionary-developmental analysis of adaptive semantics. In T. Givon & B. Malle (Eds.), *The evolution of language out of prelanguage.* Amsterdam: John Benjamins.

Tucker, D. M., Derryberry, D., & Luu, P. (2000). Anatomy and physiology of human emotion: Vertical integration of brain stem, limbic, and cortical systems. In J. Borod (Ed.), *Handbook of the neuropsychology of emotion* (pp. 56–79). New York: Oxford.

Wiener, N. (1948). *Cybernetics, or control and communication in the animal and the machine.* Cambridge, MA: MIT Press.

Chapter 2. Structures of Intelligence

The goal of this chapter is the basic one of relating brain structure to psychological function. Because anatomical foundations are so important, I encourage introductory reading that is rooted in the brain's evolved anatomy. John Hughlings Jackson's (1879, 1958) classic works remain required reading for neuropsychologists. Yakovlev's (1948) analysis of the basis of movement clearly outlines the continuity of elementary with complex neural control systems. Mesulam's (2000) introductory chapter weaves a functional description of memory, emotion, and attention with the corticolimbic network architecture that has emerged from modern anatomical studies. Swanson's (2000) review article and introductory book (2003) build on the highly integrative studies he and his students have conducted, illustrating the subcortical, as well as cortical, basis of behavior.

With specific reference to brain lateralization, Bogen and Bogen's (1969) series of papers on split-brain patients provides an excellent introduction to the profound sense of the significance of hemispheric specialization when it was "rediscovered" in these studies. This is a significance that thoughtful students today can rediscover with the same level of excitement. Although it appeared in the same year, Semmes's (1968) important article reflected a long series of studies on lateralized brain lesions in the 1950s, rather than the "new" split-brain studies of the 1960s. Nonetheless, the implications for the neural structures of the mind were easily as significant as those of the split-brain studies and were framed in an even more explicit hypothesis.

A good deal of literature on brain lateralization arose in the later decades of the 20th century. However, other than the remarkable dissociations seen in the split-brain studies, there were few fundamental characterizations of cognitive processes that differed from those emerging from the careful neuropsychological testing of brain-lesioned patients in the 1950s. A good overview of this later work is found in Davidson and Hugdahl (1994). One theoretical approach to learning that I found integrative is the Goldberg and Costa (1981) model. My own initial attempts to understand hemispheric specialization for emotion (Tucker, 1981) led to questions about cortical-limbic interaction that shaped my research and theory for many years.

The anterior-posterior dimension of brain organization has been studied from many perspectives. Luria's (1973) neuropsychological studies of brain lesions were formative for modern concepts of the frontal lobe regulation of behavior. Allport (1985) provided a particularly clear statement of the need for focusing attention in the anterior (frontal) brain, compared to the breadth of attention allowed by the posterior brain's parallel perceptual processing. In modern cognitive neuroscience models, Posner's research and thinking have shown how attention can take different forms in frontal and posterior cortical networks (e.g., Posner, Peterson, Fox, & Raichle, 1988; Posner & Dehaene, 1994).

Several early theoretical models, such as Yakovlev's (1948), anticipated the core-shell organization of corticolimbic networks. But it was only in the modern anatomical studies of the 1970s, particularly those by Pandya and his associates at Boston university, that we came to understand the connection architecture of the mammalian cerebral hemisphere, with its limbic core and neocortical sensory and motor networks (Pandya & Yeterian, 1984, 1996). Barbas (2000) has continued this tradition of anatomical studies with obvious and significant implications for understanding brain function. As Pandya and associates have repeatedly emphasized, the modern evidence has consistently supported Sanides's (1970) speculations on the evolution of the cortex in a core-to-shell direction, beginning with the emergence of the limbic cortex and followed by the association cortex and only later by the differentiation of specialized sensory and motor neocortices.

Allport, A. (1985). Selection for action: Some behavioral and neurophysiological considerations of attention and action. In H. Heuer & A. F. Sanders (Eds.), *Issues in perception and action.* Hillsdale, NJ: Erlbaum.

Barbas, H. (2000). Connections underlying the synthesis of cognition, memory, and emotion in primate prefrontal cortices. *Brain Research Bulletin, 52*(5), 319–330.

Bogen, J. E., & Bogen, G. M. (1969). The other side of the brain: III. The corpus callosum and creativity. *Bulletin of the Los Angeles Neurological Society, 34,* 191–220.

Davidson, R. J., & Hugdahl, K. (Eds.) (1994). *Human brain laterality.* New York: Oxford University Press.

Goldberg, E., & Costa, L. D. (1981). Hemisphere differences in the acquisition and use of descriptive systems. *Brain and Language, 14,* 144–173.

Jackson, J. H. (1879). On affections of speech from diseases of the brain. *Brain, 2,* 203–222.

Jackson, J. H. (1958). Selected writings. In D. Taylor (Ed.), *Selected writings of John Hughlings Jackson.* New York: Basic Books.

Luria, A. R. (1973). *The working brain: An introduction to neuropsychology.* New York: Basic Books.

Mesulam, M.-M. (2000). Behavioral neuroanatomy: Large-scale networks, association, cortex, frontal syndromes, the limbic system, and hemispheric specializations. In M.-M. Mesulam (Ed.), *Principles of behavioral and cognitive neurology* (2d ed., pp. 1–120). New York: Oxford University Press.

Pandya, D. N., & Yeterian, E. H. (1984). Proposed neural circuitry for spatial memory in the primate brain. *Neuropsychologia, 22*(2), 109–122.

Pandya, D. N., & Yeterian, E. H. (1996). Comparison of prefrontal architecture and connections. *Philosophical Transactions of the Royal Society of London. Series B, Biological Sciences, 351*(1346), 1423–1432.

Posner, M. I., & Dehaene, S. (1994). Attentional networks. *Trends in Neurosciences, 17*(2), 75–79.

Posner, M. I., Petersen, S. E., Fox, P. T., & Raichle, M. E. (1988). Localization of cognitive operations in the human brain. *Science, 240,* 1627–1631.

Sanides, F. (1970). Functional architecture of motor and sensory cortices in primates in the light of a new concept of neocortex evolution. In C. R. Noback & W. Montagna (Eds.), *The primate brain: Advances in primatology: Vol. 1* (pp. 137–208). New York: Appleton-Century-Crofts.

Semmes, J. (1968). Hemispheric specialization: A possible clue to mechanism. *Neuropsychologia, 6,* 11–26.

Swanson, L. W. (2000). Cerebral hemisphere regulation of motivated behavior. *Brain Research, 886*(1–2), 113–164.

Swanson, L. W. (2003). *Brain architecture: Understanding the basic plan.* New York: Oxford University Press.

Tucker, D. M. (1981). Lateral brain function, emotion, and conceptualization. *Psychological Bulletin, 89*(1), 19–46.

Yakovlev, P. I. (1948). Motility, behavior, and the brain. *Journal of Nervous and Mental Disease, 107,* 313–335.

Chapter 3. Principles of Representation and Control

[1] The "connectionist," or parallel-distributed, processing advances of the 1980s offered explicit new models for the distributed processing of brain networks that are implied by the heavily interconnected cortical anatomy. Although much of the research reasoned from the computational model to psychology (Rumelhart & McClelland, 1986), some theorists such as Grossberg (1980, 1984) recognized the importance of distributed representational models for interpreting neurophysiological evidence.

As important as behaviorist learning theory was to 20th-century psychology and to modern society generally, the real progress in theoretical models of animal learning was made by specialists in the field and known largely only to them, at a time when the "cognitive revolution" had led academic and popular attention away from learning theory. As I outline in Chapter 3, Kamin's (1965) observations that his rats learned only when they were surprised constituted a turning point and were soon formulated in a new systematic learning theory by Rescorla and Wagner (1972). Research by Amsel and Stanton (1979, 1980) and Bitterman and associates (Tennent & Bitterman, 1975; Papini & Bitterman, 1990) then clarified the specific principles of learning reviewed in Chapter 3. Continuing this fundamental theoretical development, the elegant studies and formulations of Papini (2002, 2003) have shown that rats are above all cognitive animals.

Phan Luu and I have attempted to integrate the implications of modern learning theory with current ideas of corticolimbic mechanisms of self-regulation (Luu & Tucker, 2003; Tucker & Luu, in press). We have been strongly influenced in this effort by the neurophysiological evidence and theory organized by Gabriel and his associates (e.g., Gabriel, Sparenborg, & Stolar, 1987; Gabriel, Sparenborg, & Kubota, 1989). Although the neurophysiological mechanisms actually lead in a direction somewhat different from the summary ideas developed in Chapter 3, it seems clear that modern learning theory should continue to have a key role in modern theoretical neuroscience.

The speculations on epistemology concluding Chapter 3 are certainly no more than speculations, and it may seem pretentious to claim they follow in the tradition of Piaget's (1971) *Genetic Epistemology.* Still, following Kuhn's (1996) insightful sociological analysis, there can be little doubt of the existence of a motivated basis for the development of ideas, even in the most formal statements of scientific theory.

Amsel, A. (1979). The ontogeny of appetitive learning and persistence in the rat. In N. E. Spear & B. A. Campbell (Eds.), *Ontogeny of learning and memory* (pp. 189–224). Hillsdale, N. J.: Erlbaum.

Amsel, A., & Stanton, M. (1980). Ontogeny and phylogeny of paradoxical reward effects. *Advances in the study of behavior, 11*, 227–274.

Gabriel, M., Sparenborg, S., & Kubota, Y. (1989). Anterior and medial thalamic lesions, discriminative avoidance learning, and cingulate cortical neuronal activity in rabbits. *Experimental Brain Research, 76*(2), 441–457.

Gabriel, M., Sparenborg, S. P., & Stolar, N. (1987). Hippocampal control of cingulate cortical and anterior thalamic information processing during learning in rabbits. *Exp Brain Res, 67*(1), 131–152.

Grossberg, S. (1980). How does a brain build a cognitive code? *Psychological Review, 87*, 1–51.

Grossberg, S. (1984). Some psychophysiological and pharmacological correlates of a developmental, cognitive, and motivational theory. In R. Karrer, J. Cohen, & P. Tueting (Eds.), *Brain and information: Event-related potentials: Vol. 425* (pp. 54–82). New York: New York Academy of Sciences.

Kamin, L. J. (1965). Temporal and intensity characteristics of the conditioned stimulus. In W. F. Prokasy (Ed.), *Classical conditioning: A symposium* (pp. 118–147). New York: Appleton-Century-Crofts.

Kuhn, T. (1996). *The structure of scientific revolutions* (3rd. ed). Chicago: University of Chicago Press.

Luu, P., & Tucker, D. M. (2003). Self-regulation and the executive functions: Electrophysiological clues. In A. Zani & A. M. Preverbio (Eds.), *The cognitive electrophysiology of mind and brain* (pp. 199–223). San Diego: Academic Press.

Papini, M. R. (2002). Pattern and process in the evolution of learning. *Psychological Review, 109*(1), 186–201.

Papini, M. R. (2003). Comparative psychology of surprising nonreward. *Brain Behavior and Evolution, 62*(2), 83–95.

Papini, M. R., & Bitterman, M. E. (1990). The role of contingency in classical conditioning. *Psychological Review, 97*(3), 396–403.

Piaget, J. (1971). *Genetic epistemology.* New York: Norton.

Rescorla, R. A., & Wagner, A. R. (1972). A theory of pavlovian conditioning: Variations in the effectiveness of reinforcement and nonreinforcement. In A. H. Black & W. F. Prokasy (Eds.), *Classical conditioning II: Current research and theory* (pp. 65–99). New York: Appleton-Century-Crofts.

Rumelhart, D. E., & McClelland, J. L. (1986). *Parallel distributed processing: Explorations in the microstructure of cognition: Vol. 1. Foundations.* Cambridge, MA: MIT Press.

Tennant, W. A., & Bitterman, M. E. (1975). Blocking and overshadowing in two species of fish. *Journal of Experimental Psychology—Animal Behavior Processes, 1*(1), 22–29.

Tucker, D. M., & Luu, P. (2006). Adaptive binding. In H. Zimmer, A. Mecklinger, & U. Lindenberger (Eds.), *Binding in human memory: A neurocognitive approach.* New York: Oxford University Press.

Chapter 4. Motivated Experience

One important line of research showing the limitations of our common assumptions about conscious knowledge of cognitive processes is often described as the judgment and heuristics literature (e.g., Kahneman, Slovic, & Tversky, 1982). Social psychological studies have pointed to substantial evidence of the inadequacies of simple introspection (Nisbett & Wilson, 1977).

The view of the mind from the beginning of the 20th century is an interesting perspective as we consider the science of the mind in the early 21st. Two examples of the optimism for what would be gained in the 20th century are James's *Principles of Psychology* in 1890 and Freud's *Interpretation of Dreams* in 1900.

In the Introduction I cited some of the classical work on cybernetics and control theory. Although much of modern cognitive science and cognitive neuroscience has been preoccupied with representation (and not attending well to the fundamental problems of control), there are important exceptions in which control theory has become complex and well reasoned in relation to brain physiology and function. One example is the work of Kelso and his associates (Kelso, 1995). Another is the seminal neurophysiological research and provocative theory of Walter Freeman (1995).

The idea that emotions are caused by thoughts (and do not act to shape cognition actively) can be seen as a long-standing assumption of cognitivist thinking. William James is often credited with introducing this nearsightedness and with his notion that our actions (such as running when you see a bear in the woods) are physiological reactions, whereas our emotions (the psychological experience) only occur with reflection on our actions. What was a creative idea in the 19th century became a way for psychologists to use cognitive explanations instead of neuropsychological ones in the 20th century (e.g., Schacter & Singer, 1962). On the other hand, a much more interesting development of Jamesian thinking appeared when the complex and fragmentary nature of neural representations of bodily processes and attitudes was considered, such as in Damasio's (1996) somatic marker hypothesis.

Doug Derryberry, Phan Luu, and I have worked together for many years to understand motivational influences on experience, beginning with primitive limbic influences (Derryberry & Tucker, 1991), then incorporating subcortical mechanisms of vertical integration (Tucker, Derryberry, & Luu, 2000), and more recently considering primitive bases of complex human values (Tucker, Luu, & Derryberry, 2005). The work of Jürgens and Ploog has been particularly important in showing the influence of vertical integration of the brain stem, the midbrain, and the diencephalic, limbic, and cortical systems in emotion (Jürgens & Ploog, 1970; Ploog, 1981, 1992).

More recently, the analysis of the hierarchic integration of the multiple, vertically organized brain system has been brought very effectively to psychological theory by the work of Lewis (2005) and Lewis and Todd (2005). The combination of developmental reasoning and neuroanatomical analysis has allowed them to gain powerful insights into traditionally central mechanisms such as the cognitive appraisal of emotional events.

Earlier I pointed out the central role of Pandya's studies of primate corticolimbic anatomy in my theorizing (e.g., Pandya & Yeterian, 1984). To relate this anatomy to memory, the excellent studies and reviews by Squire (e.g., 1987) and Schacter (e.g., 1997) show the essential role of corticolimbic consolidation in the formation of memory. Remaining to be elucidated is the Pandyan architecture that will achieve the results of consolidation through neurophysiological mechanisms.

Finally, the active influence of memory mechanisms in the process of perception has been demonstrated and clearly understood by Shepard (1984). An important example of electrophysiological evidence of the two-way interactions of memory and perception has been illustrated by the studies of Naya, Yoshida, and Miyashita (2001).

Damasio, A. R. (1996). The somatic marker hypothesis and the possible functions of the prefrontal cortex. *Philosophical Transactions of the Royal Society of London. Series B, Biological Sciences, 351*(1346), 1413–1420.

Derryberry, D., & Tucker, D. M. (1991). The adaptive base of the neural hierarchy: Elementary motivational controls of network function. In A. Dienstbier (Ed.), *Nebraska symposium on motivation* (pp. 289–342). Lincoln: University of Nebraska Press.

Freeman, W. (1995). *Societies of brains: A study in the neuroscience of love and hate.* Hillsdale, NJ: Erlbaum.

Freud, S. (1953). *The interpretation of dreams.* London: Hogarth Press. (Original work published 1900)

James, W. (1953). *The principles of psychology.* New York: Dover. (Original work published 1890)

Jürgens, U., & Ploog, D. (1970). Cerebral representation of vocalization in the squirrel monkey. *Experimental Brain Research, 10,* 532–554.

Kahneman, D., Slovic, P., & Tversky, A. (1982). *Judgment under uncertainty: Heuristics and biases.* New York: Cambridge University Press.

Kelso, J. A. S. (1995). *Dynamic patterns: The complex organization of brain and behavior.* Cambridge, MA: MIT Press.

Lewis, M. D. (2005). Bridging emotion theory and neurobiology through dynamic systems modeling. *Behavioral and Brain Sciences, 28*(2), 169–194; discussion 194–245.

Lewis, M. D., & Todd, R. M. (2005). Getting emotional: A neural perspective on emotion, intention, and consciousness. *Journal of Consciousness Studies, 12*(8), 210–235.

Naya, Y., Yoshida, M., & Miyashita, Y. (2001). Backward spreading of memory-retrieval signal in the primate temporal cortex. *Science, 291,* 661–664.

Nisbett, R. E., & Wilson, T. D. (1977). Telling more than we can know: Verbal reports on mental processes. *Psychological Review, 84,* 231–259.

Pandya, D. N., & Yeterian, E. H. (1984). Proposed neural circuitry for spatial memory in the primate brain. *Neuropsychologia, 22*(2), 109–122.

Ploog, D. W. (1981). Neurobiology of primate audio-vocal behavior. *Brain Research Reviews, 3,* 35–61.

Ploog, D. W. (1992). Neuroethological perspectives on the human brain: From the expression of emotions to intentional signing and speech. In A. Harrington

(Ed.), *So human a brain: Knowledge and values in the neurosciences* (pp. 3–13). Boston: Birkhauser.

Schacter, D. L. (1997). *Searching for memory: The brain, the mind, and the past.* New York: Basic Books.

Schacter, F., & Singer, J. E. (1962). Cognitive social and physiological determinants of emotional states. *Psychological Review, 69,* 379–399.

Shepard, R. N. (1984). Ecological constraints on internal representation: Resonant kinematics of perceiving, imagining, thinking, and dreaming. *Psychological Review, 91*(4), 417–447.

Squire, L. R. (1987). *Memory and brain.* New York: Oxford University Press.

Tucker, D. M., Derryberry, D., & Luu, P. (2000). Anatomy and physiology of human emotion: Vertical integration of brain stem, limbic, and cortical systems. In J. Borod (Ed.), *The neuropsychology of emotion* (pp. 56–79). New York: Oxford University Press.

Tucker, D. M., Luu, P., & Derryberry, D. (2005). Love hurts: The evolution of empathic concern through the encephalization of nociceptive capacity. *Development and Psychopathology, 17*(3), 699–713.

Chapter 5. Visceral and Somatic Frames of Mind

Most neuroscientists view their discipline as informed only by the publications of the last few years. Certainly there are technologies available now, in neuroimaging and genetics for example, that are yielding important new interpretations. However, a broader theoretical understanding of the brain was often approached very effectively by researchers many years ago. Moreover, the insights of one era of brain research are often lost during the next. Coincidentally, several publications of 1948 framed the key evidence for the ideas in Chapter 5.

We saw above that the concepts of cybernetics were integrated across the boundaries of homeostatic principles and electronic circuit design by Wiener (1948). In addition, an elegant—and quite complete—statement of the emergence of motive from the visceral core of the brain was formulated by Yakovlev that same year. Action is an exteriorization from this interior base. Although Yakovlev may not have separated the visceral and somatic divisions of the nervous system in the way I have developed the reasoning in Chapter 5, his description fully anticipates the material there, as he captures the developmental progression from urge to actualization in each instance of behavior.

The anatomical groundwork for understanding the emergence of functional dualism between visceral and somatic divisions of the higher (telencephalic) central nervous system was laid out in C. Judson Herrick's (1948) publication, *The Brain of the Tiger Salamander.* This relatively large, neotenic amphibian enabled anatomical study that illustrated the way in which the major circuits of the telencephalon are ordered by their evolutionary roots in the somatic and visceral divisions of the diencephalon (thalamus for external, somatic, sensory, and motor communication; hypothalamus for the two-way regulation of the internal visceral milieus).

This progression or "microgenesis" of each act and thought was formulated in psychological developmental terms by Heinz Werner (1957/1948). Werner's "comparative psychology" allowed an understanding of the generality of organizational processes—differentiation and integration—that are formative of the mind in exactly the same way they are formative of embryogenesis. For Werner, complex cognitive representations are achieved by the hierarchic integration of the now differentiated mental contents, a theme I have attempted to instantiate directly within the framework of corticolimbic (core-shell) anatomy.

Moving to more recent work, the microgenetic analysis of psychological processes was grounded in neurophysiological mechanisms by Jason Brown (1977, 1988). Like Herrick, Yakovlev, and other traditional scholars of the brain since Jackson, Brown recognized that the human brain's more recently elaborated features can function only when fully supported by the more primitive substrate of evolved forms. As a student of developmental psychology and psychoanalysis, Brown was able to relate the brain's hierarchic anatomy to the developmental progression from diffuse to articulated organization required for each thought and action. In our efforts to relate psychological development and self-regulation to neural mechanisms, Derryberry, Luu, and I have tried to weave the theme of vertical integration throughout our work, thanks largely to Brown's influence.

The theoretical goal of Chapter 5 is to explain the organization of abstract concepts from the informatic mechanisms emerging from the concrete boundaries of both internal visceral and external sensorimotor constraints. Understanding abstract thought has been a traditional goal of both psychological and neuropsychological research. Many of the traditional psychological approaches to abstraction have had a strong developmental influence (Pikas, 1966). The "loss of the abstract attitude" was one of the important general effects recognized in the early psychological studies of brain damage (Goldstein, 1952).

One of the most integrative theoretical studies of psychological mechanisms of abstract thinking is that by Harvey, Hunt, and Schroder (1961). They drew on not only developmental studies but also philosophical models that emphasize the effectiveness of dialectical opposition in forming hierarchic patterns that subordinate at a new level ideas that were incompatible at a lower level. I have applied this kind of reasoning to the structural patterns forming (in a kind of opposition) within limbic (visceral) and neocortical (somatic) networks.

Brown, J. W. (1977). *Mind, brain, and consciousness: The neuropsychology of cognition.* New York: Academic Press.

Brown, J. W. (1988). *The life of the mind: Selected papers.* Hillsdale: Erlbaum.

Cicchetti, D., & Tucker, D. M. (1994). Development of self-regulatory structures of the mind. *Development and Psychopathology, 6,* 533–549.

Goldstein, K. (1952). The effect of brain damage on the personality. *Psychiatry, 15,* 245–260.

Harvey, O. J., Hunt, D. E., & Schroder, H. M. (1961). *Conceptual systems and personality organization.* New York: Wiley.

Herrick, C. J. (1948). *The brain of the tiger salamander*. Chicago: University of Chicago Press.

Pikas, A. (1966). *Abstraction and concept formation: An interpretive investigation into a group of psychological frames of reference*. Cambridge, MA: Harvard University Press.

Werner, H. (1957). *The comparative psychology of mental development*. New York: Harper. (Original work published 1948)

Yakovlev, P. I. (1948). Motility, behavior, and the brain. *Journal of Nervous and Mental Disease, 107*, 313–335.

Chapter 6. Subjective Intelligence

Chapter 5 introduced a new theme; Chapter 6 exercises it in relation to familiar themes in both neurological and psychological domains. With the notion of abstraction in corticolimbic networks, we revisit the structural dimensions of the brain's organization.

Perhaps the most difficult conceptual orientation for Chapter 6 is that required to see the brain across various dimensions. We carve out the left/right, front/back, and core/shell dimensions, as we did in Chapter 2, and it is natural to see these as orthogonal, that is, as intersecting at right angles. In fact, however, these dimensions are not independent like a good coordinate system but are inherently correlated: The core-shell dimension seems to be organized differently, for example, for the left and right hemispheres, and this presents a challenge for both the writing and reading of the chapter.

The psychological goal of Chapter 6 is to consider how the bodily mechanisms of abstraction operate in subjective experience. I begin developing a neuropsychological analysis of the embodiment of mind, complementing the embodiment movement that has been influential in philosophy in recent years (Johnson, in press; Johnson & Lakoff, 1999).

The use of stories of adolescents coping with daily affairs serves as a vehicle for exploring the subjective perspective that may arise from embodied neural mechanisms. Although we can gain a fresh view of the subjective process by seeing how it can emerge from fundamentally unconscious biological mechanisms, new views of objectivity may occur as we come to understand the fundamental nature of subjective biases.

Johnson, M. (in press). *The meaning of the body*. Chicago: University of Chicago Press.

Johnson, M., & Lakoff, G. (1999). *Philosophy in the flesh: The embodied mind and its challenge for Western thought*. New York: Basic Books.

Chapter 7. Objective Experience

Research on the neural mechanisms of social cognition is growing rapidly at present, and it offers an important challenge in bringing the new tools of neuroscience to human problems. In this chapter I contend that our capacity to represent the perspectives of others provides the essential objective basis of the mind. Many intriguing findings in the current literature suggest new insights into the representation of both personal and interpersonal perspectives (e.g., Farrer & Frith, 2002). At the time of this writing, a brief survey of the current cognitive neuroscience literature shows the interesting results of the analytic, mechanistic approach of neuroscience with the complex, integrative questions of research on the mind in a social context.

Farrer, C., & Frith, C. D. (2002). Experiencing oneself vs. another person as being the cause of an action: The neural correlates of the experience of agency. *Neuroimage*, 15(3), 596–603.

Chapter 8. Information in the Age of Mind

To look to the future, we take up the perspective of history. Darwin's analysis of evolutionary mechanisms occurred in the undeniably political context of his times. People today are more sophisticated in many ways, but many still have a hard time accepting evolution as the context of personal history.

My thoughts on the continuing evolution of an informatic shell are of course just speculations, in the tradition of Sagan or Kurzweil. But the significance of our evolving computational technology for the future of mind does gain meaning, I believe, by the study of the neural mechanisms of experience. Once we recognize that the mind is embodied within its peculiar wetware, it becomes less difficult to envision how it could become embodied in other physical forms.

Darwin, C. (1872). *Expressions of the emotions in man and animals.* London: John Murray.

Index

Page numbers in boldface indicate figures.

abstract thought. *See also* concept formation; core-shell theory
  adolescent mind, 195, 200
  and the body, 25, 204
  brain stem arousal mechanism, 277
  corticolimbic networks, 244
  Einstein genius, 218–219
  encephalization of visceral function, 208
  and experience, 28–29, 220–221
  exteriorization, 240
  and fantasy, 215
  genetic epistemology, 279
  hierarchic structure of, 25, 226, 254–255, 294–295
  hypothesis of encephalized conceptual structure, 222–223
  and information, 297
  and intelligence, 204–205, 219, 227–230, 298
  intersubjective reasoning, 255–256
  IQ tests as measure, 32–33
  language, 223
  memory consolidation, 210, 298–300
  motivational control, 212, 229
  and perception, 247
  recursive processing, 223–226
  representation, 221–222
  self-actualization, 298–300
  sensorimotor networks, 210, 219
  and values, 248
accommodation, 134, 138–139
action
  attentional control, 69
  and cognition, 216–217, 252
  core-shell theory, 27–28, 181
  and creativity, 16
  ethical decisions, 264
  formation in child development, 182
  frontal lobe, 180–186, 249
  and perception, 126–127
  visceral nervous system, 248
activity-dependent specification, 161
adolescent mind. *See also* Andrew; Jared; Kim
  and abstract thought, 195, 200
  "cutting," 259–260
  heterochrony, 202–203
  infatuation, 267–268
  self-consciousness, 18
adrenalin, 151
aesthetic sense
  encephalization, 244–245
  and perception, 264
  posterior brain, 28, 76–77, 244
affordances, 251–252, 285–286
agency, 144–145
Age of Mind
  conscious self-actualization, 298–300
  evolution of the mind, 281
  and the Information Age, 295–298
  neural networks, 284
  neurocybernetics, 291–294
allocentrism, 238, 267–268
Allport, Alan, 69, 102
Alzheimer's disease, 193
amnesia, 84–85, 191–192, 193
Amsel, Abram, 125
amygdala, 38–39, 167, 172, 183
analogical representation, 95–96, 241–242

Andrew
  aesthetic sense, 245
  considers conversation with Jared, 233–234, 236–239, 255–256
  contemplates the night, 247–248
  lunch with Kim and Jared, 279–280
  offers transmission to Jared, 86–89, 105
  self-awareness, 252–253
  talking to Jared at the party, 163
anhedonia, 123
animal learning theory. See also mammalian brain; rat brain
  expectations, 20–21, 140, 191
  motivational control, 113
  Pavlovian principles, 114
  representation, 148
  rewards, 232
anterior (front) brain
  action control, 59, **61**, 252
  association areas, 83
  brain anatomy, **36**
  cognition and limited-capacity resources, 69
  divisions, 41, **62**
  ethical decisions, 28
  information and action, 102, 243
  motor systems, 77, 248–250
  perception and action, 126
  visual cortex, 172
anterior attention system, 72–73
anterior cingulate cortex, 72–73, 181, 257, 261
anterograde amnesia, 84–85
anxiety. See also uncertainty
  Andrew considers conversation with Jared, 236–239
  and depression, 263–264
  and emotions, 163–165
  and experience, 221
  and motivation, 183
  motor system of anterior brain, 77
  pain and experience, 295
apraxia, 75
archicortex, 38
architecture, 100–102, 129, 136, 179–180. See also SPD (stability-plasticity dilemma)
arousal, 151
artificial intelligence, 7–8, 20
artificial neural networks, 7–8, 13, 95, 113
assimilation, 134, 139
association areas
  abstract thought, 219
  limbic networks, 90
  and mind, 83
  motivational control, 90
  ontogenetic recapitulation, 200–201
  sensorimotor networks, 90
  sensory cortex, 60
  Wernicke's aphasia, 245–246
attention, 68–70, 72–73
auditory cortex, 59, **62**, 88, 173. See also hearing
automatization, 241, 250, 285–286, 289
avoidance reflex, 76
awareness
  evolution of, 271, 284
  and memory, 84
  and motor control, 71
  self-awareness, 94–95, 252–253
  vertical integration, 275–276
  working memory, 70
axon fiber tracts, 19

back projections
  brain anatomy, 83
  limbifugal, **174**–176, 182, 217, 246
  memory and perception, 178–179, 252
back propagation, 104
Barbas, Helen, 173, 202
basal ganglia
  brain stem, 124
  olfactory circuits, 39
  subcortical structures, 40
  telencephalon, 38, 167
Bateson, Gregory, 151
behaviorism, 114–115
bipedalism, 36, 64–65
bipolar disorder, 127–128
birds, 205–206
Bitterman, M. E., 124–125
blindsight, 206
blocking effect, 119–121
body dialectic, 219
Bogen, Joseph, 45, 48
Brady, James, 34
brain anatomy. See also individual components; spinal cord
  architecture of, 13–14
  association areas, 60, **62**, 83
  cerebral hemispheres, 58–59
  complexity and expectancy, 140
  dorsal, 36
  enchanted loom, 190–191
  hemispheres, **36**
  hemispheric specialization, 26–28, 43
  and information, 16–17

mapping, 50–53
mirroring, 82
naïve subjectivity, 15–16
neoteny, 201, 223
network architecture of, 102
organization of, 19
phylogenesis, 281
radial dimension, 78
Society for Neuroscience, 295
split-brain observations, 49
ventral, 36
brain damage
   blindsight, 206
   brain mapping, 50–53
   Broca's aphasia, 250–251
   cognitive development in children, 55
   hemispheric deficits, 42–43
   hemispheric specialization, 48
   and intelligence, 33–34
   loss of abstract thought, 205
   pseudobulbar palsy, 265–266
   social interaction and, 44
   subcortex damage, 40–41
   Wernicke's aphasia, 245–246
brain stem
   and abstract thought, 277
   definition, 19
   diencephalon, **165**
   limbus, **42**, 78
   motivational control, 90
   primary sensory cortex, 82
   primordial brain, 37, 164
   pseudobulbar palsy, 266
   spinal cord, **156**
   subcortical structures, 40
   terminal additions, 124
   vertical integration, 37, 186
Broca's aphasia, 250–251
Brown, Jason, 245–246, 277

cats, 40
central sulcus, **36**, 59–**61**, **64**, **156**
cerebellum, 37–**38**
cerebral hemispheres, **38**, **156**
certainty, 296–298
children
   action formation in, 182
   back propagation training, 104
   cognitive development, 54–55
   depression in, 263
   embryogenesis, 281
   hemispheric specialization, 235
   intelligence and differentiation, 101
   language acquisition, 202–203, 224, 242

myelination during maturation, 200
and the objective self, 268–269
Piaget learning theory, 134
social bonding and relationships, 258–259
theory of mind, 269
cingulate cortex, **64**
cingulate gyrus, **64**, **156**, 168, 172
civilization, 29
closed loop, 154
cognition
   attentional control, 68–69
   biological substrates of, 52
   bodily control of, 59
   and brain architecture, 14
   critical appraisal, 245–246
   delayed, 91
   and emotions, 148, 152–153, 239, 284, 292
   encephalization of visceral function, 207–208, 232
   evolution in the mammalian brain, 65
   expectancy system, 131
   exteriorization, 251
   hemispheric specialization and, 56–57
   and human values, 264, 294
   and language, 284–286
   memory consolidation, 148
   mood-dependent, 127–128
   and motivation, 112–113, 157
   network architecture, 136
   and observation, 247
   primary process and fantasy, 214, 231
   in rats, 115–116
   secondary process cognition, 215–217
   sensorimotor networks, 90–93
   social relationships, 258
   and speech, 250–251
   spreading activation, 106–107
cognitive behavior therapy, 148
cognitive development, 134
cognitive neuroscience, 10
cognitive representations, 112
cognitive science, 10, 147–148
cognitive theory, 151–152
commissures, 45, 49, 136
common sense, 177
computational neuroscience, 7–8
computational simulation, 289–290
computers, 5, 7, 20, 95–96
concept formation. *See also* abstract thought
   core-shell theory, 211–212, 238
   corticolimbic networks, 180
   dialectics of intelligence, 231, 284

connectionist networks
  architecture of, 179–180
  artificial neural networks, 7–8, 20
  cross-level connections, **81**,
    177–178, 218, 221–223
  distributed representation, 96–97, 130,
    139–141, 173–**174**, 293
  layers, 100
  nerve nets, 97
  organization of experience, 13
  pattern completion, 107
  and representation, 97–99, 113, 129,
    136, 239
  SPD dilemma, 99–100, 136
  training, 104, 130
consciousness
  anterior attention system, 72–73
  core-shell theory of abstract thought, 232
  corticolimbic networks, 272
  dogma of immaculate perception, 132
  expectancy system, 131
  expert performers, 31
  exteriorization, 240
  kindling reaction, 198
  and language, 250–251
  limitations of, 129
  and memory, 84, 142, 188
  motive blindness, 239–240, 247, 299
  pseudobulbar palsy, 272
  and subjectivity, 22, 232, 269–271
  thin ice of, 146
  visceral nervous system, 275–276
contextual self-concept, 168
control process, 112
control theory. *See also* feedback control;
    feedforward control
  and behaviorists, 114–115
  computational neuroscience, 8
  cybernetics, 150–151
  economic forecasting, 122
  mood-based, 127–128
  and neural processes, 95
  representation and adaptation, 105
  self-regulation, 262
  stimulus-response, 118–119
Copernican revolution, 138
core-shell theory. *See also* abstract thought;
    corticolimbic networks
  brain architecture, 19
  consciousness, 232
  dialectics of intelligence, 27–28,
    224–225, 251, 279
  emotional evaluation, 238
  epistemology, 278–279
  ethical experience, 248
  exteriorization, 240, 251
  hemispheric specialization, 27–28, 236,
    266–267
  and intelligence, 204–205, 211,
    224–225
  Jared contemplating Andrew's offer,
    195–198
  memory consolidation, 182, 218,
    223, 297
  ontogenesis, 290–291
  pain perception, 256
  perception and action networks, 243
  radial dimension, 78
  subjective-objective dimension, 254
corpus callosum, **165**
cortex
  cerebral hemispheres, **38**, 167
  corticolimbic networks, **174**
  definition, 19
  divisions, 41, **62**
  gyrus, 36
  hemiplegia, 41
  input/output architecture, 59–**62**
  kindling, 188–189
  levels of, 78–**79**
  and limbic system, 13
  maturation, 201–202
  sulcus, 36
  visceral nervous system, 14
corticolimbic networks. *See also*
    core-shell theory
  and abstract thought, 219, 244
  architecture, 129, 171–172,
    176–177, 179–180
  brain anatomy, 19
  and cognition, 246–247
  connections, 80–82
  consciousness, 272
  distributed information representation,
    173–**174**
  encephalization of visceral function,
    208, 223
  and fantasy, 214–215
  heteromodal association networks,
    180–182
  and information, 16–17
  and intelligence, 207–208,
    211–212
  and language, 242
  lateralization, 234–235
  long-term memory, 89
  memory consolidation, 86, 124,
    174–176, 194–195
  memory representation, 60, 183, 186
  pain and bodily integrity, 257

316 Index

and perception, **165**, 180, 252
schematic, **81**
senses, 78–**79**, **80**
Costa, Louis, 55
creativity
 and action, 16
 and fantasy, 214, 217–218
 nature of, 145–146, 229
 right hemisphere, 56
critical appraisal, 245–246
cross tolerance, 259
cultural schism, 17–18
cultural transmission, 285–286
"cutting," 259–260
cybernetics
 control theory, 150–151
 distributed networks, 20
 and mind, 22
 negative feedback, 262
 Norbert Wiener, 115

Darwin, Charles, 283–284
data, 3–6, 12, 20
deconstructionists, 11–12, 14
decussation, 41–42
delayed cognition, 91–92
delayed gratification, 182
depression
 in children, 263
 ECT therapy, 193
 and emotions, 152–153, 263–264
 and experience, 23
 learned helplessness, 123–124
 mood swing, 127–128, 179, 299
 negative feedback, 262–263
 psychological pain, 261–262
Derryberry, Douglas, 261
The Descent of Man, 283
despair, 263
developmental epistemology, 134
dialectical balance, 22
dialectics, 12, 132, 138–139, 141
dialectics of intelligence. *See also* intelligence
 concept formation, 231, 284
 core-shell theory, 27–28, 224–225, 251, 279
 exteriorization of consciousness, 240
 hemispheric specialization, 236
 intersubjective reasoning, 255–256
 learning, 134–135
 SPD dilemma, 139
diencephalic thalamus, 81–82
diencephalon (interbrain)
 brain anatomy, **38**, **165**–167
 hypothalamic division, 23–24

memory consolidation, 171
primate vision, 205–206
and telencephalon, 172, 299
thalamic division, 24
vertical integration, 37, **159**–160, 271–272, 281–282, 288
differentiated specificity, 25
differentiation, 140
digestion, 133–134
distributed networks, 100–102
distributed representation. *See also* parallel-distributed representation; representation
 artificial neural networks, 95
 connectionist networks, 95–97, 130, 139–141, 173–**174**, 293
 definition, 20
 functional sculpting, 162
 holistic systems, 22
 Jared, 105
 and learning, 138
 left hemisphere, 234
 motivational control, 92–93
 stimulus-response and learning, 137–138
 training, 98
dogma of immaculate perception, 132, 176
domain knowledge, 89, 112
dorsal cortex, **64**–65
Dragons of Eden, 6
dual brain theory, 43–46
The Duality of the Mind, 43
dynamical systems theory, 54

economics, 121–122, 260, 263, 292
ECT (electroconvulsive therapy), 193–194
education, 298–300
EEG (electroencephalogram), 292
egocentrism
 Andrew's conversation with Jared, 238
 ego halo, 249
 and embodied mind, 28–29
 and emotional contagion, 268
 and experience, 199
 Jared's conversation with Andrew, 141, 196, 198
 Jared's infatuation with Kim, 267
 motive controls, 205
 motive memories, 247
 perception and need, 244
 regression in service of the ego, 227–229
 subjective basis for, 254
ego halo, 249, 267
Einstein, Albert, 214, 218–219, 275

electroconvulsive therapy (ECT), 193–194
electroencephalogram (EEG), 292
embodied mind
  cognition, 67, 110, 252
  connectionist networks, 83
  and egocentrism, 28
  and emotions, 152
  and intelligence, 92
  Johnson, 293
  Kim takes Jared's keys, 186
  neural basis of, 16
  subjectivity, 254
embryogenesis
  cellular organization, 54
  connectionist networks, 136, 179
  core-shell theory, 182
  development and differentiation, 101
  and intelligence, 161–162, 281
  mechanisms, 299
  microgenesis, 277
  neural tube, 64
  and psychological development, 160–161
  self-actualization, 299–300
  telencephalon, 41
  terminal additions, 124
  vertical integration, 149, 158–**159**, 160, 169–170
emotional contagion, 267–268
emotions
  and cognition, 148, 239, 284
  as cognitive construct, 151–153
  and decision making, 16
  The Expression of the Emotions in Man and Animals, 283–284
  hemispheric damage and, 49–50
  and intelligence, 273
  kindling reaction, 189
  neural basis of, 123–124
  pain perception, 257, 266
  and rational cognition, 239, 292
  right hemisphere interpretations, 237–239
  syncretic holism, 50
  visceral nervous system, 14
  vocalization, 266
empathy
  Andrew's conversation with Jared, 238
  emotional contagion, 267–268
  and experience, 293
  sensory representation, 91–92
  and social interaction, 198–199
encephalization. *See also* vertical integration
  aesthetic sense, 244–245
  definition, 19–20, 40

hypothesis of encephalized conceptual structure, 222–223, 229–230, 266
  motive control, 182
  natural selection, 272
  pain and bodily integrity, 256
  self-control, 288
  somatic networks, 205–207
  visceral nervous system, 207–208, 232
endbrain. *See* telencephalon (endbrain)
endogenous opiates, 257–258
epilepsy, 45, 189, 191
epistemology, 278–279
ethical decisions, 28, 76–77, 264
ethical experience, 248–250
evolution
  behavioral principles, 114
  decussation, 42
  delayed gratification, 182
  intelligence and cultural transmission, 6
  and language, 3–4
  mammalian brain, 160
  of mind, 281
  natural selection, 283–284
  order of brain levels, 37–39
  sensorimotor networks, 63–**64**
  stability and plasticity, 133–134
  vertebrate brains, 19, 34–35
  vertical integration, 207
executive control, 14
exoencephalon, 288, 290–291
expectancies, 244, 247, 251, 259
expectancy-based learning, 124–125, 140
expectancy system, 203
expectations
  and hope, 118
  and information, 132–133, 225–226
  Jared's date with Kim, 128
  in mammals, 21, 126–127, 130–131, 191
  memory consolidation, 149
  memory representation, 186
  motivation, 137, 148, 181
  transferences, 236
experience. *See also* memory consolidation; motivational control
  and abstract thought, 28–29, 220–221
  definitions, 187–188
  as developmental process, 276–278
  distributed representation, 293–294
  domain knowledge, 112
  ethical and aesthetic, 76–77
  expert performers, 31
  functional differentiation, 161–162
  historical self, 132
  and information, 13

and interpersonal relations, 24–25, 199
intuitions, 25–26
language and grammar, 241–242
and memory, 83–86
mood states, 158
motive blindness, 232
motive memories, 23–24, 231, 300
neurophysiology of, 190
and perception, 89, 175–176, 181, 243
philosophical approach, 17–18
representation and control, 129
and the senses, 70–72
sensorimotor networks, 119, 180
and social interaction, 195
structure and process of, 53–58
unconscious nature of, 141, 191
values and empathy, 293
visceral basis of memory, 22–24
visceral nervous system, 237–239
expert performers, 30–32, 220–221, 226–227
The Expression of the Emotions in Man and Animals, 283
exteriorization, 240, 244, 246, 251
extinction rate, 116–117, 125
eyeblink conditioning, 125

face recognition, 7
family values, 263–265
fantasy
  and abstract thought, 215
  and cognition, 214–215, 231
  and creativity, 217–218
  cybernetic, 289, 296
  and intelligence, 213, 219, 225
  primary process cognition, 251
  unconscious mind, 209
feedback control
  artificial neural networks, 113
  control theory, 115
  corticolimbic networks, **174**–175, 182
  in humans and machines, 9
  mammalian information, 21
  and pain, 261
feedforward control, 21, 115, 126–127
fire together, wire together, 97, 161
fMRI (functional magnetic resonance imaging), 10, 40, 198–199, 292
forebrain. *See* prosencephalon (forebrain)
The Formation of Vegetable Mould Through the Actions of Worms With Reflections on Their Habits, 283
fornix, **165**
free will, 74, 144–145
Freud, Sigmund

adolescent infatuation, 267–268
biological objectivity and mind, 15
development of the mind, 277
ego development, 250
failure of psychoanalysis, 147
Freudian child development theory, 28
hedonic cathexis, 209
motive memories, 191
overdetermined, 274
primary process cognition, 214, 227–228
secondary process cognition, 215–217
frog brain, 271–272
frontal lobe
  action formation, 180–186, 249
  association areas, 60, **62**
  brain anatomy, **36**
  damage to, 75–76, 205, 250–251
  higher-order association cortex, 79–**80**
  and limbic system, 76–77
  obsessive-compulsive, 267
  speech interpretation, 88
frontal networks, 27–28, 137–138
functional differentiation, 161–162
functional magnetic resonance imaging (fMRI). *See* fMRI (functional magnetic resonance imaging)
functional sculpting, 161–162, 281

gender discrimination, 103–104
general intelligence factor, 33
generality, 176–177
genetic epistemology, 279
Gestalt psychology
  analogical representation, 95–96
  isomorphism, 70, 244
  Jared contemplating Andrew's offer, 112
  pattern formation in perception, 9
  pattern recognition, 98
ghost in the machine, 204
Gibson, James, 251–252
global precedence, 49
Goldberg, Elkhanon, 55
graceful degradation, 98
grammar. *See also* language
  left hemisphere, 109
  and motor control, 68
  symbolic representation, 241–242, 285
  and vocalization, 289
gut-level experience
  and abstract thought, 295
  and intuition, 25–26
  Kim worried about Jared, 163
  and mind, 238–239
  and perception, 178
gyrus, 36

Index 319

hallucination, 175
handedness, 43, 50
hearing. *See also* auditory cortex
   corticolimbic networks, **62**, 172
   posterior brain, 59, **61**
   primary sensory cortex, 79–**80**
   somatic networks, 63
Heath, R. G., 157
Hebb, Donald, 97
Hecaen, Henri, 47–48
hedonic cathexis, 209, 260, 262, 264
Hegelian analysis, 135
hemiplegia, 40–41
hemispheric dominance, 43
hemispheric specialization
   architecture, 129
   asymmetry of mental content, 46–48, **51**, 56–57
   bifurcation of intelligence, 45
   brain anatomy, 19
   child development, 242
   connectionist networks, 136
   consolidation of memory, 26–28
   core-shell theory, 266–267
   dialectic organization of mind, 239–240
   divisions, 41
   embryogenesis, 41
   and emotion, 237–239
   and intelligence, 27–28, 58, 231–232, 240
   language, 107–109
   lateralization, 75, 234–235
   parallel-distributed representation, 110
   and social interaction, 264
   split-brain, 48–49
   structure and process, 70
Heraclitus, 135
Herrick, C. Judson, 167, 172
heterochrony, 201, 202–203
heteromodal association cortex, 78–**79**, **81**
heteromodal association networks
   corticolimbic networks, 171–172, **174**, 180–182
   differentiation, 201
   memory consolidation, 86, 210, 223
   perception, 83
   speech interpretation, 88
   Wernicke's aphasia, 245–246
hierarchic structure
   abstract thought, 25, 229–230, 254–255
   brain anatomy evolution, 39, 206–207
   hemispheric specialization, 27–28
   integration, 103
   mammalian brain, 60, 125

higher-order association cortex, 79–**80**
hindbrain. *See* rhombencephalon (hindbrain)
hippocampus, 38, 85, 167, 172
historical self, 132, 236
holistic memory, 66–67
holistic systems
   abstract thought, 25, 279
   distributed representation, 20, 22
   dynamic interaction, 54
   hemispheric specialization, 26–28
   intuitions, 25–26
   SPD dilemma, 102–103
homeostasis
   embryogenesis, 54
   encephalization of visceral function, 207–208
   experience and behavior, 12–13
   hypothalamic projections, 172
   Jared at the diner, 66
   and memory consolidation, 148
   and need states, 151
   neurophysiology of, 154
   temperature, 150
   visceral nervous system, 63–**64**, 166
hope
   and depression, 127
   and economics, 122, 260
   and expectation, 118
   learned helplessness, 123–124
   and learning, 138
   negative feedback, 262
   and perception, 180
   in rodent cognition, 115–117
   and truth, 12
hostile cathexis, 209
humanities
   cultural schism, 17–18
   dialectic dead-ends, 12
   naïve subjectivity, 15
   and psychology, 9–10
   schism with science, 232
   syncretism of experience, 293–295
   "two cultures" lecture, 11
hunger, 153–154
hypothalamus
   diencephalon, 23, **165**–166
   function, **38**
   limbic cortex, 172
   and motivation, 23, 177, 181, 186, 209
   and telencephalon, 167–168, 179, 299
   visceral nervous system, 167–168, 214, 237

ideation, 182
increasing differentiation, 99
incubation, 146
infatuation, 267–268
informatics, 284–289, 291, 297–298
information
    blocking effect, 119–121
    brain anatomy, 14, 102–103
    and certainty, 296–298
    and change, 130
    and civilization, 29
    conscious learning, 129
    cultural transmission of, 285
    and expectation, 132–133, 225–226
    and fantasy, 214–215
    in holistic systems, 22
    informatics, 287–288
    mammalian, 21
    meaning and bodily function, 8, 12–13
    memory consolidation, 148
    nature of in humans, 21–22
    neural networks, 16–17, 20
    paradox, 3–4
    parallel channels, 65–67
    relational nature, 99, 229–230, 247
    in scientific theory, 139
    self-regulation in mammals, 21
    sensation and need, 169
    serial channels, 67–68
    shell-to-core function, 243–244
    SPD dilemma, 139
    and uncertainty, 130–131, 140, 218
    and value, 10, 122
Information Age
    and the Age of Mind, 295–298
    computational simulation, 290
    and data, 3–4, 16
    evolution of the mind, 281
    and learning, 133
    Moore's law, 5–6
infragranular layer, 173–**174**
inhibitory control, 60, 249
instantiation, 279
insula, **156**, 172
insular cortex, **64**, 257
integration, 140
Intel Corp, 5
intelligence. *See also* dialectics of intelligence
    and abstract thought, 204–205, 219, 227–230, 298
    analogical representation, 241
    and artificial intelligence, 6–7
    artificial neural networks, 7
    and attention, 72–73
    automatization, 289

bifurcation, 45
brain anatomy, 14
and brain injury, 33–34
and cultural transmission, 6, 285–286
embryogenesis, 281
emotions and abstract concepts, 273
evolution of, 286–287
and expectation, 126–127
expert mind, 30–32, 220–221, 226–227
and fantasy, 213–214, 225
hemispheric asymmetry, 55
hemispheric specialization, 27–28, 58, 231–232, 240
information paradox, 4
IQ tests as measure, 32–33
and machine intelligence, 20, 96, 287, 289–290
memory consolidation, 26–28, 170, 226
objective, 28–29
Piaget learning theory, 134
radical neoteny, 282
representation and control, 105
sensorimotor networks, 63–**64**, 92, 161–162
and social interaction, 92, 255
SPD dilemma, 100
stimulus-response, 119
structure of, 30–32, 211–212, 274
subjective nature of, 15–16, 26–28, 231–233, 254–255, 276–278
symbolic representation, 242
visceral nervous system, 207–208
interbrain. *See* diencephalon (interbrain)
interpersonal relations, 24–25
intersubjective reasoning, 255–256, 259, 269
introspection, 57
intuitions, 25–26
IQ (intelligence quotient), 32–33, 44, 46–48
isomorphism, 70, 179, 244

Jackson, John Hughlings
    bodily control of mind, 15, 59
    brain injury imperception, 75
    hemispheric specialization, 44
    neuroanatomy, 157–158
    sensorimotor machine of the mind, 62
James, William, 9, 73–74, 151, 293
Jared
    Andrew offers transmission, 86–89, 91–92
    asks Kim out, 90–91, 197
    car trouble, 52–53

Jared (*continued*)
  contemplates Andrew's offer, 111–113, 195–198, 277–278
  crush on Kim, 128
  date with Kim, 162–164
  distributed representation, 105
  driving home, 126–127
  hurt Andrew's feelings, 141–142
  infatuation with Kim, 267–268
  insensitivity toward Andrew, 198, 203
  introspection and mind, 57
  Kim takes his keys, 185, 216–217
  lateral view of his brain, **35**–36
  lunch at the diner, 66–67
  lunch with Kim and Andrew, 279–280
  medial view of right hemisphere, **37**
  self-awareness, 94–95
  spreading activation, 106–107
  talks to Andrew at the party, 163
  visual fixation, **42**
Johnson, Mark, 293
Jürgens, Uwe, 265

Kamin, L. J., 120
Keele, Steve, 262
Kim
  agrees to date Jared, 90–91, 128
  date with Jared, 162–164
  fantasy about Jared, 209
  history term paper, 212–213, 225–226, 278
  kindling reaction, 189–190
  lateral view of left hemisphere, **163**
  lunch with Jared and Andrew, 279–280
  pact with Tara, 162, 168₇169, 185
  takes Jared's keys, 185, 216–217, 273–275
  vertical integration, 164, 166–170, 182–186
kindling
  definition, 188
  and fantasy, 215
  and intelligence, 211
  Jared's insensitivity to Andrew, 203
  stress response, 189–190
  and vertical integration, 198
knowledge, 276–279
Kuhn, Thomas, 138–139, 141
Kurzweil, Ray, 6

language. *See also* grammar
  and abstract thought, 223
  acquisition of, 202–203, 224
  and cognition, 250–251, 284–286
  and culture, 3, 6
  evolution of, 3–4
  hemispheric specialization, 107–109, 235
  left hemisphere and meaning, 236
  left hemisphere as seat of mind, 43–44
  and memory, 84
  and mental content, 47
  and motor control, 67–68
  Wernicke's aphasia, 245–246
lateralization
  brain structures, 109, 232
  left and right hemispheres, 234–236
  and perception, 49
  sensory systems, 74–75
learned helplessness, 123–124, 153
learning. *See also* information
  and change, 130, 141
  conscious, 129
  distributed representation, 137–138
  extended neoteny, 282
  heterochrony, 201
  mammalian brain, 126, 130–131, 297–298
  and motivation, 143
  negative sculpting, 135–137
  network plasticity, 200
  Piaget theory, 133–134
  SPD dilemma, 133
learning theory, 262, 292
left-handers, 50
left hemisphere. *See also* hemispheric specialization
  analytic perception, 107, 234
  Andrew's conversation with Jared, 238–239
  brain mapping, **51**
  cognitive development in children, 55
  decussation, 41–**42**
  Jared contemplating Andrew's offer, 111–112
  Kim's, **163**
  language meaning, 236
  motor system of anterior brain, 75
  SPD dilemma, 136
  speech, 43, 88
  spreading activation, 109
  symbolic representation, 240–241
  visuospatial abilities, 47
Lévi-Strauss, Claude, 275
limbic cortex
  aphasia, 245–246
  brain anatomy, **38**
  corticolimbic networks, **174**–175, 232
  hypothalamic projections, 172
  opiate receptors, 257
  visceral functions, 63–**64**

limbic networks
   and abstract thought, 277
   association areas, 90
   connections, 177
   and fantasy, 209
   kindling, 189
   memory consolidation, 83, 186, 209, 292
   and plasticity, 200
   self-regulation, 260–263
   vocalization, 265–266
limbic nuclei, 40
limbic system
   bodily control, 13–14
   brain anatomy, 78
   and empathy, 199, 267
   executive control, 14
   frontal lobe connections, 76
   function, 82
   hemispheric specialization, 26–28
   hypothalamic projections, **64**, 168
   informatics, 291
   memory consolidation, 23–24, 170–171, 186
   memory damage, 85
   motivation and cognition, 157
   and neocortex, 41
   pain perception, 256–257
   spinal cord, **156**
limbifugal processing. *See also* core-shell theory
   action control, 181, 251–252
   back projections, 246
   cortex procession, 175–176, 211
   primate brains, 222
   semantic integration, 224
   visceral control centers, 180
limbipetal processing. *See also* shell-to-core function
   and cognition, 246–247
   memory consolidation, 176
   perception, 243
   semantic integration, 224
limbus
   brain stem, **42**, 78
   cerebral hemispheres, **38**, 167
   definition, 13
long-term memory, 65, 89, 193–194.
   See also memory consolidation
love, 265–268
Luu, Phan, 260–262

machine intelligence, 20, 96, 287, 289–290
mammalian brain

analogical representation, 241–242
cortex evolution, 98
corticolimbic networks, 80, 124
cultural transmission in, 285
dimensions of, 58
evolution of, 160, 282
expectancy system, 119, 126–127, 137, 140
expectations, 21, 119, 137, 140
hierarchic structure, 125
input/output architecture, 60
kindling, 188
learning, 126, 130–131, 297–298
limbic networks, 208
memory consolidation, 65
neoteny, 201
positive expectations, 262
representation and learning, 130, 232
social bonding, 258
uncertainty and learning, 130–131
mania, 153
Marx, Karl, 135
mathematics, 242, 285–286, 289
maturation, 201–202, 290
medulla, 37–**38**
memories, 299
memory consolidation. *See also* corticolimbic networks; experience
   abstract thought, 210, 279, 298–300
   and amnesia, 191–192
   association areas, 60
   basis for perception, 175–176
   bodily control of, 59
   and cognition, 148
   consciousness, 84, 142, 188
   core-shell theory, 182, 218, 223, 297
   corticolimbic networks, 194–195
   distributed information representation, 162
   and emotions, 284
   epilepsy, 191
   experience, 83–86, 95, 194
   hemispheric specialization, 236
   hierarchic structure of, 26, 125, 226
   homeostasis, 148
   kindling reaction, 190
   limbic networks, 83, 186, 209, 292
   limbic system, 13, 23–25
   long-term memory, 89, 193–194
   mammalian brain, 65
   neocortex storage, 85–86
   network architecture, 170–171, 294
   neural plasticity, 149, 201–202
   and perception, 246, 249
   personal significance and, 28

memory consolidation (*continued*)
  recursive processing, 224
  and representation, 122, 178–179, 209, 222
  right hemisphere, 236
  sensorimotor networks and experience, 119, 275
  and sleep, 190–191
  spreading activation, 106–108
  stimulus-response, 114, 118, 158, 161
  vertical integration, 149
memory representation, 60, 179–180
mental disease, 202–203
mesencephalon (midbrain)
  divisions, 37–**38**
  reticulum, 164–165
  spinal cord, **156**
  vertical integration, **159**–160, 271–272, 281, 288
metencephalon, 37–**38**
microgenesis, 277
midbrain. *See* mesencephalon (midbrain)
mirroring
  Andrew's conversation with Jared, 238
  brain anatomy, **61**, 82, 172–173
  and evolution, 178
  motor cortex, 92
  motor systems, 198–199, 267
mood states, 127–128, 153, 158, 164
mood swings, 127–128, 153–**155**, 179, 299
Moore's law, 5–6
morphogenesis
  decussation, 41–42
  and intelligence, 161–162
  negative sculpting, 135–137
  and phylogenesis, 281–282
  Piaget learning theory, 134
motivation. *See also* mood states; motivational control; motive blindness; motor control
  anhedonia, 123
  and cognition, 112–113, 157
  cognitive behavior therapy, 148
  and cognitive science, 147–148
  and emotions, 152, 284
  empathy, 267
  endogenous opiates, 257
  and expectations, 137, 148, 181
  and experience, 13, 23–24, 143–144
  hypothalamus, 177
  mechanisms, 150–151
  neuroeconomics, 292
  and representation, 252
  vertical integration, 155–**156**

motivational control. *See also* hypothalamus; motivation
  and abstract thought, 212, 222–223, 229
  association areas, 60, 90
  and creativity, 145–146
  distributed representation, 92–93, 139
  empathy, 199
  functional sculpting, 162
  hope, 122
  hostile cathexis, 209
  and learning, 143
  limbic networks, 23
  overdetermined, 274
  and uncertainty, 140–141
  working memory, 154, 275
motive blindness. *See also* motivation; motive controls
  automatization, 250
  and consciousness, 239–240, 247, 299
  and creativity, 145–146
  and emotions, 151–152
  experience, 190, 204, 232
  mood swings, 153
  primary process cognition, 215
motive controls. *See also* motive blindness
  core-shell processing, 243
  encephalization, 182
  frontal lobe, 181
  functional differentiation, 161–162
  hypothesis of encephalized conceptual structure, 222–223
  and memory, 23, 186, 210
  uncertainty, 140–141
  vertical integration, 149
motive expectancies, 180
motive memories
  egocentrism, 247
  and experience, 23–24, 231, 300
  Sigmund Freud, 191
motor activity, 161
motor control
  anterior (front) brain, 59–60, **61**, 69
  hemispheric specialization, 75
  hillside walk, 71–72
  information flow, 74
  Jared at the diner, 67
motor systems, 74–75, 198–199, 248–250
mutations, 288, 290–291
myelencephalon, 37–**38**
myelination, 200

natural selection, 272, 283–284
needs. *See also* uncertainty
  definition, 12

and expectancy, 244
and homeostasis, 151
information and data, 16–17, 197
mammalian, 21, 210
motivational impulses, 13
negative contrast effect, 117–118
negative feedback, 262
negative sculpting, 135–137. *See also*
subtractive elimination
neglect syndrome, **42**, 68, 76
neocortex
brain anatomy, 38
and limbic system, 41, 174–175
and memory, 85–86, 171
myelination, 200
somatic networks, 14, **165**–166
thalamic projections, 172
vocalization, 266
neostriatum, 124
neoteny, 162, 201, 282, 284
nerve nets, 96–97
network architecture, 135–137
neural networks
brain mapping, 52
hillside walk, 71
and information, 16–17
information organization in machines, 20
and intelligence, 92–93
layers, 97–98
memory consolidation, 149
plasticity, 149, 287
neural tube, 64, **159**–160
neuraxis
and evolution, 160–161
and intelligence, 273
morphogenesis, 41
and motivational control, 154–**155**
spinal cord, 154–**155**
vertical integration, 157–158, 169–170
visceral/somatic pathways, 63–**64**
neuroanatomy, 32, 129, 157–158
neurocybernetics, 272–273, 291–294
neuroeconomics, 263, 292–293
neuroimaging technologies, 10, 40. *See also* fMRI (functional magnetic resonance imaging)
neuronal model, 66
neurons, 200
neuropharmacology, 259
neurophilosophy, 77
neurophysiology, 277
neuroscience, 7, 10, 12, 62
nonverbal IQ, 44
nonverbal skills, 47, 49–50, 88

objective intelligence, 28–29
object relations, 194
obsessive-compulsive, 267
occipital lobe, **36**
"Oh shit!" wave, 260–261
Old Dog Problem, 99–100
olfactory, 39
ontogenesis, 160, 200–201, 290
opiate receptors, 257, 264
orbitofrontal cortex, 172
orienting response, 68
Origin of Species, 283
orthogenetic principle, 103
overdetermined, 274

pain
anxiety and experience, 295
and empathy, 293
feedback control, 261
and love, 268
neural building block of self-control, 263
and self-control, 263–264
and social bonding, 256–258
in social interactions, 258–260
vocalization, 266
paleocortex, 38–39
paleostriatum, 124
Pandya, Deepak, 80, 173
panic attack, 164
Panksepp, Jaak, 257–259
Papez, James, 38
paralimbic cortex, 78–**79**, **80**, 83, 171
parallel-distributed representation. *See also* distributed representation
awareness, 68
hemispheric specialization, 110
Jared and Andrew, 105, 107
memory consolidation, 192
nerve nets, 96–97
parietal lobe, **36**, 42, 60–**62**, 68
partial reinforcement extinction effect, 116–117
pattern completion, 98, 107
Pavlovian principles, 66, 114, 190
perception
and abstract thought, 247
and action, 126–127
aesthetic sense, 264
computational simulation, 290
core-shell theory, 27–28, 243–245
corticolimbic pathways, **165**, **174**, 246, 252
dogma of immaculate perception, 131–132
and experience, 89, 181, 243

Index 325

perception (*continued*)
Gestalt psychology, 9
global precedence, 49
holistic integration, 67, 180
isomorphism, 70, 244
memory consolidation, 86, 148,
175–176, 178–179, 249
paralimbic cortex, 82–83
performance IQ, 44
PET scanner, 10, 40
phenomenology, 70, 120, 132
phylogenesis
and brain anatomy, 281
definition, 160
microgenesis, 277
ontogenesis, 200–201, 282, 290
Piaget, Jean
dialectical analysis, 135
genetic epistemology, 279
intelligence development in the child, 9, 277
learning theory, 133–134
place avoidance, 120
plasticity. *See also* SPD (stability-plasticity dilemma)
feedback control in mammals, 21
holistic systems, 22
and mental disease, 202–203
neural networks, 149, 287
radical neoteny, 162, 284
sensorimotor network maturation, 200, 223
Plato, 135
Ploog, Detlev, 265
political correctness, 11–12, 14
pons, 37–**38**
Posner, Michael, 72–73, 106, 109
postcentral gyrus, 41, **62**, **64**, **156**
posterior (back) brain
aesthetic sense, 244
association areas, 83
brain anatomy, **36**
critical appraisal, 245–246
damage to, 75–76
divisions, 41, **62**
and limbic system, 76–77
limbipetal processing, 246–247
parallel channels of information, 68–69
perception and action, 126, 243–244
senses, 59, **61**, 180
SPD dilemma, 102
posterior attention system, 72
posterior networks, 27–28
precentral gyrus, 60, **62**, **64**, **156**

premotor cortex, 79–**80**
Pribram, Karl, 7, 76, 244, 248–249
primary auditory cortex, 88
primary motor cortex, 79–**80**, **81**, 201
primary process cognition
and creativity, 218
and ego, 249–250, 254
Einstein's genius, 219
fantasy, 214, 251
unconscious mind, 215–216
visceral nervous system, 218, 227–228
primary sensory cortex
concept formation, 231
corticolimbic networks, 171–172, **174**
differentiation, 201
senses, 79–**80**
and the thalamus, 82
unimodal association cortex, 176–177
primate brains
commissurotomy operations, 45
cortex and subcortex, 40
corticolimbic networks, 80
limbifugal activity, 222
vision, 205–206
vocalization, 265
problem solving, 146
prosencephalon (forebrain), **159**–160
prosody, 88
pseudobulbar palsy, 265–266, 272
psychoanalysis, 147, 151
psychologists, fear of, 146
psychology, 9–10, 17–18
psychomotor retardation, 264
psychosurgery, 257

radical neoteny, 162, 282, 284
rat brain
blocking effect, 119–121
economic forecasting, 121–122
expectations and learning, 135, 180
experiments, 114–116
learned helplessness, 123–124
negative contrast effect, 117–118
partial reinforcement extinction effect, 116–117
working memory, 119
rationality, 27
recursive processing
abstract thought, 297
affordances, 252
experience, 276–278
memory consolidation, 243
representation, 223–226
Reed, Marjorie, 261

reflexes
    and memory, 158
    Pavlovian principles, 114
    spinal cord, 154–**155**
    stimulus-response, 118, 161
reinforcement, 115–116, 118
relational knowledge, 140
representation. *See also* distributed
    representation; memory consolidation
    abstract thought, 207, 219, 221–222
    animal learning, 148
    cognitive behavior therapy, 148
    connectionist networks, 97–99, 113,
        129, 136, 239
    control-biased, 127, 130
    definition, 94
    and depression, 124
    economic forecasting, 122
    Einstein's fantasy, 214
    expert performers, 220–221
    hemispheric specialization, 235–236
    intelligence and control, 105
    mammalian brain, 130, 232
    and motivation, 252
    principles, 139–141
    recursive processing, 223–226
    stimulus-response, 126
    telencephalon network, 167, 172–175
    and vertical integration, 169
reptiles, 124–125
Rescorla, R. A., 120
reticulum, 137, 164–165
retinotopic map, 173
retrograde amnesia, 191–192
rewards
    animal learning, 148, 232
    economic forecasting, 121–122
    Kim at party, 168
    learned helplessness, 123
    partial reinforcement extinction effect,
        116–117
    and reinforcement, 118
rhombencephalon (hindbrain)
    brain anatomy, 37–**38**
    vertical integration, **159**–160, 281, 288
    visceral control centers, 164
right hemisphere. *See also* hemispheric
    specialization
    brain injury, 107–108
    brain mapping, **51**, 234
    cognitive development in children, 55
    decussation, 41–**42**
    emotions and damage, 49–50
    holistic systems, 107–108, 136,
        235, 239
    intelligence and creativity, 56
    liberation movement, 43–46
    memory consolidation, 236
    prosody, 88
    sensorimotor networks, **81**
    spreading activation, 109
    visceromotor functions, **64**
right-handers, 43, 50

Sagan, Carl, 6
schizophrenia, 144–145
science, 11–12, 15, 17–18, 232
scientific revolution, 138–139
secondary process cognition
    consciousness, 215–217
    and creativity, 218
    and ego, 249–250
    Einstein's genius, 219
    sensorimotor networks, 218, 227–228,
        251, 254
self-actualization, 199, 298–300
self-awareness
    adolescent mind, 195
    Andrew, 252–253
    bodily basis for, 169–170
    in children, 104
    Jared's afternoon, 94–95
self-consciousness, 18, 196, 245–246
self-control, 161, 262–264, 288
self-organizing, 161, 282, 287
self-regulation
    control theory, 262
    in depression, 263
    homeostasis, 154
    Jared's insensitivity to Andrew, 203
    Kim takes Jared's keys, 274–275
    motive expectancies, 180
    pain, hope and information, 260–263
    and plasticity, 202–203
self-representation, 168–169, 202–203
semantic activation, 107, 224
Semmes, Josephine, 50–53, 109–110,
    136, 234
senses, **61**, 70–72, 78–**80**, 172–173
sensitization effect, 188
sensorimotor cortex, 82
sensorimotor networks
    abstract thought, 210, 219
    bipedalism, 64–65
    cognition, 90–93, 286
    cybernetic fantasies, 289, 296
    differentiated specificity, 25
    evolution of, 63–**64**
    and experience, 180
    heterochrony, 201

sensorimotor networks (*continued*)
  and intelligence, 92, 161–162
  language acquisition, 202
  maturation, 200
  memory consolidation, 119, 149, 223
  and the mind, 62
  mirroring, 199
  posterior attention system, 72
  secondary process cognition, 218, 254
  spinal cord, **155–156**
sensory systems, 74–75, 241
septal nuclei, 167
septum, 38
sexual arousal, 157
sexual maturation, 137
Shannon, Claude, 3–4, 113
shell-to-core function. *See also* limbipetal processing
  abstract thought, 211, 224–225
  network architecture, **81**, 221–222
  and perception, 27, 243–244
Shepard, Roger, 175–176
Sherrington, Charles, 190–191
Skinner, B. F., 116
sleep, 190–191
smell/taste, 59, **61**, 66–67
Snow, C. P., 11
social bonding
  family values, 263–265
  love, 265–268
  and pain, 256–258
  and subjective intelligence, 265
  transferences, 258–259
social interaction
  and depression, 262, 264
  and empathy, 198
  and experience, 195
  hemispheric specialization, 264
  and intelligence, 92, **255**
  and opiates, 264
  rejection and pain, 258–260
  right hemisphere damage, 44
Society for Neuroscience, 295
Socrates, 134–135
solipsism, 12
somatic networks
  encephalization, 20, 205–207, 223
  neocortex, 14
  pain reflexes, 256
  secondary process cognition, 227–228
  and the thalamus, 167–168
somatic pathways, 63–**64**
somatic/visceral function, 167
somatosensory cortex, **61–62**, 172
SPD (stability-plasticity dilemma)
  architecture of, 100–102
  connectionist networks, 99–100, 136
  dialectics of intelligence, 139
  expectations and learning, 133
  hierarchic integration, 103
  holistic systems, 102–103
  identity and learning, 135
  information and learning, 121
  memory consolidation, 192
specificity, 176–177
speech, 43, 88, 250–251
Sperry, Roger, 45, 48
spinal cord. *See also* brain anatomy
  and brain stem, 37
  levels of, **156**
  neural tube, 64
  pain reflexes, 256
  reflexes, 154–**155**, 158
  sensorimotor networks, 63–**64**, **155–156**
Spitz, René, 263
split-brain, 45, 48–49
spreading activation, 106–108, 112
stability-plasticity dilemma (SPD). *See* SPD (stability-plasticity dilemma)
stimulus-response
  control theory, 118–119
  encephalization, 207–208
  evolutionary principles, 114
  expectancy system, 132
  functional sculpting, 161
  and learning, 113–114, 137–138
  and memory, 158
  partial reinforcement extinction effect, 116–117
  reinforcement, 115–116
  and representation, 126
  reptiles and fish, 130
  spinal cord, 154–**155**
  vertical integration, 169–170
stimulus-response behaviorism, 147
strychnine neuronography, 78
subcortical structures, 40–41
subjectivity
  and deconstructionists, 11–12
  expanding consciousness, 232, 269–271
  and human values, 264, 294
  and intelligence, 15–16, 26–28, 231–232, 254–255, 276–278
  intuitions, 25–26
  naïve introspection, 21–22
subtractive elimination, 160–161, 281
sulcus, 36
supragranular layer, 173–**174**, 175

symbolic representation, 240–241
symmetry, 41
syncretic holism, 25, 50, 179–180, 221–223

Tara, 162, 169, 184
telencephalon (endbrain)
  crying, 266
  definition, 19
  and the diencephalon, 172, 299
  divisions, **38**
  embryogenesis, 41
  and hypothalamus, 167–168, 179, 299
  memory consolidation, 170–171
  reticulum, 137
  self-representation, 168, 271
  sensorimotor networks, 64
  vertical integration, **159**–160, 166–168, 179, 281–282, 288
  visceral nervous system, 291
temperature, 150
temporal lobe, **36**, 84–85, 189
temporal-parietal cortex, 88
terminal additions, 124, 288, 290–291
thalamus
  archicortex, 38
  corticolimbic networks, **174**
  diencephalon, **38**, **165**–167, 299
  neocortex, 172
  primary sensory cortex, 82, 167–168
theory of mind, 269, 299
Thorndike's law of effect, 114, 116
tonotopic, 173
touch, 59, **61**
transferences, 236, 258–259
truth, 12
Turing machines, 5
"two cultures" lecture, 11

uncertainty
  and creativity, 145–146
  information and data, 16, 130–131
  information and need, 197, 218, 297
  mammalian, 21, 140–141, 143
unconscious mind
  affordances, 286
  automatization, 250, 285–286
  and awareness, 276
  experience, 141, 270
  and fantasy, 209
  and human evolution, 283–284
  Jared contemplating Andrew's offer, 197–198
  memory consolidation, 190–191, 211
  motive controls, 23–24, 232

neuropsychological processes of, 273
primary process cognition, 215
unimodal association cortex
  corticolimbic networks, 171–172, **174**
  differentiation, 201
  memory consolidation, 210, 223
  perception, 83
  primary sensory cortex, 176–177
  schematic, **81**
  senses, 79–**80**
  speech interpretation, 88

values
  and abstract thought, 248
  and cognition, 294
  and ethics, 28
  and experience, 293
  family values, 263–265
  and information, 10, 122
  opiate drugs, 257
vasocongestion, 157
vasodilation, 157
ventral (bottom) cortex, **64**–65, 172
verbal skills, 46
vertebrates
  brain anatomy shared with humans, 34
  brain evolution, 19, 37–39, 94
  decussation, 41–**42**
  pain and vocalization, 264–265
  symmetry in, 41
  terminal additions, 124
  vertical integration of nervous system, 36–37
vertical integration. *See also* encephalization
  awareness, 275–276
  definition, 19–20, 157
  and embryology, 158–**159**, 160, 182
  evolved mind, 282–284
  functional differentiation, 160–161
  Kim at party, 164, 166–170, 182–186
  kindling reaction, 190, 198
  memory consolidation, 149
  spinal cord, **155**–**156**
  stimulus-response, 169–170
  subcortical structures, 40–41
  subjective experience, 264
  telencephalon, 179
  theory of mind, 299
  vertebrate nervous system, 36–37, 206, 281
visceral nervous system
  and action, 248
  and consciousness, 275–276
  encephalization, 20, 207–208
  exoencephalon, 291

Index 329

visceral nervous system (*continued*)
   and the hypothalamus, 167–168, 214, 237
   limbic system, 13–14
   pain reflexes, 256
   and perception, 181
   primary process cognition, 218, 227–228
   rhombencephalon, 164
   and right hemisphere elaboration, 237
   spinal cord, 154–**155**, **156**
   telencephalon (endbrain), 291
visceral pathways, 63–**64**
vision
   encephalization, 271–272
   network architecture, 136, 171–172
   parietal lobe, 60, **62**
   posterior brain, 59, **61**
   primary sensory cortex, 79–**80**
   somatic networks, 63
   thalamus, 82
visual cortex, 172, **174**, 188
visual field, **42**, 108, 173, 272
visuospatial abilities, 47
vocabulary, 32–33
vocalization, 264–266, 289
von Bertalanffy, Karl Ludwig, 54–55

Wagner, A. R., 120
Watson, John B., 114
Werner, Heinz, 54, 103, 277
Wernicke's aphasia, 245–246
wetware, 7–8, 15–16
Wiener, Norbert, 115, 150. *See also* cybernetics
Wigan, A. L., 43–45
Wolfram, Stephan, 289–290
working memory
   Andrew considers conversation with Jared, 237
   definition, 70
   evolution in the mammalian brain, 65
   expert performers, 220–221
   Kim at party, 166
   Kim's conversation with Tara, 162
   and motivation, 148
   motivational control, 154, 275
   rat brain, 119
   relational knowledge, 140
   sensory networks, 65–66
writing, 4

Yakovlev, P. I., 200, 240, 251–252